Environmental Division 2021

AA002866

Held at the 2021 AIChE Annual Meeting

Boston, Massachusetts, USA and Online
7 - 11 November and 15 - 19 November 2021

ISBN: 978-1-7138-5705-1

Printed from e-media with permission by:

Curran Associates, Inc.
57 Morehouse Lane
Red Hook, NY 12571

CURRAN ASSOCIATES INC.
proceedings.com

Some format issues inherent in the e-media version may also appear in this print version.

Copyright© (2021) by AIChE
All rights reserved.

Printed with permission by Curran Associates, Inc. (2022)

For permission requests, please contact AIChE
at the address below.

AIChE
120 Wall Street, FL 23
New York, NY 10005-4020

Phone: (800) 242-4363
Fax: (203) 775-5177

www.aiche.org

Additional copies of this publication are available from:

Curran Associates, Inc.
57 Morehouse Lane
Red Hook, NY 12571 USA
Phone: 845-758-0400
Fax: 845-758-2633
Email: curran@proceedings.com
Web: www.proceedings.com

TABLE OF CONTENTS

625621 An Optimization Framework for the Evaluation of Energy-Water Systems for Agricultural Applications

Abstract:

The food-energy-water nexus (FEWN) dictates that food, energy and water are interconnected resources, and consequently, when decisions focusing on one resource are investigated, the other ones need to be considered. Therefore, the upcoming challenges of an evergrowing population, decreasing water sources, together with the impact of climate change can be tackled sustainably by the FEWN [1]. Further, arid, and semiarid regions are traditionally characterized by water scarcity. Here, technology-based water supply systems, i.e., water purification via RO desalination, can pose opportunities to harness saline water sources by utilizing varying energy supply systems for various applications [2, 3]. Additionally, there is an essential need to ensure the sustainable supply of food for a growing population [4]. However, FEWN studies have been conducted for various objectives, yet they have not been widely extended to irrigation planning [5].

Therefore, this work investigates the suitability and sustainability of energy-water options to meet varying water demands in agriculture systems. To do so, renewable, and non-renewable based energy sources, as well as varying sources of water are considered. Applicable water sources consist not only of ground or surface water, but also of purified water via RO desalination. A surrogate model is developed to incorporate all aforementioned alternatives, resulting in a mixed-integer non-linear (MINLP) optimization program. The MINLP can be optimized for an array of objectives, not only minimizing operational cost, but also satisfying a specified water demand as defined by the agricultural system. Additionally, the process costs of agricultural systems for maximization of the water utilization can be compared. Further, alternative water supply scenarios can be analyzed to identify trade-offs between competing irrigation and agriculture strategies. Thus, this work formulates a decision framework regarding energy-water systems, to fulfill agricultural scenarios, with minimal process costs, as well as maximum water utilization.

[1] R. Cory Allen, Yaling Nie, Styliani Avraamidou, Efstratios N. Pistikopoulos. Infrastructure Planning and Operational Scheduling for Power Generating Systems: An Energy-Water Nexus Approach, Computer Aided Chemical Engineering, 47, 233-238, 2019. https://doi.org/10.1016/B978-0-12-818597-1.50037-0

[2] Marcello Di Martino, Styliani Avraamidou, Julie Cook, Efstratios N. Pistikopoulos. An Optimization Framework for the Design of Reverse Osmosis Desalination Plants under Food-Energy-Water Nexus Considerations, Desalination, 503, 2021. https://doi.org/10.1016/j.desal.2021.114937

[3] Marcello Di Martino, Styliani Avraamidou, Efstratios N. Pistikopoulos. Superstructure Optimization for the Design of a Desalination Plant to Tackle the Water Scarcity in Texas (USA), Computer Aided Chemical Engineering, 48, 763-768, 2020. https://doi.org/10.1016/B978-0-12-823377-1.50128-2

[4] Sarah Namany, Tareq Al-Ansari, Rajesh Govindan. Optimisation of the energy, water, and food nexus for food security scenarios, Computers and Chemical Engineering, 129, 2019, https://doi.org/10.1016/j.compchemeng.2019.106513

[5] Fatima-zahra Lahlou, Hamish R. Mackey, Gordon McKay, Udeogu Onwusogh, Tareq Al-Ansari.

Water planning framework for alfalfa fields using treated wastewater fertigation in Qatar: An energy-water-food nexus approach, Computers and Chemical Engineering, 141, 2020, https://doi.org/10.1016/j.compchemeng.2020.106999

Copyright © American Institute of Chemical Engineers. All rights reserved.

631639 Practical Considerations of a Moringa Seed Coated Filter for Water Purification

Abstract:

Moringa seeds offer a local and sustainable solution for water treatment processes in areas where waterborne illnesses are prevalent. In this process, the cationic proteins found in the seeds adsorb onto negatively-charged surfaces (e.g. sand) and capture bacteria and viruses from water. Previous work showed that lab scale filters can remove 8-log bacteria and 7-log viruses using small sand sizes. The purpose of this talk is to discuss practical considerations for the scale-up of these filters to be used in the field such as modeling the scale-up, removing organic matter that will block available sites and using available sand and other packing material.

Copyright © American Institute of Chemical Engineers. All rights reserved.

632762 Fully Functional Large-Area Nanoporous Single Layer Graphene Membranes for Desalination and Water Purification

Abstract:

An ideal desalination membrane should exhibit minimum thickness to maximize water permeance and narrow
pore size distribution for efficient ionic /molecular separations. Monolayer graphene with uniform distribution of
high-density of sub-nanometer pores has been considered as the perfect material offering ultra-fast water
permeance and high solute rejection. However, scalable production of graphene membranes with control over
sub-nanometer pores remains challenging. Here, we report the fabrication of large-area nanoporous graphene
membrane for high-efficiency water desalination. The density of defects that manifest as small nanopores in
graphene is significantly increased via facile and scalable oxidative etching of the graphene lattice. Novel
interfacial polymerization is used to seal tears and unwanted defects in graphene. The resulting centimeterscale
monolayer graphene membranes with high-density of sub-nanometer nanopores not only allow high
water permeance (upto 23 times higher than conventional membranes), but also show an excellent rejection
of salt ions and organic molecules (~100% rejection). Our work provides a facile and scalable approach for
atomically thin desalination and nanofiltration membranes.

References:
Cheng et al. NanoLetters 2020
Kidambi et al. Adv. Mat. 2018
Kidambi et al. Adv. Mat. 2018
Kidambi et al. ACS App. Mat. Int. 2018
Kidambi et al. Adv. Mat. 2017
Kidambi et al. Adv. Mat. 2017
Kidambi et al. Nanoscale 2017

Copyright © American Institute of Chemical Engineers. All rights reserved.

631523 The Investigation of Industrial Crop Production in Arid Regions

Abstract:

Arid regions cover 35% of the land area in the world and is known by water scarcity, high temperatures, low vegetation and high salinity soil (Tchakerian, 2015). This extreme environment make any agriculture activity difficult. Moreover, arid regions tend to use desalination for water production and rely on synthetic fertilizer for soil enhancement, which creates intense industrial power and chemical production that result in large carbon dioxide footprint. This work explores the synergy between industrial waste resources such as wastewater, bio-sludge to assess the possibility of crop production. The focus is on industrial non-food crops that are typically used in the manufacturing industry, such as cotton, rubber, etc. Industrial crops give flexibility to use industrial byproducts and can aid land reclamation. This work falls under the circular economy by creating integration to eliminate waste and create value added products.

Tchakerian, V.P., 2015, Hydrology, Floods and Droughts-Deserts and Desertification, Encyclopedia of Atmospheric Sciences, 184-192. DOI: 10.1016/B978-0-12-382225-3.00035-9

Copyright © American Institute of Chemical Engineers. All rights reserved.

626038 Cobalt (II) Activated Sodium Percarbonate for the Degradation of Tetracycline Antibiotics By Using Reactive Oxygen and High Valent-Cobalt Oxo Species.

Abstract:

The present study is one of the alluring advance oxidation processes (AOPs) involving the reaction of sodium percarbonate (SPC) and cobalt (II) generating reactive oxygen species (ROSs) i.e., hydroxyl radicals, superoxide radicals and high-valent cobalt oxo used for the destruction of chlorinated organic solvents (COSs). The contribution and influence of these different ROSs in the process of degrading the pollutants is unclear and unanswered till now. The production and existence of the cobalt oxo was confirmed in acidic media as significant intermediately oxidant. Methyl phenyl sulfoxide was used as probe compound. The scavengers tests were also performed to evaluate the presence of hydroxyl and superoxide radicals whereas, most of the ROSs converted from cobalt oxo to reactive oxygen radicals, thus increased the pH values of the solution and molar ratio of the SPC/Co (II). The degradation of the pollutants was found highly dependent on the substrate reactivity and the availability of the present number of ROSs in the system. Cobalt oxo has a higher rate of reaction affinity with organic substrates having least ionization potentials. On the other hand, contrary to the free radicals' oxidation, cobalt oxo showed high resistance to humic acid and other ions in the process. This research can open new horizons for the degradation of chlorinated organic pollutants (COSs) using cobalt (II) activated SPC System.

Copyright © American Institute of Chemical Engineers. All rights reserved.

The emergence of novel viruses, including SARS-CoV-2, and prevalence of difficult-to-culture and non-culturable viruses in the environment has increased interest in developing predictive models for viral inactivation during disinfection. To develop these models, a mechanistic understanding of how disinfectants inactivate viruses, and which disinfectant species drive inactivation, is desirable. Chlorination is used widely to inactivate viruses in water, and chlorine inactivates viruses by oxidizing viral components, including nucleic acids. To date, molecular biologists have assumed that hypochlorous acid/hypochlorite HOCl/OCl- is the chlorinating agent that initiates nucleic acid damage, and chlorination rates determined for viral genomes are grounded in that assumption. However, recent studies have underscored that minor constituents of chlorine, molecular chlorine (Cl_2), and dichlorine monoxide (Cl_2O), which exist in equilibrium with HOCl/OCl-, can influence chlorination rates of organic compounds, especially at elevated chloride concentrations (> 3 mM) in which they are favored to form. Here, we show that the first-order chlorination rates of dsDNA and ssDNA extracted from representative bacteriophages are enhanced in the presence of chloride. To our knowledge, this is the first study to show that chlorination rates of biological macromolecules are affected by chloride concentrations present at levels relevant to water disinfection. We further show that these enhanced chlorination reaction rates can be explained by the presence of Cl_2 and Cl_2O in solution. The results of this study will inform kinetic models for viral genome chlorination, and bring us one step closer toward accurate viral inactivation predictive models.

628362 Catalytic Capacitive Deionization for Adsorption and Reduction of Aqueous Nitrate

Abstract:

The growing global population is driving up demand for fresh water, meanwhile scarce drinking water sources are increasingly being polluted with anthropogenic discharges from agricultural waste and industrial processes. Capacitive deionization (CDI) is an environmental remediation technology that can remove wastewater contaminants through physical separation, but it produces concentrated waste streams that necessitate further processing and increase operational complexity. Heterogenous catalysis is a promising opportunity to degrade toxic wastewater contaminants, but the kinetics for degradation can be slow as it is limited by mass transport. We report a hybrid approach called catalytic capacitive deionization (CCDI) that combines the advantages of capacitive deionization and heterogeneous catalysis in a single continuous flow process. Porous electrodes that contain catalytic nanoparticles are used in a flow-through CCDI device, in which ionic contaminants are electrosorbed onto the electrode and subsequently degraded to harmless by-products. We demonstrate CCDI through the removal and reduction of nitrate in water towards innocuous nitrogen gas. A feed stream containing 50 mg-N/L nitrate was continuously fed to the CCDI reactor at 1 mL/min. The nitrate was removed and treated through a 20-minute combined adsorption/catalytic cycles. Treated effluents at voltages above 1.3 V contained 2 mg-N/L nitrate along with trace amounts of nitrite and ammonium by-products below maximum concentration levels defined for drinking purpose. We found that our CCDI system could reduce a significant fraction of the nitrate in the feed stream, and at voltages of 1.3 V or greater the major byproduct was N_2. For example, at a cell potential of 1.5 V, the CCDI system removed 92% of the nitrate in the feed, and 91% was selectively transformed to produce innocuous N_2. An energy efficiency analysis demonstrates that this technology is more competitive than other water treatment methods. This work demonstrates that catalytic CDI can potentially be used as a novel approach to treat a wide-range of aqueous ionic contaminants, significantly advancing the next generation of CDI systems.

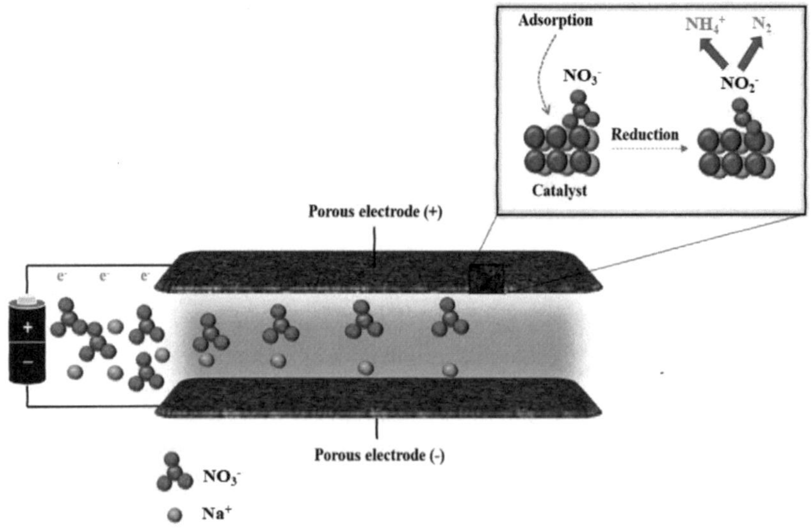

Copyright © American Institute of Chemical Engineers. All rights reserved.

Effective Municipal Wastewater Treatment at Low-Cost Using Coagulation/Precipitation Followed by Nano-Disinfection

Mohamed K. Mostafa[1], Ahmed S. Mahmoud[2], Mohamed T. Kamar[3], Maha M. Elshfai[4], M. Moussa[3,5], and Robert W. Peters[6]

[1]Faculty of Engineering and Technology, Badr University in Cairo (BUC), Cairo, Egypt.

[2]Scientific Research Development unit, Egyptian Russian University (ERU), Egypt

[3]Environmental Engineering Program, Zewail City of Science and Technology, Giza, Egypt

[4]Sanitary and Environmental Institute (SEI), Housing and Building National Research Center (HBRC), Egypt

[5]Environmental Engineering, Faculty of Engineering Mataria, Helwan University, Egypt

[6]Department of Civil, Construction and Environmental Engineering, University of Alabama at Birmingham, Birmingham, AL, USA

Abstract

Conventional wastewater treatment plants suffering from several industrial effluents overlap, trihalomethane (byproduct after free chlorine disinfection), and high construction cost. This study successfully prepared and compared a novel mixing of coagulants ferric/ferrous chloride (50:50) with other traditional individual coagulants such as ferric chloride, ferrous chloride, and aluminum chloride for domestic wastewater treatment. The results obtained indicated that the mixture of ferric/ferrous chloride is more efficient than other coagulants to meet the requirements of the Egyptian code for reuse of treated wastewater in different agricultural applications. Also, this study successfully replaced traditional chlorine addition by low-cost green synthesis of nano-Zero Valent Iron (GT-nZVI) for tertiary wastewater treatment. A continuous flow system was successfully built to achieve the treatment requirements in an economical and fast way. Finally, this study strongly supported using the proposed system rather than traditional wastewater treatment technologies in decentralized wastewater treatment units to overcome the combination of industrial effluents to the municipal wastewater networks and to reduce hazardous material produced after using traditional chlorine disinfectant.

1. Introduction

Wastewater is the most observable environmental problem among many in urban areas (Ngoc and Schnitzer, 2009, Mahmoud et al., 2021b). The increasing world population increases water consumption and produces a huge amount of wastewater every day. Nowadays about 2.4 billion people in urban and rural areas don't have contact with adequate sanitation services (Daily and Ehrlich, 1992, Gleick, 2000, Rijsberman, 2006, Mostafa and Peters, 2015). There is an expectation of increasing the demand for sanitation by 2 billion will live in towns and cities, mainly in developing countries within 20 years (Langergraber and Muellegger, 2005, Mostafa and Peters, 2017, Mostafa and Peters, 2016b). More than 90% of developing country's sewage is discharged untreated and polluting lakes, rivers, and coastal areas (Islam and Tanaka, 2004, Liu and Diamond, 2005, M Abdeldayem et al., 2020). Conventional sanitation techniques, based on water-wasting technology, flush toilets, aren't a suitable solution for industrial and developing countries (Weng, 2002, Del Porto, 2005, Diaz and Barkdoll, 2006, Jason A. Heberling, 2018, Hamdy et al., 2019). The sanitation systems were built and designed on the premise that human excreta are a waste suitable for disposal into the environment and it can assimilate this waste (Esrey and Andersson, 1999, Langergraber and Muellegger, 2005). Sanitation systems that provide Ecological Sanitation (EcoSan) is a cycle of a sustainable closed system, can close the gap between sanitation and agriculture (Flores et al., 2009, Morsy et al., 2020). The EcoSan strategy is resource control and represents a complete concept towards ecological and economically stable sanitation (Winblad et al., 2004, Simha and Ganesapillai, 2017, Mostafa and Peters, 2016b). The main point is to decrease nutrient and water circles with less expense in material and energy to provide sustainable development (Brown and Lugo, 1994, Pruss-Ustun and Organization, 2008). So, the EcoSan system is an alternative technique to circumvent the disadvantages of conventional wastewater systems (Langergraber and Muellegger, 2005, Mostafa et al., 2017).

This study attempts to explain and determine the wastewater treatment by using different coagulation techniques followed by nano adsorption and disinfection techniques. The effects of coagulant selection were studied by using the most common coagulants used for wastewater treatment proposes. Also, the selected coagulant was optimized by studying the effect of pH and dose. Finally, advanced nano adsorption and disinfection techniques were studied by using low-cost GT-nZVI sorbent material and the effect of operating conditions was optimized by studying the effect of pH, dose, time, and stirring rate.

2. Material and Methods

2.1. Chemicals and Reagents

The following chemicals were used in this study: Ferric chloride ($FeCl_3.6H_2O$, Extra pure, Loba Chemie), soft black tea, Ethanol (C_2H_6O, 99% pure, Chem lab), Sodium hydroxide (NaOH, 99% pure, Oxford Co.), Sulfuric acid (H_2SO_4, 96%, Fisher chemical), Ferrous chloride anhydrous ($FeCl_3$, Extra pure, Loba Chemie), Zinc sulfate ($ZnSO_4$, 99.9, Winlab) and standard 5% sodium hypochlorite solution.

2.2. Samples collection

the samples were collected from a definite location at 30.12707, 31.70291, Badr University, Cairo, Egypt. 48 composite samples were collected on 7&8th April 2021. The collected samples were introduced in the chemistry lab for examination of water and wastewater, Housing and Building National Research Centre (HBRC), Egypt to be examined and treated using the selected materials according to the author's request.

2.3. Preparation of GT-nZVI

The extraction of reducing agents from soft black tea as well as reduction of metal cation from its soluble salt was adapted from previous studies (Mahmoud et al., 2019a, Mahmoud et al., 2020a).

2.4. Batch studies

In the first experiment, the coagulation process was done after mixing wastewater sample with seven coagulants namely $AlCl_3$, $Al_2(SO_4)_3$, $FeCl_3$, $FeCl_2$, $FeCl_3$/ $FeCl_2/H_2O_2$, $FeCl_3$/ $FeCl_2/Cl_2$ and $ZnSO_4$. The coagulation process was conditioned at 3 min in the jar test with stirring speed 150 rpm and settled for an hour. After complete precipitation, the adsorption process was proceeded by using GT-nZVI. The operating conditions were designated based on earlier studies (Mahmoud et al., 2020a, Mahmoud and Mahmoud, 2021, Rabie S. Farag and Peters, 2019). Table 1 summarizes the batch experiments at different operating conditions pH, dose(g), contact time (min), and stirring rate (rpm). A known selected weight of GT-nZVI 0.5g was equilibrated with 1000 mL of partially treated wastewater sample in Erlenmeyer flasks and shacked for a known period at room temperature. After equilibration, the suspension of the adsorbent was separated using Whatman filter paper No. 2, and the remaining (COD, BOD, TSS, TN, and TP) concentrations were measured according to standard methods for the examination of water and wastewater 23rd edition (Baird, 2017). The percentage of removal efficiency was calculated using Eq. 1 (Mahmoud et al., 2021a, Rabie S. Farag and Peters, 2018). The amount of sorbed contaminants were calculated using Eq. 2 (Maha M. Elshfai and Elsaid, 2021, Mahmoud et al., 2021a, Rabie S. Farag and Peters, 2018). All experiments were performed in triplicate and all results were within lab measurements (Mahmoud et al., 2018).

$$Sorption\ (\%) = \left(\frac{C_0 - C_e}{C_0}\right) \times 100 \qquad (1)$$

where C_o is the initial concentration (mg/L) and C_e is the equilibrium concentration in solution (mg/L).

$$Q_e(mg/g) = \frac{(C_0 - C_e)V}{m} \qquad (2)$$

where q_e is the equilibrium adsorption capacity (mg/g), V is the volume of aqueous solution (L), and m is the dry weight of the adsorbent (mg) (Mahmoud et al., 2020b, Mahmoud et al., 2019b).

Table 1. Batch experiments at different materials and different operating parameters.

Effect	Dose (g)	Contact time (min)	pH	Stirring rate (rpm)	Temperature (°C)
$AlCl_3$, $Al_2(SO_4)_3$, $FeCl_3$, $FeCl_2$, $FeCl_3$/ $FeCl_2$/H_2O_2, $FeCl_3$/ $FeCl_2$/Cl_2 and $ZnSO_4$					
Material selection	Dose (1g/L)	3min mixing/20min flocs formation/60min settling	pH (7.12)	----	25
$FeCl_3$/ $FeCl_2$/H_2O_2					
Effect of pH	0.5	3min mixing/20min flocs formation/60min settling	3-10	-----	25
Effect of adsorbent dose	0.1-0.8	3min mixing/20min flocs formation/60min settling	7	-----	25
GT-nZVI					
Effect of pH	0.7	60	6-9.5	250	25
Effect of adsorbent dose	0.1-1.0	60	8	250	25
Effect of contact time	0.7	10-80	8	250	25
Effect of stirring rate	0.7	60	8	100-400	25

3. Result and dissections

3.1. Characterization

The XRD pattern as shown in Figure (1a) shows two peaks at $2\theta = 44.58°$ and $64.99°$ indicated the formation of alpha iron structure 110 and 200 planes (Mahmoud et al., 2017). The XRD analysis also verified that there are other iron oxides, carbide, iron sodium phosphate, magnesium iron hydride, iron-chromium oxide, iron (III) borate, and ferrimolybdite formations. Figure (1b) shows the SEM image of the prepared GT-nZVI particles before any treatment process. The GT-nZVI showed an irregular surface structure with an average size of about 83nm. Also, many pores were observed, which help to mass transfer of soluble low molecular weight organic compounds to the inner iron nanoparticles. Figure (1c) shows the EDAX analysis for the selected nanoparticles at 20 mm, indicated the formation of nano-iron particles with outer oxide and carbon and oxygen layers and other elements which come from the extraction solution (Mostafa and Peters, 2016a).

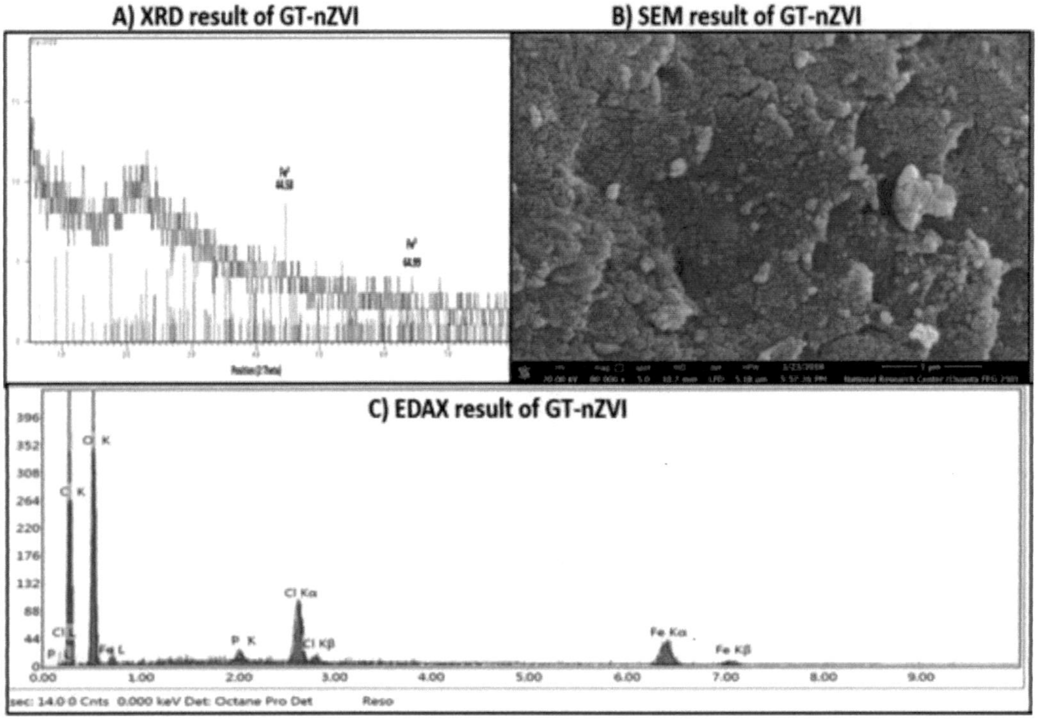

Figure 1: Characterization of GT-nZVI A) XRD of the freshly prepared GT-ZVI, B) SEM of GT-nZVI and C) EDAX results of GT-nZVI.

3.2. Selecting the coagulant

Different chemical treatment techniqes were tested for wastewater treatment by using traditional chemical coagulants and fenton coagulants. The obtained results indicated that coagulation techniqe was effective for contaminants remediation. The results indicated that the $AlCl_3$, $Al_2(SO_4)_3$, $FeCl_3$, $FeCl_2$, $FeCl_3/FeCl_2/H_2O_2$, $FeCl_3/FeCl_2/Cl_2$ and $ZnSO_4$ was able to reduce the carbonaceous and nitrogenous organic contaminants (COD, BOD, TSS, Oils, and TN), phosphorous contaminants (TP), and sulfur (H_2S) contaminants. Figure 1a shows the reduction of COD, BOD, TSS, and TN concentrations by a percent of 57.2, 69.1, 75.4, 61.6, 77.6, 73.1, and 50.1% for COD, 60.1, 71.2, 66.2, 65.1, 79.0, 75.1, and 50.9% for BOD, 59.8, 67.8, 61.1, 65.1, 79.1, 71.8, and 51.5% for TSS and 30.2, 27.7, 42.9, 40.4, 60.7, 56.6, and 18.4% for TN after using $AlCl_3$, $Al_2(SO_4)_3$, $FeCl_3$, $FeCl_2$, $FeCl_3/FeCl_2/H_2O_2$, $FeCl_3/FeCl_2/Cl_2$ and $ZnSO_4$, respectively. Figure 1b shows the reduction of TP, oils and sulfides by a percent of 26.2, 36.7, 35.7, 38.4, 71.8, 55.4, and 3.3% for TP, 85.4, 89.3, 92.5, 94.2, 97.3, 96.7, and 15.0% for Oils and 39.1, 51.3, 56.3, 53.8, 51.8, 76.1, and 7.6% for H_2S after using $AlCl_3$, $Al_2(SO_4)_3$, $FeCl_3$, $FeCl_2$, $FeCl_3/FeCl_2/H_2O_2$, $FeCl_3/FeCl_2/Cl_2$ and $ZnSO_4$, respectively. Finally, the obtained results indicated that $FeCl_3/FeCl_2/H_2O_2$ is able to degrade high level of carbon, nitrogen, phosphorus and sulfur contaminants to fulfill the Egyptian laws.

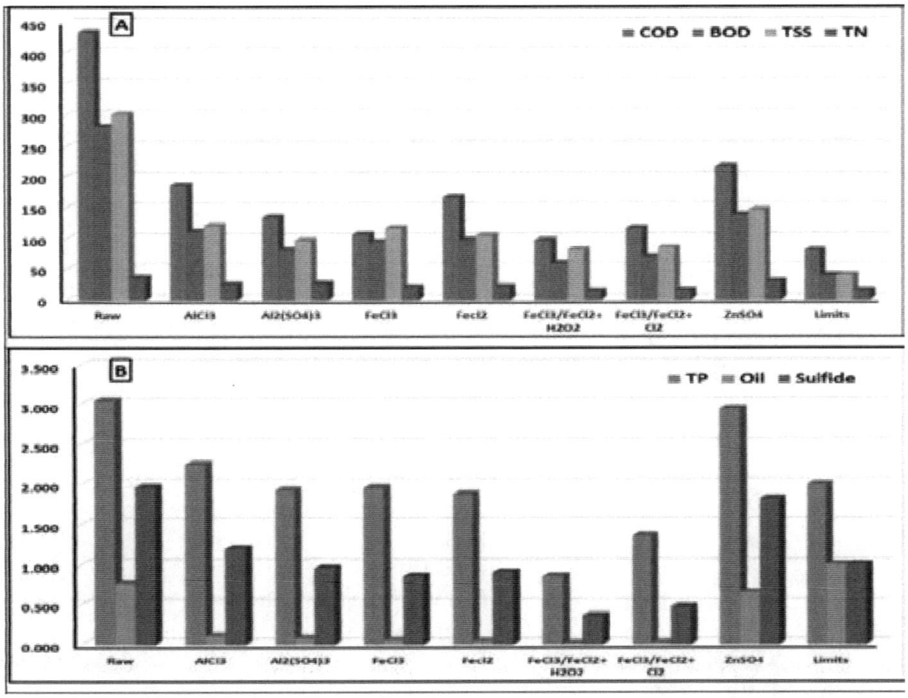

Figure 2: Effect of different coagulants on wastewater treatment A) Reduction of COD, BOD, TSS, and TN concentrations, and B) Reduction of TP, oils, and total sulfide concentrations.

3.3. Selecting optimum conditions for the coagulation process

3.3.1. Effect of pH

The effect of pH was studied at different pH values ranged between 3 and 10 for wastewater treatment by using $FeCl_3/FeCl_2/H_2O_2$ Fenton coagulation process. Table 1 and Figure 3 show the wastewater treatment removal %. The obtained results indicated that $FeCl_3/FeCl_2/H_2O_2$ was able to degrade and precipitate a wide range of wastewater contaminants at neutral pH between 7 and 8. Also, results indicated that the coagulation process by $FeCl_3/FeCl_2/H_2O_2$ was able to precipitate and degrade a wide range of organic contaminants at acidic and alkaline media. Finally, the selection of pH values was dependent on both, the flocs formation and best removal efficiency.

Table 1: Effect of pH on wastewater treatment by using $FeCl_3/FeCl_2/H_2O_2$

pH	3	4	5	6	7	8	9	10
COD removal %	71.3	73.6	74.5	75.6	77.6	77.2	76.1	74.5
BOD removal %	74.4	75.1	76.2	77.6	79.0	79.0	77.9	76.5
TSS removal %	73.4	74.4	76.1	78.1	79.1	78.4	77.4	77.4
Fe removal %	87.7	88.6	88.7	90.4	90.6	90.6	91.4	91.4
Al removal %	100.0	100.0	100.0	100.0	100.0	100.0	100.0	100.0
TP removal %	55.4	63.6	66.4	68.9	71.8	70.2	67.4	65.2
TN removal %	52.7	50.8	52.2	52.7	60.7	57.1	56.6	55.5
Zn removal %	68.8	72.9	72.9	75.0	83.3	81.3	81.3	77.1
Phenol removal %	100.0	100.0	100.0	100.0	100.0	100.0	100.0	100.0
CN removal %	52.6	57.9	68.4	68.4	78.9	89.5	89.5	89.5
Oil removal %	94.5	94.5	95.4	96.2	97.3	97.0	96.2	95.0
Sulfide removal %	37.6	37.1	39.5	43.1	51.8	51.8	48.7	45.7

An et al.2014 studied orthophosphate removal by using chemical precipitation by using ferrous chloride and the obtained results indicated that the maximum removal percentage was 99% occurred at pH 8 showing agreement with the obtained results (An et al., 2014). Mei et al., 1998 studied control of continuous flow wastewater treatment by using coagulation technique by ferrous chloride and hydrogen peroxide and the results indicated that slightly acidic media pH 5 can perform faster coagulation (Syu and Chen, 1998). Quasim et al., 1992 studied the coagulation process of ferric ions for TOC and turbidity removal, and the obtained results showed effective removal at alkaline pH (Qasim et al., 1992)

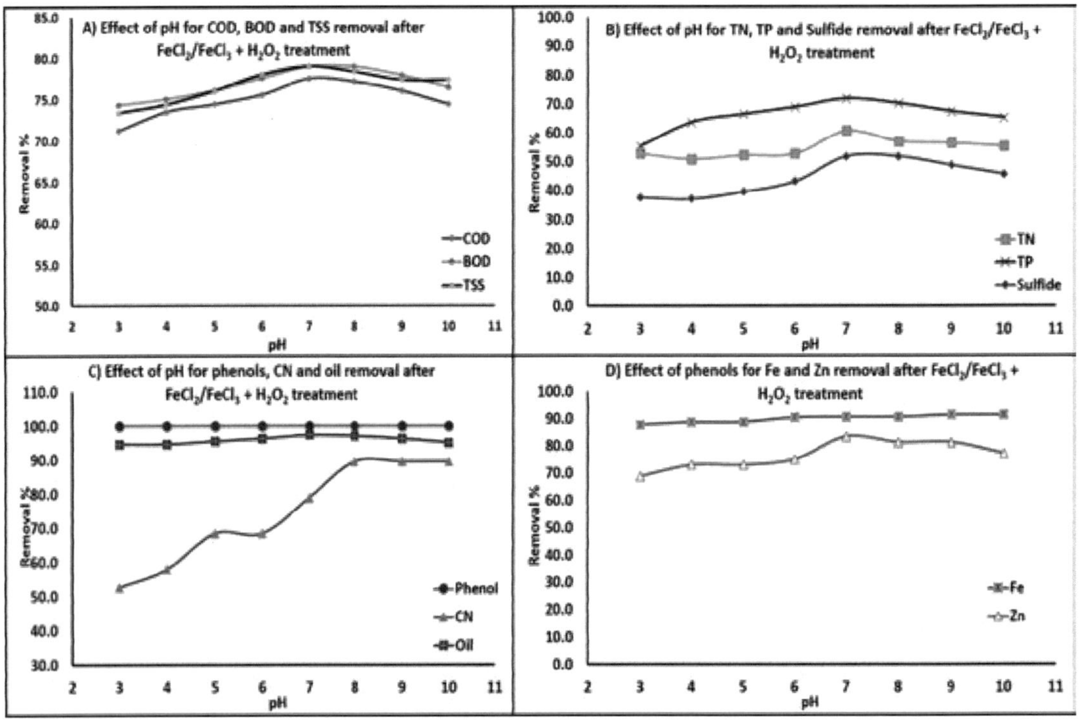

Figure 3. Effect of pH on wastewater treatment by coagulation technique using $FeCl_3/FeCl_2/H_2O_2$ A) Reduction of COD, BOD, and TSS concentrations, B) Reduction of TP, TN, and total sulfide concentrations, C) Reduction of Phenols, Cyanides, and oils concentrations, and D) Removal of Iron and Zinc concentrations.

3.3.2. Effect of coagulation dose

The effect of dose was studied by using different mixed 50:50:1 from $FeCl_3/FeCl_2/H_2O_2$ using mixed doses namely 0.1, 0.2, 0.3, 0.4, 0.5, 0.6, 0.7 and 0.8. The obtained results indicated that the minimum effective dose was 0.5g/L from mixed coagulant. Figure 4a describes the removal of the most common carbonaceous organic contaminants represented in COD, BOD, and TSS showed effective removal of 77.6, 79, and 79.1%, respectively. Also, nitrogen, phosphorous, and sulfur contaminants showed accepted removal percentages were 60.7, 71.8, and 51.8%, respectively as shown in figure 4b. Finally, small concentrations of oil, CN, phenols, Iron (Fe), and Zinc (Zn) can be founded in the university wastewater and can be eliminated by the minimum coagulant dose as shown in Figures 4c and 4d.

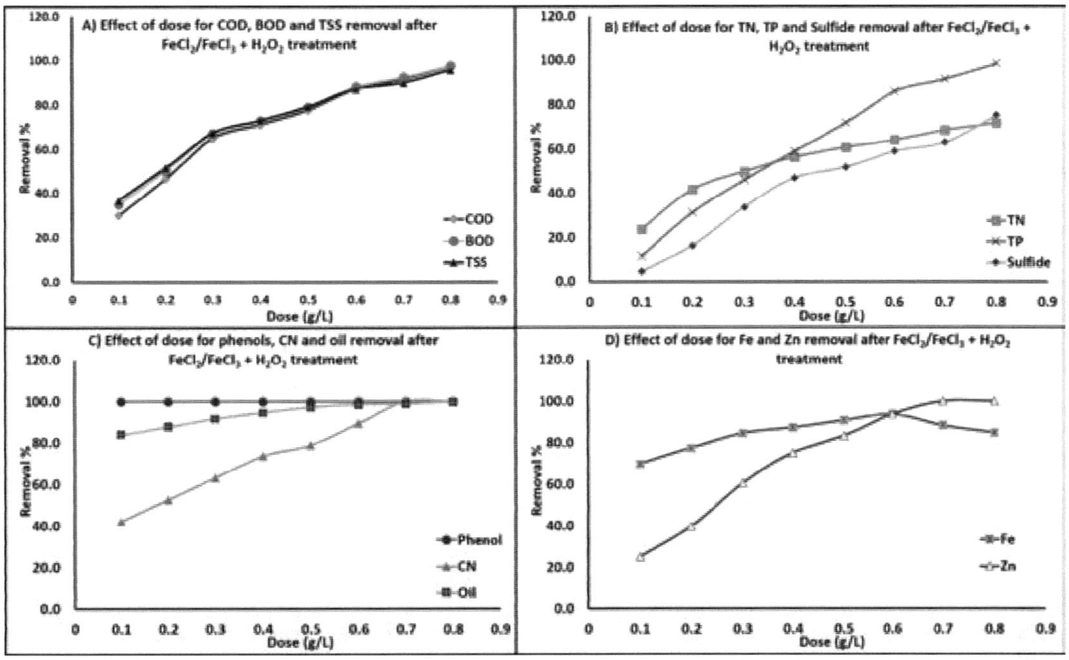

Figure 4. Effect of coagulant dose on wastewater treatment A) Reduction of COD, BOD, and TSS concentrations, B) Reduction of TP, TN, and total sulfide concentrations, C) Reduction of Phenols, Cyanides, and oils concentrations, and D) Removal of Iron and Zinc concentrations.

3.4. Selecting optimum operating conditions for the GT-nZVI treatment process

Previously chemically prepared nZVI and GT-nZVI showed effective removal for different contaminants removal at specific optimum conditions and can be applied for advanced wastewater treatment techniques instead of other traditionally disinfectants (Mahmoud and Mahmoud, 2021, Mahmoud et al., 2021a, Rabie S. Farag and Peters, 2019, Farag and Peters, 2018, Farag et al., 2018). The most effective operating conditions are pH, dose, time, and stirring rate effects (Maha M. Elshfai and Elsaid, 2021, Mahmoud et al., 2019a). The effect of temperature is not an effective parameter for wastewater treatment and the effect of concentrations can be applied only for contaminants removal in synthetic water (Mahmoud et al., 2018, SaryEl-deen et al., 2017).

3.4.1. Effect of pH

The effect of pH was studied at allowable final drain pH as shown in Figure 5a. The obtained results the most effective pH occurred at pH 8. The effective removal percentages that occurred for COD, BOD, TSS, TN, and TP organic contaminants were 68.2, 72.9, 73.0, 32.9, and 65%, respectively. The previous studies showed effective removal percentages at pH ranging between 7 and 8, showing agreement with the obtained results (Mahmoud et al., 2020a, Mahmoud et al., 2018, Mostafa et al., 2017, El-Shafei et al., 2016).

3.4.2. Effect of dose

The effect of GT-nZVI doses was studied using different doses as shown in Figure 5b. The results indicated that the removal efficiency increased with dose and the effective dose was 0.7g/L. As dose increase, the vacant sites also increase which enable to increase the adsorption mechanism into these vacant sites. Finally, the selected GT-nZVI dose showed agreement with the previous studies. Farag et al., 2018 studied aqueous phosphate chemisorption onto GT-nZVI and the obtained results indicated that the effective dose to decrease 66% from orthophosphate 10mg/L solution was 0.6g/L (Rabie S. Farag and Peters, 2018).

3.4.3. Effect of contact time

The effect of contact time was studied using different times namely at 10, 20, 30, 40, 50, 60, 70, and 80min as shown in Figure 5c. The results indicated that the equilibrated time was 60 min showing agreement with the previous studies. Mahmoud et al., 2019 studied the reduction of organic matter represent in COD and BOD using GT-nZVI at low cost and the obtained results indicated that effective removal efficiency was equilibrated at 60min showing agreement with the obtained results.

3.4.4. Effect of stirring rate

The effect of stirring rate was studied at 100, 150, 200, 250, 300, 350, and 400 rpm as shown in Figure 5d. the obtained results indicated that the minimum effective stirring rate was 250 rpm. The prepared GT-nZVI showed accumulation after formation. These accumulations need to separate to increase the surface area of nanoparticles and increase the adsorption and degradation process. Several studies were conducted for wastewater treatment by adsorption process and the obtained results indicated that the stirring rate effect ranged between 150 and 250 rpm showing agreement with the obtained results.

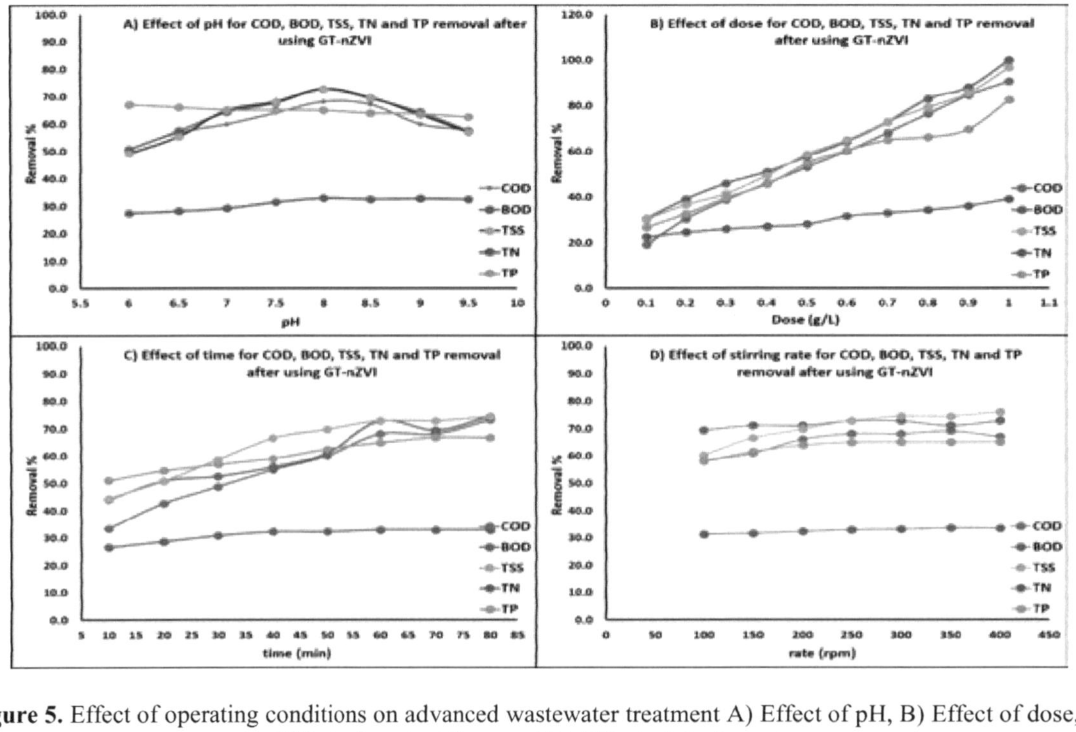

Figure 5. Effect of operating conditions on advanced wastewater treatment A) Effect of pH, B) Effect of dose, C) Effect of contact time, and D) Effect of stirring rate.

Cost estimation for the proposed sytems

The operating expenses (OPEX) was calculated based on four main parameters: (1) coagulant and adsorbent materials (Scenario 1: Al_2SO_4 and aluminum NPs) or (Scenario 2: ferric/ferrous chloride (50:50) and green synthesis of nano-Zero Valent Iron (GT-nZVI)), (2) energy need to operate the system, (3) maintenance costs, and (4) labour cost. The energy cost was computed as \$0.00054/m^3 using an electricity pricing of \$0.10/kWh. The labour cost was computed as \$0.1/m^3, while the maintenance cost for the whole system was estimated as \$0.0037/m^3. The total OPEX was calculated as \$2.3/m^3, in case of applying first scenario one time (Table 1). However, the aluminum NPs can be recovered and reused for at least 3 times, which is expected to reduce the cost to about \$0.8/m^3. The total OPEX was calculated as \$0.6/m^3, in case of applying second scenario one time (Table 1).

Table 1. OPEX of the proposed treatment systems.

Raw materials	Amount	Price (\$)	1m^3	Cost (\$) for first scenario	Cost (\$) for second scenario
Operating Cost					
Al_2SO_4 or ferric/ferrous chloride (50:50)	1 Ton	200	630g	0.126	0.126
C_2H_5OH (Pure life Co.)	1000 mL	0.65	300mL	0.097	0.097
Sodium Boron Hydride	100 g	2	100 g	2	X
Green tea	100 g	0.3	100 g	X	0.3
Lind Berg hotplate (400°C)	2400 Watt	0.1488	2 h at 200°C	0.0744	0.0744
Dryer (EHRET) (300°C)	1400 Watt	0.0434	1h at 150 °C	0.0217	0.0217
Wisestir Magnetic stirrer	200 Watt	0.0062	1h	0.0062	0.0062
Centrifuge PLC-012E	1200 Watt	0.0372	1h	0.037	0.037
GLT 6 jar test (300 rpm)	19 Watt	0.00158	1h	0.00158	0.00158
Total Dollars for each m^3				**2.3**	**0.6**

Conclusion

This study successfully studied advanced wastewater treatment using traditional Fe (III) coagulant, Advanced Oxidation Process (AOP) by using Fenton coagulant together followed by nanotechnology on a bench scale. The selection of nanomaterial was performed according to cost, efficiency, and ease of preparation. This study successfully prepared and characterized GT-nZVI for advanced wastewater treatment. The effect of $AlCl_3$, $Al_2(SO_4)_3$, $FeCl_3$, $FeCl_2$, $FeCl_3/FeCl_2/H_2O_2$, $FeCl_3/FeCl_2/Cl_2$ and $ZnSO_4$ coagulants individually were studied. The effective coagulant that was observed to fulfill the Egyptian code of reuse was $FeCl_3/FeCl_2/H_2O_2$. The effect of coagulation pH and dose were studied and the obtained results indicated that 0.5 g of coagulant at pH 7 can degrade and precipitate more than 77% for carbonaceous organics, more than 60 and 70% from phosphorous and nitrogen organic loads, respectively, and more than 50% from the sulfur organic load. The advanced level of the treatment process was proceeded by adding 0.7 g/L of freshly prepared GT-nZVI at pH 8, contact time 60 min, and stirring rate 250 rpm. The selected optimum conditions were done by studying the effect of the operating parameter (pH, dose, time, and stirring rate). Finally, this model can degrade, precipitate and adsorb a wide range of wastewater contaminants to fulfill advanced treatment requirements.

Acknowledgements

The authors gratefully acknowledge the financial support given by the Joint Academy of Scientific Research and Technology (ASRT)/the Bibliotheca Alexandrina (BA) Research Grants (Grant Number: 1469).

References

AN, J.-S., BACK, Y.-J., KIM, K.-C., CHA, R., JEONG, T.-Y. & CHUNG, H.-K. 2014. Optimization for the removal of orthophosphate from aqueous solution by chemical precipitation using ferrous chloride. *Environmental technology,* 35, 1668-1675.

BAIRD, R. B. 2017. Standard methods for the examination of water and wastewater, 23rd. Water Environment Federation, American Public Health Association, American ....

BROWN, S. & LUGO, A. E. 1994. Rehabilitation of tropical lands: a key to sustaining development. *Restoration Ecology,* 2, 97-111.

DAILY, G. C. & EHRLICH, P. R. 1992. Population, sustainability, and Earth's carrying capacity. *BioScience,* 42, 761-771.

DEL PORTO, D. 2005. Urban and industrial watersheds and ecological sanitation: two sustainable strategies for on-site urban water management. *Water Crisis: Myth or Reality?*, 285.

DIAZ, J. & BARKDOLL, B. Comparison of wastewater treatment in developed and developing countries. World Environmental and Water Resource Congress 2006: Examining the Confluence of Environmental and Water Concerns, 2006. 1-10.

EL-SHAFEI, M., MAHMOUD, A., MOSTAFA, M. & PETERS, R. Effects of entrapped nZVI in alginate polymer on BTEX removal. AIChE Annual Meeting. San Francisco, CA, 2016. 13-18.

ESREY, S. & ANDERSSON, I. Environmental Sanitation from an Eco-Systems Approach. Proceedings of the Workshop on Ecological Sanitation. Mexico, 1999.

FARAG, R. S., ELSHFAI, M. M. & MAHMOUD, A. S. 2018. Adsorption and kinetic studies using nano zero valent iron (nZVI) in the removal of chemical oxygen demand from aqueous solution with response surface methodology and artificial neural network approach. *Journal of Environment & Biotechnology Research,* 7, 12-22.

FARAG, R. S., ELSHFAI, M. M., MAHMOUD, A. S., MOSTAFA, M. K., & PETERS, R. W. Green synthesis of nano iron carbide: Preparation, characterization and application for aqueous phosphate removal. aper presented at the Environmental Division 2018 - Core Programming Area at the 2018 AIChE Annual Meeting, 307-317, 2018.

FLORES, A., BUCKLEY, C. & FENNER, R. 2009. Selecting sanitation systems for sustainability in developing countries. *Water science and Technology,* 60, 2973-2982.

GLEICK, P. H. 2000. *The world's water 2000-2001: the biennial report on freshwater resources*, Island Press.

HAMDY, A., MOSTAFA, M. K. & NASR, M. 2019. Techno-economic estimation of electroplating wastewater treatment using zero-valent iron nanoparticles: batch optimization, continuous feed, and scaling up studies. *Environmental Science and Pollution Research,* 26, 25372-25385.

ISLAM, M. S. & TANAKA, M. 2004. Impacts of pollution on coastal and marine ecosystems including coastal and marine fisheries and approach for management: a review and synthesis. *Marine pollution bulletin,* 48, 624-649.

JASON A. HEBERLING, Y. G. A., AHMED S. MAHMOUD, MOHAMED K. MOSTAFA, ROBERT W. PETERS. AOP Performance at Wastewater Treatment Plants. ". Proc. 2018 Annual AIChE Meeting, Pittsburgh, PA, (October 28 – November 2). 2018.

LANGERGRABER, G. & MUELLEGGER, E. 2005. Ecological Sanitation—a way to solve global sanitation problems? *Environment international,* 31, 433-444.

LIU, J. & DIAMOND, J. 2005. China's environment in a globalizing world. *Nature,* 435, 1179.

M ABDELDAYEM, O., ELDAGHAR, O., K MOSTAFA, M., M HABASHY, M., HASSAN, A. A., MAHMOUD, H., M MORSY, K., ABDELRADY, A. & PETERS, R. W. 2020. Mitigation plan and water harvesting of flashflood in arid rural communities using modelling approach: A case study in Afouna village, Egypt. *Water,* 12**,** 2565.

MAHA M. ELSHFAI, A. S. M. & ELSAID, M. A. Comparative Studies of using nZVI and Entrapped nZVI in Alginate Biopolymer (Ag/ nZVI) for Aqueous Phosphate Removal. 12th international conference on nano-technology for green and sustainable construction, 2021. https://www.researchgate.net/publication …, 21.

MAHMOUD, A. S., FARAG, R. S. & ELSHFAI, M. M. 2020a. Reduction of organic matter from municipal wastewater at low cost using green synthesis nano iron extracted from black tea: Artificial intelligence with regression analysis. *Egyptian Journal of Petroleum,* 29**,** 9-20.

MAHMOUD, A. S., FARAG, R. S., ELSHFAI, M. M., MOHAMED, L. A. & RAGHEB, S. M. 2019a. Nano zero-valent aluminum (nZVAl) preparation, characterization, and application for the removal of soluble organic matter with artificial intelligence, isotherm study, and kinetic analysis. *Air, Soil and Water Research,* 12**,** 1178622119878707.

MAHMOUD, A. S., ISMAIL, A., MOSTAFA, M. K., MAHMOUD, M., ALI, W. & SHAWKY, A. M. 2020b. Isotherm and kinetic studies for heptachlor removal from aqueous solution using Fe/Cu nanoparticles, artificial intelligence, and regression analysis. *Separation Science and Technology,* 55**,** 684-696.

MAHMOUD, A. S., MOHAMED, N. Y., MOSTAFA, M. K. & MAHMOUD, M. S. 2021a. Effective Chromium Adsorption From Aqueous Solutions and Tannery Wastewater Using Bimetallic Fe/Cu Nanoparticles: Response Surface Methodology and Artificial Neural Network. *Air, Soil and Water Research,* 14**,** 11786221211028162.

MAHMOUD, A. S., MOSTAFA, M. K. & ABDEL-GAWAD, S. A. 2018. Artificial intelligence for the removal of benzene, toluene, ethyl benzene and xylene (BTEX) from aqueous solutions using iron nanoparticles. *Water Supply,* 18**,** 1650-1663.

MAHMOUD, A. S., MOSTAFA, M. K. & NASR, M. 2019b. Regression model, artificial intelligence, and cost estimation for phosphate adsorption using encapsulated nanoscale zero-valent iron. *Separation Science and Technology,* 54**,** 13-26.

MAHMOUD, A. S., MOSTAFA, M. K. & PETERS, R. W. 2021b. A prototype of textile wastewater treatment using coagulation and adsorption by Fe/Cu nanoparticles: Techno-economic and scaling-up studies. *Nanomaterials and Nanotechnology,* 11**,** 18479804211041181.

MAHMOUD, A. S., SARYEL-DEEN, R. A., MOSTAFA, M. K. & PETERS, R. W. Artificial Intelligence for Organochlorine Pesticides Removal from Aqueous Solutions using Entrapped nZVI in Alginate Biopolymer. 2017 Annual AIChE Meeting; Minneapolis, MN, October 29 - November 3, 2017, 2017.

MAHMOUD, M. & MAHMOUD, A. S. 2021. Wastewater treatment using nano bimetallic iron/copper, adsorption isotherm, kinetic studies, and artificial intelligence neural networks. *Emergent Materials***,** 1-9.

MORSY, K. M., MOSTAFA, M. K., ABDALLA, K. Z. & GALAL, M. M. 2020. Life cycle assessment of upgrading primary wastewater treatment plants to secondary treatment including a circular economy approach. *Air, Soil and Water Research,* 13**,** 1178622120935857.

MOSTAFA, M. & PETERS, R. 2017. Reuse paint wastewater in the manufacture of cement bricks and tiles. *Journal of Material Cycles & Waste Management,* 19.

MOSTAFA, M. & PETERS, R. W. 2016a. A comprehensive assessment of water quality at the Rosetta branch of the Nile River, Egypt. *Journal of Civil Engineering and Architecture,* 10**,** 513-529.

MOSTAFA, M. K., MAHMOUD, A. S., SARYEL-DEEN, R. A. & PETERS, R. W. Application of Entrapped Nano Zero Valent Iron into Cellulose Acetate Membranes for Domestic Wastewater Treatment. 2017 Annual AIChE Meeting; Minneapolis, MN, October 29 - November 3, 2017, 2017.

MOSTAFA, M. K. & PETERS, R. W. 2015. Use river pollutant modeling to simulate and predict the change in the Damietta Branch water quality before and after construction of the Ethiopian Dam. *Journal of Environmental Protection,* 6, 935.

MOSTAFA, M. K. & PETERS, R. W. 2016b. Improve effluent water quality at Abu-Rawash wastewater treatment plant with the application of coagulants. *Water and Environment Journal,* 30, 88-95.

NGOC, U. N. & SCHNITZER, H. 2009. Sustainable solutions for solid waste management in Southeast Asian countries. *Waste management,* 29, 1982-1995.

PRUSS-USTUN, A. & ORGANIZATION, W. H. 2008. Safer water, better health: costs, benefits and sustainability of interventions to protect and promote health.

QASIM, S., HASHSHAM, S. & ANSARI, N. 1992. TOC removal by coagulation and softening. *Journal of Environmental Engineering,* 118, 432-437.

RABIE S. FARAG, M. M. E.-S., AHMED S. MAHMOUD, MOHAMED K. MOSTAFA, & PETERS, R. W. Green Synthesis of Nano Iron Carbide: Preparation, Characterization and Application for Removal of Phosphate from Aqueous Solutions. Proc. 2018 Annual AIChE Meeting, Pittsburgh, PA, (October 28 – November 2). 2018.

RABIE S. FARAG, M. M. E., AHMED S. MAHMOUD, MOHAMED K. MOSTAFA, AHMED KARAM, & PETERS, R. W. Study the Degradation and Adsorption Processes of Organic Matters from Domestic Wastewater using Chemically Prepared and Green Synthesized Nano Zero-Valent Iron. 2019 Annual AIChE Meeting; Orlando, FL, November 10 - 15, 2019, 2019. https://www.researchgate.net/publication ….

RIJSBERMAN, F. R. 2006. Water scarcity: fact or fiction? *Agricultural water management,* 80, 5-22.

SARYEL-DEEN, R. A., MAHMOUD, A. S., MAHMOUD, M., MOSTAFA, M. K. & PETERS, R. W. Adsorption and Kinetic Studies of using Entrapped Sewage Sludge Ash in the Removal of Chemical Oxygen Demand from Domestic Wastewater, with Artificial Intelligence Approach. 2017 Annual AIChE Meeting, 2017.

SIMHA, P. & GANESAPILLAI, M. 2017. Ecological Sanitation and nutrient recovery from human urine: How far have we come? A review. *Sustainable Environment Research,* 27, 107-116.

SYU, M.-J. & CHEN, B.-C. 1998. Back-propagation neural network adaptive control of a continuous wastewater treatment process. *Industrial & engineering chemistry research,* 37, 3625-3630.

WENG, C. N. A critical review of Malaysia's accomplishment on water resources management under agenda 21. Proceedings of International Conference on Environmental Management, Centre for Graduate Studies, National University of Malaysia (UKM), Bangi, 2002. 315-331.

WINBLAD, U., SIMPSON-HÉBERT, M., CALVERT, P., MORGAN, P., ROSEMARIN, A., SAWYER, R., XIAO, J., RIDDERSTOLPE, P., ESREY, S. A. & GOUGH, J. 2004. *Ecological sanitation.*

622142 Degradation of Phenol Derivatives Using Advanced Oxidation Process in Batch and Continuous Film Reactors

Abstract:

Abstract

Advanced oxidation processes (AOPs) constitute a promising technology for the treatment of wastewaters containing refractory organics like nitrophenols which form a group of special interest due to their high toxicity and low biodegradability. The AOP techniques investigated are (i) Photolysis by UV irradiation (ii) UV assisted photo-oxidation [UV/H_2O_2], (iii) UV assisted photocatalysis [UV/ TiO_2 /H_2O_2] (iv) UV assisted photocatalytic oxidation [UV/TiO_2/H_2O_2] and (iv) Photo-Fenton [UV/H_2O_2/Fe^{2+}] oxidation. These processes were investigated in a batch photoreactor and continuous flow cascade film reactor. The relative efficiency of these processes and the influence of process parameters like pH, pollutant concentration and photocatalyst dosing on the overall degradation of the substrate are evaluated.

The obtained results show that the photo-Fenton process has maximum efficiency in oxidative degradation of nitrophenols followed by UV assisted photo-oxidation, UV assisted photocatalytic oxidation, UV assisted photocatalysis and photolysis. The photo-Fenton oxidation process gave the maximum degradation efficiency of greater than 95% in both batch and continuous reactors. The obtained results show that the studied AOPs had maximum efficiency degradation at pH = 4.6. The pollutants concentration decay profile has been used to determine the kinetic parameters which indicates that the processes follows first order process.

Copyright © American Institute of Chemical Engineers. All rights reserved.

623094 Bioelectrochemical Sulfate Reduction Enhanced Nitrogen Removal from Industrial Wastewater Containing Ammonia and Sulfate

Abstract:

Many industries discharge wastewater containing sulfate (SO_4^{2-}) and ammonium (NH_4^+). Low organic carbon/nitrogen ratio and existing SO_4^{2-} in industrial wastewater limited nitrogen removal efficiency in traditional wastewater treatment processes. Sulfide (S^{2-}) autotrophic denitrification coupled with heterotrophic sulfate reduction provided an ideal strategy. However, limited electron donors still hindered efficient nitrogen removal. Herein, a microbial-electrochemical sulfate reduction was coupled with sulfide autotrophic denitrification and heterotrophic sulfate reduction to achieve high nitrogen removal efficiency. Benefitting from the microbial-electrochemical sulfate reduction with a biocathode, the total nitrogen (TN) removal efficiency was increased from ~25% to the maximum of ~ 85% with inputting -45 mA electric current. With the help of supplying electrons to denitrification through SO_4^{2-} reduction, Columbic efficiency was also improved to 61.5%. Through controlling the input current and recirculation rate to increase the turnover numbers (TONs) of SO_4^{2-}-S, high TN removal efficiency can be obtained at various concentration of SO_4^{2-}. Furthermore, experimental investigations revealed that coupling microbial-electrochemical sulfate reduction could relieve S^0 deposition and enhance the sulfur recovery, which also contributed to improve TN removal efficiency. Moreover, microbial community analysis showed the main functional bacteria of each process in the coupled system were *Desulfatirhabdium, Thiobacillus, Nitrospira and Desulfovibrio* respectively. Treating real industrial wastewater, the coupled system also achieved high TN removal efficiency (~75%). Overall, the coupled system provides electrons for the denitrification process through microbial-electrochemical sulfate reduction to obtain a superior TN removal performance, which offered a new approach for efficient nitrogen removal of industrial wastewater containing SO_4^{2-}.

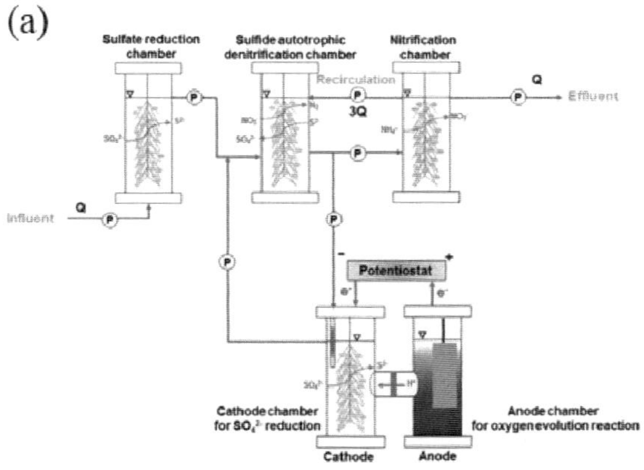

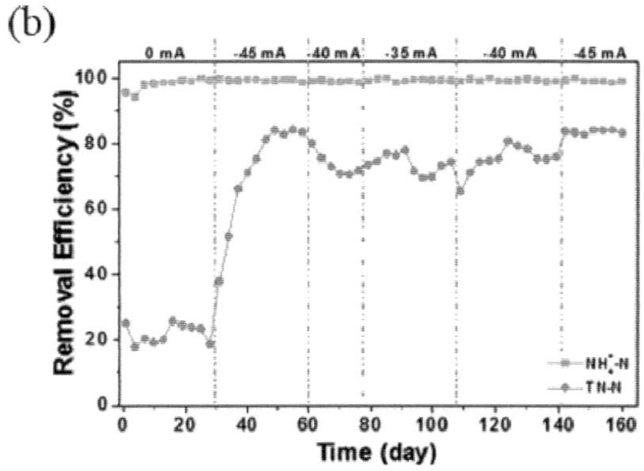

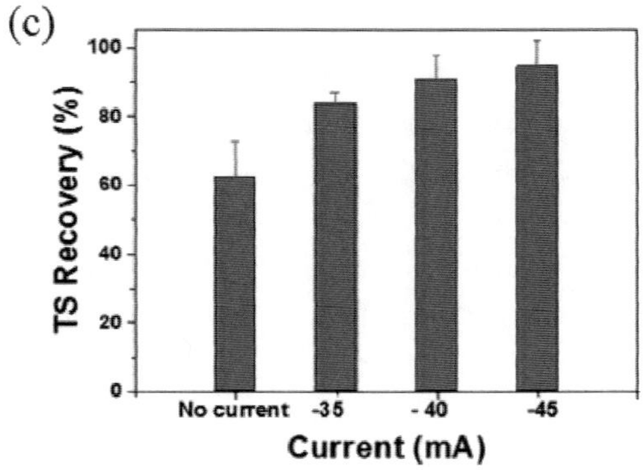

(a) Schematic diagram of the coupled system. (Q: flow.)

(b) NH_4^+-N and TN removal efficiency with different input currents (HRT of cathode chamber = 18 h).

(c) TS recovery with different input currents (HRT of cathode chamber = 18 h).

Copyright © American Institute of Chemical Engineers. All rights reserved.

623245 Metal Single-Atom Catalysts Driven Heterogeneous Electro-Fenton Technology for Fast Elimination of Refractory Contaminants

Abstract:

Fenton reaction represents the promising technology to tackle severe water environment pollution issues via super-active ·OH production, due to the high reaction rate, non-toxic reagents, and environmentally friendly process. However, it suffers from the low atomic efficiency, poor pH adaptability, and catalyst non-reusability in a homogeneous or heterogeneous system. Herein, metal single-atom catalysis is introduced to drive heterogeneous electro-Fenton for ·OH evolution utilizing earth-abundant oxygen and water as raw materials. An iron single-atom settled by defective three-dimensional porous carbon (FeSA/3DPC) with unique unsaturated coordination is designed to tune multi-electron oxygen reduction processes, enabling O_2-to-·OH conversion. The isolation feature of iron atoms and the unique unsaturated coordination are identified via atom resolved scanning transmission electron microscopy and extended X-ray absorption fine structure analysis. The prepared FeSA/3DPC catalyst can dramatically catalyze O_2 or H_2O_2 into ·OH in circumneutral solution, delivering almost two-orders of magnitude higher mass activity in ·OH productivity compared to nanoparticular iron oxide catalyst. The FeSA/3DPC electro-Fenton system exhibits fast elimination of refractory toxic pollutants(

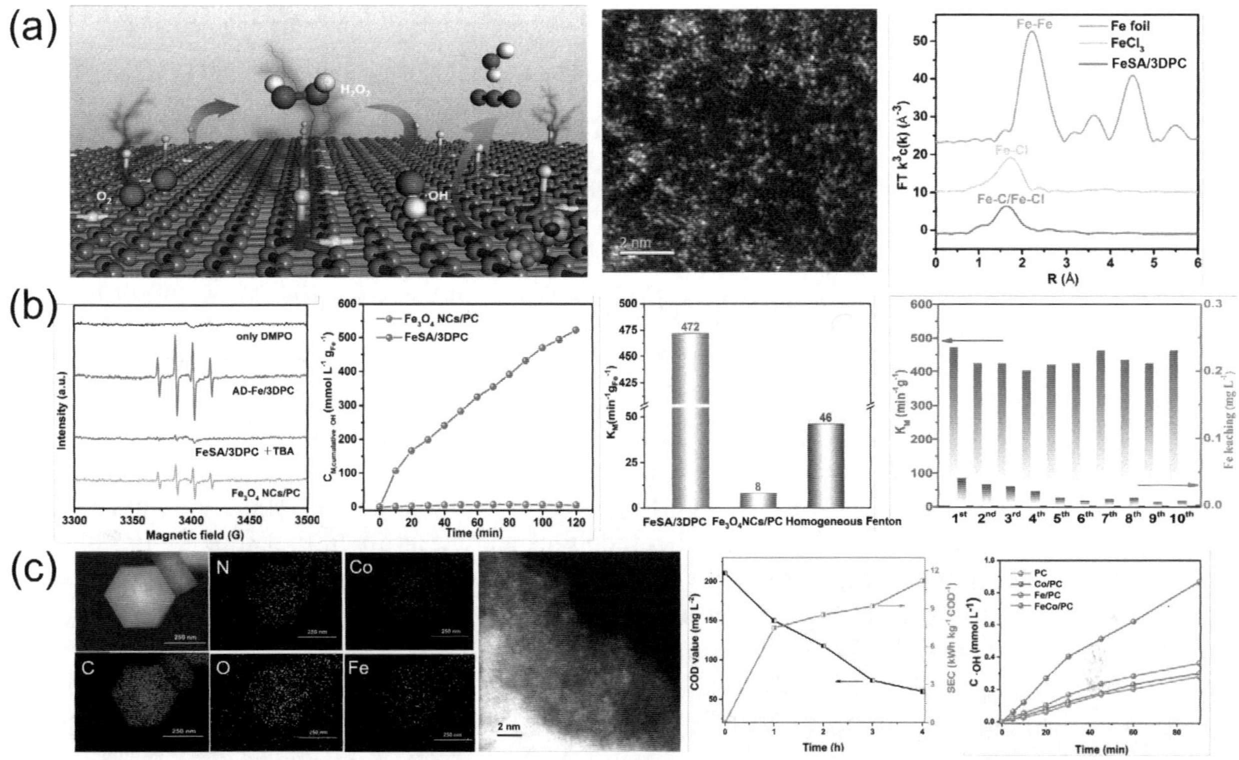

Figure 1. Fe single-atom catalyst: (a) Schematic diagram of electrochemical ·OH evolution from O_2 over monodispersed C, Cl dual-coordinated iron atom and the confirm of single-atom iron atom by atom-resolved scanning transmission electron microscopy and extended X-ray absorption fine structure analysis; (b) The qualitative and quantitative detection of ·OH by spin-trapping ESR spectra and probe reaction using p-HBA; the comparison of mass activities in kinetic rate of SMX degradation; reusability tests for SMX degradation and Fe leaching during successive 10 times runs. (c) Fe, Co dual-atom catalyst: elements mapping of C, N, O, Fe, Co from atom-resolved scanning transmission electron microscopy; the treatment for actual refinery wastewater and electricity consumption; the quantitative detection of ·OH.

Copyright © American Institute of Chemical Engineers. All rights reserved.

626089 Ultra-Fine Heterogeneous Reduced Graphene Oxide Supported Zero Valent Iron-Copper Bimetallic Composite for the Efficient Degradation of Antibiotic Sulfamethoxazole

Abstract:

In this study, a novel catalyst i.e., ultra-fine heterogeneous reduced graphene oxide supported zero valent iron-copper bimetallic composite ZVI-Cu@rGO was prepared using wet ion exchange process for the activation of sodium percarbonate (SPC) and significant degradation efficiency was achieved using sulfamethoxazole (SMX) as a target pollutant. The characterization of ZVI-Cu@rGO was performed using Scanning electron microscopy (SEM), X-ray photoelectron spectroscopy (XPS), X-ray diffraction (XRF), Fourier transform infrared spectrometer (FTIR), Transmission electron microscopy (TEM) and Brunauer–Emmett–Teller (BET) to evaluate the psychochemical properties. The mass ratio of ZVI-Cu@rGO (1:0.5:1) indicated significant degradation efficiency (95.8%) of SMX in the first 60 minutes with a final pH value of 5.96 whereas, the removal of total organic carbon (TOC) was 79.5%. Electron spin resonance (ESR) was performed to confirm the presence of hydroxyl and superoxide radicals produced in SPC activated ZVI-Cu@rGO. The plausible reaction mechanism was also proposed using the outcomes of the experiments. Higher stability, reusability (even after 5 runs) and a wider pH range (3-8) makes this newly synthesized catalyst highly suitable for the efficient degradation of SMX.

Copyright © American Institute of Chemical Engineers. All rights reserved.

636455 Use of Environmental Sustainability Principles to Help Solve Environmental Problems of Concern Today

Abstract:

In 1987, the concept of sustainable development was first put forward in the Bruntland Report which was prepared for the World Commission on Environment and Development. The links between inequality, poverty, and environmental degradation were recognized; the report sought a way for a fairer world with less damage to the environment that supports communities. The report defined sustainable development as "Development that meets the needs of the present without compromising the ability of future generations to meet their own needs". As a consequence, environmental sustainability ensures that in meeting our needs for water, food, and shelter as well as engaging in activities that make our lives enjoyable, we don't damage the environment, deplete our natural resources, and protect global ecosystems. Important aspects tied in with address: (1) Behavioral economics; (2) Intergenerational equity; and (3) What to do with the developing world. As engineers, we will increasingly need to be considering these concepts.

Recent research activities aimed toward environmental sustainability have involved: condensate recovery in air handlers; improved building energy conservation; use of green roofs combined with cistern collection and rain gardens for storm water mitigation on campus buildings; modeling of pollutants fate and transport in the Nile River tributaries; use of electrochemical sensors in storm drains to characterize storm water runoff in urban roadways; use of vertical gardening systems and community gardens to address urban food deserts; lead contamination in drinking water systems physical/chemical treatment of hazardous wastes, particularly in the area of sonication (acoustic cavitation) and photocatalysts to treat organic contaminants; and waste plastics in the marine environment. Highlights of several of these research topics will be described in the context of environmental sustainability.

Copyright © American Institute of Chemical Engineers. All rights reserved.

636472 Advancing Methodologies for Sustainability assessment of transition to Low Carbon and Circular Economy via integration of process systems engineering and macroeconomics

Abstract:

Decarbonization and implementing Circular Economy are two major goals for sustainable development to mitigate the challenges of climate change and increasing anthropogenic wastes. Among numerous pathways proposed for this transition, it is important to evaluate the systems level economic and environmental impacts of these pathways for selection of the most sustainable pathways. Industrial Ecology (IE) is a field of study that focuses on developing methodologies and computational tools to assess large scale impact of these interventions. One key method in IE utilizes the macroeconomic Input-Output model for evaluating the impact of planned transitions such as shifts to renewable energy, bio-based production etc. However, the existing techniques do not efficiently integrate process engineering and thermodynamics knowledge in sustainability assessments, thus limiting the automation in evaluation of new technologies under development at scale. Further, lack of engineering integration and reliance on only monetary based model introduces uncertainty in analysis due to price variation and obsolete models along with missing feedback in considering sustainability criteria in design. In this talk, I will present the advancement in systems level sustainability assessment techniques by addressing these limitations through integrating process systems engineering and thermodynamics with macroeconomic Input-Output framework. Specifically, I will present novel methodologies and computational tools that will enable fast, and reproducible quantitative evaluation of systems level environmental and economic impacts of adopting emerging technologies for decarbonization and circular economy pathways.

Copyright © American Institute of Chemical Engineers. All rights reserved.

622295 Thermochemical Co-Conversion of Waste Polyolefins with Low-Rank Aromatic-Rich Hydrocarbons into an Intermediate of High-Quality Anisotropic Pitch

Abstract:

As primary packaging materials, polyolefin plastics (PP, HDPE, and LDPE) are the most widely used plastics in daily life. As a result, polyolefin plastics commonly have a useful life shorter than one month and account for more than 40% of the total waste plastic amount. However, current recycling and discharge methods (energy recovery, fractions recycled, and landfill) of these waste polyolefin plastics are considered neither economical nor environmentally friendly, because they have a high recycling cost and low financial output, and have led to a great waste of resources and serious secondary pollution. Therefore, a total thermochemical conversion of waste plastics into value-added chemicals and materials can be more attractive than conventional waste plastic recycling and discharge methods.

Even so, due to the composition of polyolefins, current thermochemical products derived from polyolefins are more like liquid fuels, which makes the process less competitive than petroleum-derived liquid fuels, especially when the environmental impacts are not considered. In contrast, thermochemical co-conversion of polyolefins and other low-rank aromatic-rich hydrocarbons (polystyrenes, lignite, and coals) provides a different opportunity. The co-conversion is similar to a thermochemical liquefaction process, where the melted hydrogen-rich polyolefins play the role of reactants, self-generated solvents, and possibly hydrogen sources, while the heated aromatic-rich hydrocarbons crack valuable aromatic-containing free radicals.

In the past, research on this type of co-thermochemical process was limited, because the co-conversion liquid products contained a high content of undesired heavy asphaltene and pre-asphaltene fractions, while the process was designed to produce light oil fractions. However, these heavy fractions were partially-hydrogenated yet still maintained abundant aromatic structures. Therefore, the co-conversion products can be used as an intermediate to generate a pitch material with a unique anisotropic texture. The type of pitch is named mesophase pitch and can be used as an excellent intermediate for the manufacture of value-added high-performance carbon materials such as carbon fibers, carbon foam, needle coke, and graphene supercapacitors.

In the research presented here, direct/catalytic thermochemical co-conversion of waste polyolefins (HDPE and LDPE) and aromatic-rich hydrocarbons (polystyrene, low-rank coal, metallurgical tar) under pressurized conditions was carried out in a 100 mL high-pressure stirring reactor. The effects of multiple operating parameters (types of aromatic-rich materials, reaction temperature, reaction time, and feedstock mass ratio) on the composition and yield of the targeted product intermediate were investigated. The obtained intermediates were further thermally converted into mesophase pitch, and the effects of the intermediate compositions and thermal treatment conditions on the carbonization performance and anisotropic texture development level were studied and will be discussed. The products were analyzed via solvent fractionation, FTIR, NMR, TGA, DMA (Dynamic Mechanical Analysis), and polarizing microscopy. Based on the characterization results, the proton transferability of the intermediates and the possibility of whether the polyolefins are a hydrogen vehicle or a hydrogen donor will be discussed. The co-conversion method produced a precursor with a high conversion yield and ideal composition, which demonstrated an encouraging potential for converting waste plastics and low-rank hydrocarbon resources into an anisotropic pitch with high quality for further manufacturing of high-value high-performance carbon materials. This pathway provides a bright prospect for simultaneously handling recycled waste plastics and manufacturing value-added chemicals and materials in the U.S.

Copyright © American Institute of Chemical Engineers. All rights reserved.

622718 Global Issues in Plastic Waste: Community Focused Management Approaches

Abstract:

Mismanaged plastic waste is a global environmental crisis, and the problem is particularly acute in economically developing countries and infrastructure lacking regions of the world. Unfortunately, plastic waste management approaches that work in industrialized economies are often inappropriate for developing ones. An implementation of circular economy principles has been put forward as a potentially successful process for reducing or eliminating plastic waste. However, the principles of the circular economy are primarily targeted towards regions with highly developed interconnected infrastructure. Collection, sorting, and transportation infrastructure are needed to facilitate reuse, recycling, and remanufacturing efforts. This infrastructure is often lacking or insufficient in developing economies. Therefore, to affect positive change regarding plastic waste management, approaches that empower local communities to take local control of plastic waste management without requiring sophisticated infrastructure are needed. This presentation will focus on collaborative efforts between university researchers and non-profit organizations to implement projects to assist communities in developing regions take local control of their plastic waste.

Copyright © American Institute of Chemical Engineers. All rights reserved.

624120 A Material Flow Analysis for Sustainable End-of-Life Plastic Management

Abstract:

Global plastic production was estimated to reach 360 million tons in 2018. The production phase begins with raw material, monomer synthesis, polymerization, followed by the compounding of chemical additives to create plastic for consumer use. However, over 90% of the spent plastics were either landfilled or incinerated, with both solutions being susceptible to releasing toxic substances and greenhouse gases into the environment. For instance, chemical additives may migrate out of the original plastic under certain conditions, contaminating its surroundings and harming the environment and human health. Continuous exposure to these chemicals over time will cause bioaccumulation of toxins, thus leading to unwanted consequences on human health. Despite these negative effects, chemical additives remain essential in plastic manufacturing for improving processability and properties during use. An improvement to the existing infrastructure for plastics end-of-life (EoL) management is needed to prevent and control chemical additive releases and potential risk and exposure. Therefore, this work aims to develop generic EoL processing scenarios to track and estimate the possible migration, emission, and release of plastic additives throughout the plastic EoL stage. A material flow analysis of the plastic life cycle was performed using available U.S. municipal solid waste data and published research to track plastics and chemical additives movement during the EoL stage management activities, including mechanical recycling, incineration, and landfilling. Our analysis identified the mass flow transfers, gas-phase emissions, chemical additive migration, and subsequent releases from post-consumer plastic materials. The potential releases, hazards, and risks identified in this work create an opportunity to design a safer closed-loop plastic recycling infrastructure to strategically handle chemical additives and support implementing sustainable materials management efforts to transform the U.S. plastic economy from linear to circular.

Copyright © American Institute of Chemical Engineers. All rights reserved.

626698 Use of Surfactants to Deink Plastics

Abstract:

If used polymer items are sorted into single component streams, one of the most, if not the most significant factor that reduces the quality of the product is the inclusion of non-polymeric materials into the recycled materials. One common non-polymeric materials, especially on flexible film used in film packaging, is ink. I will present our efforts on using surfactants to deink both flexible and rigid polyethylene-based materials. Our work has shown that high pH and cationic surfactants are most effective for deinking as measured by colorimetric measurements. The proposed mechanism involves both solubilization of the likely negatively charged polymer binder as well as wetting of the negatively charged particles. Other types of surfactant are effective at deinking as well. The tensile strength and elongation at break after deinking are significantly better than the material recycled without deinking, but still are slightly lower than the un-inked material. Finally, I will discuss parallels and differences of deinking with respect to washing of clothes as well as how surfactants would be effective in removing food, oil and grease.

Copyright © American Institute of Chemical Engineers. All rights reserved.

624605 Study of the Viscosity and Thermal Characteristics of Polyolefins / Solvent Mixtures: Applications for Plastic Pyrolysis

Abstract:

Plastic pollution is one of the biggest environmental problems that the world is currently facing. Pyrolysis is a frontier technique aimed at converting plastic waste back into virgin-quality resin. However, the transfer of the waste plastic feed into the pyrolysis reactor must be solved before the process can be up-scaled to a continuous process. Even though the effects of various solvents on the viscosity of polyolefin have been studied previously, an appropriate solvent that can prepare waste plastics as feed for pyrolysis has not yet been found. In this study, a new solvent that reduces the viscosity of molten plastic was introduced and characterized. The results revealed that the polymers are soluble in the ratio of up to 75 wt% plastic 25 wt% solvent. The viscosity of pure low-density polyethylene (LDPE), high-density polyethylene (HDPE), and polypropylene (PP) in the solvent was measured in different weight percentages (30% - 80% wt%) and at 160, 180, 200, 220, 240 and 260°C. The viscosity decreased with the decreasing polymer-weight percentage and increasing temperature. The viscosity of LDPE + solvent is much lower than for HDPE + solvent, and for PP + solvent. Also, the Differential Scanning Calorimetry (DSC) and Thermal Gravimetric Analysis (TGA) were applied to characterize the thermal behavior of LDPE, HDPE, and PP in the solvent in three different weight percentages (25%, 50% and 75wt%). The DSC results indicate that in the mixture of PP + solvent the melting point of PP decreases as the amount of solvent increases. Overall, these results indicate that the selected solvent is an effective agent to prepare waste plastics for pyrolysis.

Copyright © American Institute of Chemical Engineers. All rights reserved.

Waste Plastics and Microplastics in Africa: Negative Impacts and Opportunities

Gaber Abdellatif[1], Ahmed S. Mahmoud[2], Robert W. Peters[3], and Mohamed K. Mostafa[4]

[1]*Environmental Engineering Program, Zewail City of Science and Technology, Giza, Egypt*

[2]*Scientific Research Development unit, Egyptian Russian University (ERU), Badr, Egypt*

[3]*Department of Civil, Construction and Environmental Engineering, University of Alabama at Birmingham, Birmingham, AL, USA*

[4]*Faculty of Engineering and Technology, Badr University in Cairo (BUC), Cairo, Egypt*

Abstract

Waste plastics and microplastic pollution are two of the most pressing global issues during the past few decades, since the beginning of plastic mass production either for direct use in packaging and plastic products, or for indirect use for plastic dependent industries. The issue is even more problematic for Africa due to the late catch-up of the industrialization and urbanization boom for most of the African countries, which is usually accompanied with intensive production, consumption, and waste generation of plastics. Moreover, the growing population, cheaper costs of plastic products, and minor importance of Environmental conservation in developing countries in Africa are making it an easy choice to expand in plastic use and wastage. African countries have imported about 172 Mton of polymers and plastics valued between 1990 and 2017, and the per capita consumption of plastic in Africa in 2015 was 16 kg/capita/year. While plastic is very important for populations, inappropriate disposal or treatment of plastic waste can introduce hazardous and toxic materials to our environment and ecosystem like microplastics that usually consists of persistent polymers, and other toxic plastic additives like plasticizers, antioxidants, and stabilizers. Moreover, 79% of plastic waste worldwide is being discharged to landfills or open environments. That's why mixed plastic waste was listed as hazardous waste in the Basel Convention. Therefore, this paper provides a review of the life cycle of plastics in. Production and importing will be discussed, important usages will also be elaborated, and most importantly plastic waste production and current plastic waste management techniques will be evaluated and highlighted, and drawbacks of current strategies will be addressed. Finally, recommendations about better waste management strategies will be provided.

Keywords: Africa; Egypt; Kenya; Microplastics; Negative impacts; Opportunities; South Africa; Waste plastics

1. Introduction

Since 1950s, plastic production has seen an all-time surge, and that was comprehendible at a time of overall industrial expansion and growth in mass production. Plastic at this point has provided an excellent option for industries, being a cheap, light, easy to shape material that can be easily used for multiple purposes (United Nations, 2018). African Countries were late to catch the plastic boom. Some of them started their plastic boom in late 1990s. However, with the expected fast urbanization and growing population in Africa, the rest of the countries are to follow, and plastic importation is expected to double in Africa by 2030 (Babayemi et al., 2019). The main issue is that the capacity to manage the generated waste is not as developed as other continents, which has the financial capabilities to manage plastic waste, and already developed their management systems. It was estimated that between 1990 and 2017, the African continent has imported around 172 Mt of plastic, of which 126 Mt were in primary form, and 46 Mt were in products form, and this amount is expected to double by 2030 (Babayemi et al., 2019). There is no complete record of the life cycle of plastic in Africa like the one we did previously for the whole world, and this is an important alert that directs our attention to how important this is. Most of the African countries depend almost completely on import not local production (Babayemi et al., 2019).

The disturbing issue with plastic is that it doesn't degrade in the environment, but decompose into smaller particles called microplastics, that can enter our drinking water, and food chains when consumed by fish (Mostafa and Peters, 2016, Hamdy et al., 2019, Mahmoud et al., 2021). Microplastics are defined as plastic particles that is less than 5 mm long, and scientists have also introduced Nano-plastics term for particles less than 1 μm in length (WHO, 2019). Microplastics has primary sources, like products that intentionally use millimetric and sub-millimetric plastics including personal hygiene products and some medical tools, and the other primary source of microplastics is plastics industry, and raw materials used in it. The secondary source of microplastics is decomposition of microplastics into smaller particles (Pinto Da Costa et al., 2020). As shown in Figure 1, the main sectors in plastic waste generation in 2015, we will find that packaging sector is the main generator of waste with no comparison to any other sector (Geyer et al., 2017).

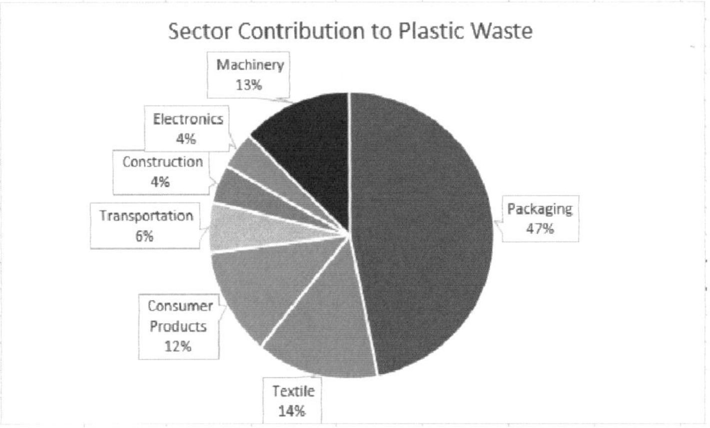

Figure 1. Plastic waste generation by sector in 2015 (Adapted from Geyer et al., 2017).

The problem with many plastic products is that a large proportion of it is designed for single use purposes then wasting, especially plastic packaging. Figure 2 shows the total production and fate of plastic from 1950 until 2015. In 2015, only 7% out of the 8.3 billion tons of plastic the world produced till that year were recycled, and 10% has been incinerated (Geyer et al., 2017). The rest of plastic waste are dumped in open areas, where they have access to the environment specially water bodies like rivers, seas and oceans. (Zaman & Newman, 2021). The exact impacts of microplastics on human health is not fully determined, yet (WHO, 2019). However, the impact of plastic waste and microplastics on our environment has been a concern as this waste end up in our oceans, polluting our water, and disturb the natural habitats of different costal and marine animals (Lebreton et al., 2017). Also, incineration in open dumps leads to air pollution and can affect local populations.

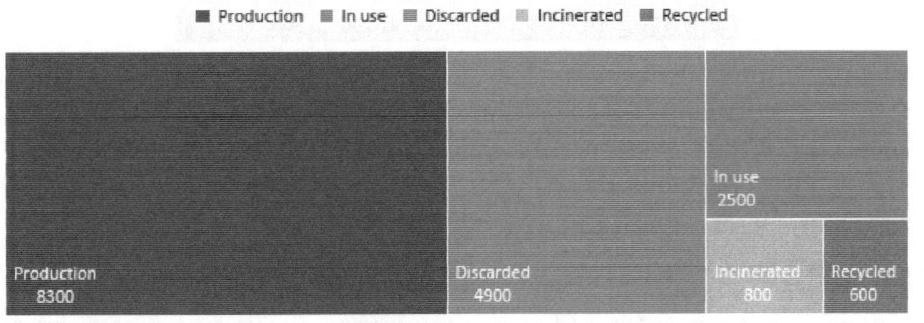

Figure 2. Total production and fate of plastic from 1950 until 2015 in million tons (Adapted from Geyer et al., 2017).

2. Methodology

This paper is meant to address the problem of waste plastics and microplastics in Africa, by conducting a thorough literature review about plastic production, main sources, life cycle and final fate in the African continent. Unfortunately, reliable data about plastic production, trading, wasting, and comparison to countries from other continents is rare to find for most of the African countries. Therefore, we have chosen three of the fastest developing countries in Africa, and of the most plastic consuming countries to be our case studies. However, we made sure to select them to be different in terms of waste management system development to show the gap, and its effect in case of plastics. These three countries are Egypt, South Africa and Kenya (Obrecht et al., 2019).

The size of the economy is important, because plastic production was found to be correlated to the GDP of the country (Babayemi et al., 2019). So, these are the countries that will have significant contribution to plastic wastes. Also, these countries have entered the market of plastic in the 90s, so they already developed a long trend in plastic production and wasting that can be used in a useful way to make reliable conclusions about their management strategies, and possible interventions.

For each country, indicators of plastic production, consumption, waste, treatment and disposal to water bodies were analyzed when available, then the current plastic waste management practices in these countries were discussed to identify the drawbacks and the gaps, and make practical recommendations about how to improve overall plastic waste management.

Multiple databases were used to find the research papers to review. The terms entered were "Micro-plastics" "waste plastics" "plastic production" and country names. The most comprehensive and recent papers, in terms of providing many indicators, and covering many countries, were then chosen to extract the data we need.

3. Analysis and Discussion

3.1. Africa

Africa is one of the fastest growing populations in the world, with a population of 1.38 billion capita, and a population growth rate of around 2.5%. At the same time, many of the African countries have achieved high economic growth rates during the past decades, and it's expected that these countries would continue growing, and that other African countries should follow due to the relatively small economies of African countries, which creates a great potential for expansion in different sectors, and the growing population that will create the demand for this growth.

Looking at the plastic situation in the African continent, we will find that the African continent depends mainly on plastic imports. Between 1990 and 2017, Africa imported 117,650 Mtons of plastic with a total of 195 M USD (Babayemi et al., 2019). The annual African consumption of plastic reached 16 kg/capita in 2015. Assuming the same value in 2017, and the population of 1.244 billion capita in the same year, the total plastic consumption of the continent would be around 20 million tons. This is only 6.7% of the global plastic consumption (D'Ambrières, 2019). Plastic waste constituted 13% of Africa's total MSW, and this is close to the global average of 12% (World Bank, 2019). However, in 2010, Sub Saharan Africa and North Africa came 3rd and 4th in the regions rank of mismanaged plastic waste contribution to the global amount, with percentages of 8.9% and 8.3%, respectively.

Table 1 summarizes plastic waste generation worldwide, in Africa and in the 3 case study countries: Egypt, South Africa, and Kenya (Khangale et al., 2020) (Sadan & de Kock, 2020) (IUCN, 2020). The data in Table 1 will be discussed in the next section.

Table 1. Plastic Waste Indicators Comparison.

Indicator	World	Africa	Egypt	South Africa	Kenya
Population in 2021 (million capita)	7900	1382	104.7	60	55.3
GDP in 2020 (Billion USD)	84540	2253	394	302	98
Plastic Consumption in 2017 (Million tons)	300	20	2.1	1.84	1.02

Annual Plastic Waste Generation 2017 (million tons)	290	-	0.97	1.2	0.5
Incinerated or Open Dumped (%)	25%	-	50%	34%	85%
Discarded or Landfilled (%)	55%	-	-	50%	-
Recycled or Reused (%)	20%	10%	50%	16%	7%
Plastic Waste contribution to total Solid Waste	12%	13%	6%	2%	20%
Annual Plastic Bags use (million bag)	500,000	-	12,000	8,000	288
Plastic Waste ending up in Water Bodies (million tons)	8	-	0.25	-	0.13

3.2. Case Studies

3.2.1. Egypt

Egypt is one of the largest producers of plastics in Africa, and this can be attributed to its large population that reached 104.7 at the time of this study (*Worldometer*, 2021). Also, Egypt has witnessed large economic activity during the last 5 year. Egypt's GDP reached 394 billion USD in 2021 with a growth rate of 3.1% (*World Bank*, 2021), and as illustrated before, this usually comes with increased demand on plastics.

In 2017, Egypt consumed 2.1 million tons of plastics, and was the largest plastic consumer in Africa. In 2016, Egypt used plastic material worth 1.6 billion USD, and the demand for plastics was predicted to increase by 7% annually, and to reach 2.3 billion USD in 2021 (Obrecht et al., 2019). However, this plastic demand comes with its costs on the environment. Nagwa El Maanawy, head of the Egyptian Plastic Technology Center, said that Egypt generates 970 thousand tons of Plastics waste, only 45% of them is recycled, 5% in reused, and 50% is incinerated. Shehab Abdel Wahab, head of the Environmental Affairs Agency, declared that Egyptians use about 12 billion plastic bags each year (*EgyptToday*, 2018). It was also reported that Egypt disposed 0.25 Mtons of plastic waste to the Mediterranean in 2019, which makes it largest contributor to plastic waste in this major sea (American Chamber of Commerce, 2020).

To solve this issue of plastic in Egypt, collaborative efforts from the government, private sector, and the public is required. Lately, multiple initiatives have been conducted by the ministry of Environment of Egypt or by key players of the private sector in the field of plastic recycling, and dependance on biodegradable and multi-use materials is needed instead of single use products, especially in packaging sector. In 2020, Egypt issued Waste Management Law #202, which creates a new regulatory authority to oversee proper waste management and recycling practices, and provides incentives to private sector and small collectors to participate in the government plan. One section of this law that grabbed much attention was about single use plastic bags reduction and finding environmentally friendly alternatives, and some coastal cities in Egypt already started banning plastic bags due to their effects on coastal areas and marine life (Business & October, 2020).

Finally, there is some substantial effort from Egypt to improve its regulatory system of SWM, providing incentives for the private sector to participate, and also in public awareness. However, more solid plans with quantitative targets needs to be made to insure continuous improvement in plastic waste management.

3.2.2. South Africa

Like Egypt, South Africa has a growing population that reached 60 million in 2021 (Worldometer, 2021) , and it's one of the largest economies in Africa with a GDP equal to 302 billion USD in 2020, despite its fall back during the last 5 years, with the growth going down to -7% last year (World Bank, 2020).

South Africa is the second largest consumer of polymer products in Africa after Egypt, with a total of 1.84 million tons of polymer products in 2017. Unfortunately, only 16% of the plastic used in South Africa is recycled, 50% is landfilled, and the rest is incinerated or dispersed with the wind, and finally end in the oceans or other water bodies. About 80% of marine waste in South Africa originates in land, and its oceans receives around 8 million tons of plastic waste annually (National Oceans and Coastal Information Management System, 2018) (Department of Environmental Affairs, 2018).

South Africa has taken many measures to control plastic pollution. These efforts included regulatory reformations and executive plans to reduce the plastic use and wasting. The Environmental Conservation act for the year 1989 included many regulations for plastic bags specifications used for different purposes, and prohibited ones (Plastic Bags Regulation, 2002). In 2020, under the South African Initiative to end plastic pollution, the Consumer Goods Council of South Africa worked on an action plan to reduce the usage of single use plastic bags. This plan had six main components: technology, innovation and design; infrastructure; bioplastics and alternatives; education and awareness in combatting litter; standards and compliance; and integrating waste pickers into the circular economy (Hanekom, 2020). Recycling has gained much focus in South Africa's SWM strategy in 2011, the strategy included the laws and regulations encouraging recycling. The strategy also detailed the institutional roles in performing recycling activities from the source point to the recycling facility (DEA, 2011). More policies have been implemented by the government to eliminate plastic waste, but from the amount of plastic waste produced currently and the modest percentage of recycled waste, we can see that the implementation of these policies faced many challenges on the ground.

3.2.3. Kenya

Kenya is one of the largest countries in Africa terms of population. It's population reached 55.3 million in 2021 (Worldometer, 2021). Starting from the 90s, Kenya has witnessed substantial economic growth. It's GDP increased from less than 10 Billion USD in 1990 to around 98 Billion USD in 2020 (World Bank, 2020).

Unfortunately, meaningful data about plastic waste in Kenya is rarely available, and this is one indicator about how primitive the situation is in Kenya in terms of plastic waste management. Kenya consumed 1.02 million tons of polymers in 2017 (Obrecht et al., 2019), and imported a total of 4.2 million tons of plastic in primary form and product form. In 2014, It was reported that around 24 million plastic bags are used monthly in Kenya, and about half of this amount is discarded to open areas, which makes the problem of plastic bags one of the top issues in waste management in Kenya (Odhiambo et al., 2014). In 2020, 20% of the total solid waste generated in Kenya was plastic waste (*World Bank*, 2020). In Mombasa, the largest coastal city in Kenya, it's estimated that 2.4 kg of plastic waste is leached to the water bodies per capita annually (*UNEP*, 2020).

This situation has evolved for several factors. First of all, industries and supermarkets produce massive amounts of plastic bags that usually overwhelm the national management system that is already weak. Also, solid waste policies are not very strict, and are not usually well implemented. Most importantly, waste transport, treatment and disposal facilities are in adequate, and the institutional capacity for waste management is very weak (Odhiambo et al., 2014). All these factors, and other factors led to the current situation in Kenya.

However, there are some positive steps that were taken lately by the Kenyan government. In 2017, single-use plastic bags were banned in a decision that grabbed the world eyes to Kenya. Moreover, a framework for plastic products production was developed between the ministry of environment and private sector to limit the overproduction of plastic packaging, and improve the quality of materials used in these products (Ministry of Environment and Forestry, 2020).

The Kenyan Solid Waste Management Strategy developed in 2018, stated that one major objective for the waste sector is to turn waste from a burden to an economic opportunity that attracts more investors, and provide more job opportunities (Ministry of Environment and Forestry, 2018). However, more effort, organization and investment in waste sector is needed to achieve the foreseen goals.

3.3. Micro-plastics in Africa

As discussed earlier, micro-plastics usually results from the decomposition of large plastic particles into microscale plastic particles that spread to our marine and terrestrial environment. Data on microplastics in Africa is also extremely limited compared to our knowledge about the global situation. This is due to the little focus from governments and researchers on the issue of micro-plastics in Africa until late in 2019 and 2020. Also, this is due to the difficulty of precisely measuring microplastic concentrations. Because, micro-plastic abundancy can be affected by other materials like sediments, or can be affected by meteorological factors like rainy seasons, when microplastic concentrations rise as reported in many places. However, other studies suggested that these correlations are not very sure, and more work needs to be done to study the real factors that affect the concentrations of micro-plastics in the environment (Alimi et al., 2021).

The concentration of microplastics in living organisms is one of the main indicators for microplastic pollution, and also for its penetration to our food systems. High concentrations in living organisms were reported in South Africa and Nigeria in insects and worms. The concentrations around 291 particle per gram wet weight of these insects. Also, high concentrations in sediments were found in the same areas, with average concentration reaching around 8000 particle per kg dry weight. The highest concentration of micro-plastic in an aquatic organism, globally, was found in Egypt's fish with a concentration of 7000 particle/fish (SH et al., 2019).

Most of the places where these samples were taken were heavily populated areas, and micro-fibers were the main components of the plastic pollution recorded in the samples taken from different places in Africa, especially Polyethylene and polypropylene. Despite, most of the studies about microplastics focus on marine life, Alimi et al. suggested that more focus in Africa needs to be given to fresh water bodies, as it receives the drainage and waste of populations around their catchments.

4. Conclusion

From the analysis made, it's clear that Africa is currently not one of the main consumers or producers of plastic products. This problem is more urgent in regions like east and south-east of Asia. However, the main concern with plastic waste in Africa is mismanagement, since Africa comes in the second place along with the Middle East in terms of mismanaged plastic waste after East and South-East Asia. However, expectation is that Africa's demand for plastic products is going to increase over time, due to the rapidly growing population, and the expected economic growth of many African countries, which is usually campaigned with a growth in the middle class creating a large market, and a surge in plastic products use. In this case, the annual mismanaged plastic waste could reach 11.5 million tons in 2025, up from 4.8 million tons in 2010.

The gap between developed economies and developing economies in East Asia and Africa in terms of mismanaged plastic waste calls our attention to the close link between the level of development of the country's economies and individuals from one side, and the level of development of its solid waste management institutions, facilities and strategies from the other side. For example, the over-burdened African countries budgets find it very hard to allocate more money to improve their waste management systems, having more pressing priorities.

Data availability about waste generation, life cycle and management is a huge problem for African countries, especially historical records. Also, there are some cases where we found large discrepancies between available sources, and some indicators were not available for the study countries in the same year to allow for comparison.

From the analysis of the 3 countries in the study, we can find that Egypt has most developed waste management system in terms of variability and comprehensiveness of efforts, while Kenya comes last due to its weak SWM system. South Africa gave special focus to plastic bags, but few efforts were achieved in the other waste sources.
Compared to the global averages, we can see that recycling and reuse in Africa in general, Kenya, and South Africa is lower than the global average. Only Egypt, has a higher recycling rate, and this number requires closer look as the percentage of mismanaged SW in Egypt is around 84% of the total SW, but plastic is a small percentage of SW, so maybe the focus in recycling in Egypt is on plastic more than other types of waste, which increased the recycled plastic to 50% of the total plastic waste in Egypt. Also, incineration and open dumping is higher in Africa and the 3 case

study countries than the global average. This confirms our previous observations about mismanagement of plastic waste. For the marine pollution, it was observed that Egypt is the main contributor to plastic waste in the Mediterranean, and Kenya with around half the population of Egypt, discard around half the amount to the ocean. Despite all of that, we observed that the ratio of plastic waste generated to plastic consumption in the 3 countries in 2017 is much lower than the global average, but to confirm this result, historical data needs to be revisited.

Africa's plastic waste problem has many aspects. First, the legal aspect, and the applicability of laws and regulations, and the ability of the government to enforce them, which in most of the cases is not achieved, or achieved in very limited places, where the institutions are strong, which takes us to the second aspect, which is institutional capacity. Last but not least, financial capability is the most important factor in building a waste management system, that's why Egypt made a good step by trying to involve the private sector and civil society in SWM, and Kenya is trying to do that, as well. This will turn the waste from a burden to an opportunity.

5. Recommendations

Our Recommendations will focus on the major problems in plastic waste management that will make some progress if solved. We will also make some suggestions that will help future research.

- African Countries, and different African Organizations should focus more on collecting plastic data, and create historical records about plastic waste generation, life cycle and management. This record can be useful to policy maker, decision makers, and researcher.
- Expectation about the progress the African countries can achieve in plastic waste management should be realistic given the financial and institutional gap.
- Since it's expected that the demand for plastic products is going to increase in Africa, more research and effort need to be done in finding alternatives to the major plastic products specially those used in packaging like single-use plastic bags. Kenya has a success story in this area that can be a model to other African countries in Africa with the same conditions.
- African governments should find a way to involve the private sector and civil society in the process of waste management, especially with the various experiences currently available in this area worldwide, to reduce the burden on the government, and create a sustainable alternative to government initiatives that faces multiple challenges over time.
- Studying the applicability of laws and regulations in the individual context of each country to be able to see a result, and gain people's respect to the law. Especially, that most of the African countries including the ones in the case studies have numerous laws that should end plastic waste, however they are struggling with the same problem for decades now.
- The connection between GDP and economic performance in General and SWM and Plastic Waste Management needs to be further analyzed to understand the capabilities and limitations of countries in this issue.

References

Alimi, O. S., Fadare, O. O., & Okoffo, E. D. (2021). Microplastics in African ecosystems: Current knowledge, abundance, associated contaminants, techniques, and research needs. *Science of the Total Environment*, *755*, 142422. https://doi.org/10.1016/j.scitotenv.2020.142422

American Chamber of Commerce, E. (2020). *IN DEPTH-Plastic in the time of pandemic - AmCham Egypt*. https://www.amcham.org.eg/publications/business-monthly/issues/295/July-2020/3975/plastic-in-the-time-of-pandemic

Babayemi, J. O., Nnorom, I. C., Osibanjo, O., & Weber, R. (2019). Ensuring sustainability in plastics use in Africa: consumption, waste generation, and projections. *Environmental Sciences Europe*, *31*(1). https://doi.org/10.1186/s12302-019-0254-5

Bank, W. (2019). *Tackling Increasing Plastic Waste*. https://datatopics.worldbank.org/what-a-waste/tackling_increasing_plastic_waste.html

Battling Kenya's Plastic Waste: Young Kenyan Woman is Transforming Waste into Sustainable and Affordable Building Materials. (n.d.). Retrieved October 3, 2021, from https://www.worldbank.org/en/news/feature/2021/03/11/battling-kenya-plastic-waste-young-kenyan-woman-transforming-waste-into-sustainable-and-affordable-building-materials

Business, L., & October, B. (2020). *Egypt ' S Waste Management Law No . 202 for the Year. 202.*

D'Ambrières, W. (2019). *Plastics recycling worldwide: current overview and desirable changes*. https://journals.openedition.org/factsreports/5102

DEA. (2011). National Waste Management Strategy. *Waste Management Document*, *1*(November), 75. https://www.environment.gov.za/sites/default/files/docs/nationalwaste_management_strategy.pdf

Egypt's initiative to eliminate plastic bags - EgyptToday. (n.d.). Retrieved September 21, 2021, from https://www.egypttoday.com/Article/1/49385/Egypt's-initiative-to-eliminate-plastic-bags

Egypt Overview: Development news, research, data | World Bank. (n.d.). Retrieved September 22, 2021, from https://www.worldbank.org/en/country/egypt/overview

Egypt Population (2021) - Worldometer. (n.d.). Retrieved September 21, 2021, from https://www.worldometers.info/world-population/egypt-population/

GDP (current US$) - Kenya | Data. (n.d.). Retrieved October 3, 2021, from https://data.worldbank.org/indicator/NY.GDP.MKTP.CD?locations=KE

GDP (current US$) - South Africa | Data. (n.d.). Retrieved September 22, 2021, from https://data.worldbank.org/indicator/NY.GDP.MKTP.CD?locations=ZA

Geyer, R., Jambeck, J. R., & Law, K. L. (2017). Production, use, and fate of all plastics ever made. *Science Advances*, *3*(7). https://doi.org/10.1126/SCIADV.1700782

Hamdy, A., Mostafa, M.K. & Nasr, M. 2019. Techno-economic estimation of electroplating wastewater treatment using zero-valent iron nanoparticles: batch optimization, continuous feed, and scaling up studies. Environmental Science and Pollution Research, 26, 25372-25385.

Hanekom, A. (2020). South African initiative to end plastic pollution in the environment. *South African Journal of Science*, *116*(5–6), 5–6. https://doi.org/10.17159/sajs.2020/8095

IUCN. (2020). National Guidance for Plastic Pollution Hotspotting and Shaping Action. *United Nations Environment Programme*, *October*, 48.

Kenya, Republic of, M. of environment and forestry. (2018). National Sustainable Waste Management Policy. *Ministry of Environement and Forestry.*

Kenya emerges as leader in fight against plastic pollution. (n.d.). Retrieved October 3, 2021, from https://www.unep.org/news-and-stories/story/kenya-emerges-leader-fight-against-plastic-pollution

Kenya Population (2021) - Worldometer. (n.d.). Retrieved October 3, 2021, from https://www.worldometers.info/world-population/kenya-population/

Khangale, U. B., Ozor, P. A., & Mbohwa, C. (2020). Plastic waste management in south africa. *Proceedings of the International Conference on Industrial Engineering and Operations Management*, *59*, 716–725.

Lebreton, L. C. M., van der Zwet, J., Damsteeg, J.-W., Slat, B., Andrady, A., & Reisser, J. (2017). River plastic emissions to the world's oceans. *Nature Communications 2017 8:1*, *8*(1), 1–10. https://doi.org/10.1038/ncomms15611

Mahmoud, A. S., Mostafa, M. K., & Peters, R. W. (2021). A prototype of textile wastewater treatment using coagulation and adsorption by Fe/Cu nanoparticles: Techno-economic and scaling-up studies. Nanomaterials and Nanotechnology, 11, 18479804211041181.

Ministry of Environment and Forestry. (2020). *REPUBLIC OF KENYA Ministry of Environment and Forestry IMPLEMENTATION PLAN FOR THE BAN OF SINGLE USE PLASTICS IN PROTECTED AREAS February 2020. February.*

Mostafa, M.K. & Peters, R.W. 2016. Improve effluent water quality at Abu-Rawash wastewater treatment plant with the application of coagulants. Water and Environment Journal, 30, 88-95.

Obrecht, M., Haddad, R. El, Elbary, R. A., Lukman, R. K., & Rosi, M. (2019). Promoting Sustainable and Circular Plastics Use in Egipt with Implementation of Ecodesign Principles. *System Safety: Human - Technical Facility - Environment*, *1*(1), 441–448. https://doi.org/10.2478/czoto-2019-0057

Odhiambo, O. R., Musalagani, A. C., Lyanda, N. J., & Ruth, S. J. (2014). the Plastic Waste Menace in Kenya : a Nairobi City Situation. *International Journal of Current Research*, *March 2002*.

Pinto Da Costa, J., Rocha Santos, T., & Duarte, A. (2020). The environmental impacts of plastics and micro-plastics use , waste and pollution: EU and national measures. *European Union*, *October*, 10–62. https://doi.org/gb5k

Plastic Bags Regulation | ELAW. (n.d.). Retrieved September 23, 2021, from https://www.elaw.org/content/south-africa-plastic-bags-regulation

Plastic Marine Pollution in South Africa – National Oceans and Coastal Information Management System. (n.d.). Retrieved September 22, 2021, from https://ocims-dev.dhcp.meraka.csir.co.za/plastic-marine-pollution-in-south-africa/

Plastic Pollution is a problem | Department of Environmental Affairs. (n.d.). Retrieved September 22, 2021, from https://www.environment.gov.za/plasticpollution

Sadan, Z., & de Kock, L. (2020). *Plastics: Facts and Futures: Moving Beyond Pollution Management Towards a Circular Plastics Economy in South Africa.* 1–134. https://wwfafrica.awsassets.panda.org/downloads/wwf_plastics_report_final_2nov2020.pdf

SH, S., M, G., & RS, M. (2019). Identification of marine microplastics in Eastern Harbor, Mediterranean Coast of Egypt, using differential scanning calorimetry. *Marine Pollution Bulletin*, *142*, 494–503. https://doi.org/10.1016/J.MARPOLBUL.2019.03.062

South Africa Population (2021) - Worldometer. (n.d.). Retrieved September 22, 2021, from https://www.worldometers.info/world-population/south-africa-population/

United Nations. (2018). The state of plastics: World Environment Day Outlook 2018. *UN Environment Programme*, 11. https://doi.org/10.1093/oxfordhb/9780199238804.003.0031

WHO. (2019). *Microplastics in drinking-water Key messages.* 1–3. https://www.who.int/water_sanitation_health/water-quality/guidelines/microplastics-in-dw-information-sheet/en/

Zaman, A., & Newman, P. (2021). Plastics: are they part of the zero-waste agenda or the toxic-waste agenda? *Sustainable Earth 2021 4:1*, *4*(1), 1–16. https://doi.org/10.1186/S42055-021-00043-8

632479 Equipment and Method for Batch Screening of Methanotroph-Microalgae Cocultures for Waste-to-Value Conversion

Abstract:

In the US, the biogas produced from anaerobic digestion (AD) of industrial, municipal and agricultural waste streams has immense potential as a renewable feedstock, with more than 7.8 million metric tons of CH_4 and over 14.4 million metric tons of CO_2 per year. However, the utilization of biogas is limited to heat and electricity production due to the presence of contaminants such as H_2S, NH_3, and volatile organic compounds (VOCs). At the same time, the digestate/wastewater from AD is rich in nitrogen, phosphorous and other contaminants. If not properly treated before released into waterways, digestate/wastewater can have detrimental impacts on the local community and environment. In fact, the excessive amount of nitrogen and phosphorus in released wastewater has caused increasingly negative consequences to our ecosystems and public health, including worsening of the greenhouse effect, reduction of the protective ozone layer, adding to smog, contributing to acid rain, and contaminating drinking water[1,2].

If biogas valorization and wastewater treatment can be integrated together through biological conversion, it will not only reduce the detrimental environmental and social impact of wastewater, but also generate revenue to offset the cost of wastewater treatment and even make the process profitable. Motivated by this idea, we have demonstrated that the methanotroph-microalgae coculture offers a highly effective platform for biogas valorization and nutrient recovery from wastewater[3,4]. This is largely attributed to the metabolic coupling of methane oxidation and photosynthesis oxygenation. In order to develop a robust and effective waste-to-value biotechnology based on the methanotroph-microalgae coculture, selection of the microbial strains that are fast growing and exhibit robust growth on wastewater would be highly desirable. To this end, we have developed a parallel screening equipment that allow us to test multiple strains simultaneously, as shown in Figure 1.

Temperature, pH, gas feed composition, and agitation affect the growth rate of microbes, so controlling these parameters is essential for screening. In addition, light intensity must be regulated for photoautotrophs. The temperature of the liquid phase was maintained at 30°C using a Type-K thermocouple connected to a heating plate. While water baths ensure a more uniform temperature distribution, the temperature gradient throughout the well-mixed solution is negligible. For pH control, most wastewater plants operate at neutral conditions. Methanotrophs produce CO_2 which decreases the pH in the liquid phase, so the addition of 1M NaOH was used to maintain a pH of 7.0 using an automated pH controller. For feed composition, each vessel was provided the same, continuous rate of feed gas throughout the experiment; methanotrophs were fed the same ratio of CH_4 and O_2 while microalgae samples were fed the same ratio of CO_2 and N_2. Agitation is necessary for a well-mixed solution, but too much agitation can disrupt optimal growth of certain cultures. The stir plates have a large stirring error of ±50 rpm. Therefore, each stir rate was chosen so that an external laser tachometer read the same stir rate for every stir plate. Although the feed gas causes uncontrollable agitation when bubbled through the liquid, the predominate source of agitation is stirring. Finally, the light intensity was consistent across all tests by using one LED light sheet set at a single intensity.

Table 1 enumerates the species of interest, which are selected based on available literature. The selected methanotrophs and microalgae species all grow readily at neutral pH conditions[5-8], which is desirable since many wastewater streams are fairly neutral. Each species will be tested in triplicate. During testing, anaerobic digestate provided by Columbus Water Works (which has AD installed) will be diluted and used for culture medium.

The efficacy of each strain was determined in three ways. The first metric was growth rate, which was calculated based on the change in optical density (OD) of the microbial solution. The optimal OD wavelength was selected based upon previous literature studies for the microbes tested. The OD was determined using ultraviolet-visible spectroscopy (UV-VIS). The second metric was to determine the extracellular biomass in the liquid phase. The extracellular polysaccharides (EPS) can be utilized as biofuel and bioplastic precursors once separated from the solid biomass. After removing cells using centrifugation, a Shimadzu total organic carbon analyzer (TOC-L) determined the concentration of EPS in the liquid phase. The last metric focuses on how effectively the microbes utilized the ammonia and phosphate present in the medium. The total nitrogen and phosphorous concentrations in the liquid phase were recorded every 24 hours.

Once the individual methanotroph and microalgae species are tested, top ranked species in terms of biomass yield and nutrient recovery performance will then be paired together and screened for synergism and biomass production. The screening results obtained from this work will provide important information future experiments, such as the choice of the coculture model systems for additional investigation, as well as for scaling up and continuous operations.

Works Cited

[1] Driscoll, C., Whitall, D., Aber, J., Boyer, E., Castro, M., Cronan, C., et al. (2003). Nitrogen pollution: sources and consequences in the U.S. northeast. *Environment* 45, 8–22. doi:10.1080/00139150309604553

[2] Galloway, J. N., Dentener, F. J., Capone, D. G., Boyer, E. W., Howarth, R. W., Seitzinger, S. P., et al. (2004). Nitrogen cycles: past, present, and future. *Biogeochemistry* 70, 153–226. doi:10.1007/s10533-004-0370-0

[3] Roberts, N., Hilliard, M., He, Q. P., & Wang, J. (2020). A Microalgae-Methanotroph Coculture is a promising platform for fuels and chemical production from wastewater. Frontiers in Energy Research, 8. doi:10.3389/fenrg.2020.563352

[4] Badr, K., Whalen, W., He, Q. P., & Wang, J. (2020). Fast and easy quantitative characterization of methanotroph-photoautotroph cocultures. *Biotech and Bioeng, 118*(2), 703-714. doi:10.22541/au.159708707.70834744

[5] AlSayed, A., Fergala, A., & Eldyasti, A. (2018). Sustainable biogas mitigation and value-added resources recovery using methanotrophs intergrated into wastewater treatment plants. *Reviews in Environmental Science and Bio/Technology, 17*(2), 351-393. doi:10.1007/s11157-018-9464-3

[6] Christenson, L., & Sims, R. (2011). Production and harvesting of microalgae for wastewater treatment, biofuels, and bioproducts. *Biotechnology Advances, 29*(6), 686-702. doi:10.1016/j.biotechadv.2011.05.015

[7] Kadir, W. N., Lam, M. K., Uemura, Y., Lim, J. W., & Lee, K. T. (2018). Harvesting and pre-treatment of microalgae cultivated in wastewater for biodiesel production: A review. *Energy Conversion and Management, 171*, 1416-1429. doi:10.1016/j.enconman.2018.06.074

[8] Zhao, W., Sun, H., Ren, Y., Wu, T., He, Y., & Chen, F. (2018). Chlorella zofingiensis as a promising strain in wastewater treatment. *Bioresource Technology, 268*, 286-291. doi:10.1016/j.biortech.2018.07.144

Figure 1: Experimental setup for the screening tests

Table 1: Microbes investigated

Methanotrophs	Microalgae
Methylococcus capsulatus	*Chlorella sorokiniana*
Methylomonas sp. LW13	*Chlorella zofingiensis*
Methylosinus trichosporium OB3b	*Chlorella vulgaris*
Methylosarcina quisquiliarum	*Chlorella kessleri*
Methylosarcina fibrata	*Scenedesmus sp.*
Methylosarcina lacus sp. LW14T	*Scenedesmus obliquus*
Methylomonas methanica S1	*Scenedesmus dimorphous*

Copyright © American Institute of Chemical Engineers. All rights reserved.

624442 Sustainability and Economic Evaluation of Tannery Wastewater Treatment Pathways Using the P-Graph Approach

Abstract:

The growth in world population has increased the pressure on freshwater demand. One of the mitigation approaches to meet this growing demand is wastewater treatment (Alasino et al., 2007; Crini and Lichtfouse, 2019; Inc et al., 2002). However, the treatment of wastewater for reuse must be given a holistic view by incorporating the techno-economic analysis and sustainability assessment simultaneously (Yenkie, 2019). The often-conflicting views of economics and sustainability associated with industrial processes can be bridged through the application of optimization methods.

The tannery industry is a major source of wastewater. It is important to treat it carefully due to the presence of high levels of contaminants, especially chromium. Treated tannery wastewater is beneficial for irrigation purposes due to the presence of nitrogen, phosphorus, potassium, and organic matter (Saxena and Bharagava, 2016). The stagewise treatment of wastewater presents a viable way of capturing all the complexities associated with the process.

The objective of this work is to provide multiple feasible pathways to tannery wastewater treatment by quantifying simultaneously the economics and sustainability. This is achieved by developing an Excel-based P-graph tool that gives the n-best feasible pathways ranked from least to highest based on cost and sustainability metrics.

In this work, treatment technologies were categorized into stages based on their efficiencies, driving forces for contaminant removal, and composition of wastewater. Further, due to the difference in efficiencies of technologies within a stage, treatment options of higher efficiency were preceded by options with lower efficiency. Through this approach, a maximal treatment structure was generated. Economic models were incorporated consisting of investment and operating costs. The sustainable process index (SPI), which has been used extensively to assess the ecological footprint of industrial processes was used as the sustainability metric (Krotscheck and Narodoslawsky, 1996; Narodoslawsky, 2015; Narodoslawsky and Krotscheck, 2004). The SPI methodology was used to evaluate the total area needed to provide a unit ($1m^3$) amount of treated wastewater. There are seven footprints associated with the SPI methodology which can be categorized into two broad areas, namely input and output. Input areas account for the resources that are consumed by the process while the output indicates that needed to embed products and emissions into the ecosphere. The P-graph approach, a graph-theoretic framework for the synthesis and optimization of process networks, was implemented by developing an Excel-based realization of software (Cabezas et al., 2018; Friedler et al., 1993; Heckl et al., 2010; Yenkie et al., 2021). The optimization problem was solved formulating a mixed-integer linear programming problem and employing the accelerated Branch-n-bound algorithm in the Excel-based P-graph tool. Figure 1 summarizes our scope of work for the tannery wastewater case study.

References

Alasino, N., Mussati, M.C., Scenna, N., 2007. Wastewater Treatment Plant Synthesis and Design. Ind. Eng. Chem. Res. 46, 7497–7512. https://doi.org/10.1021/ie0704905

Cabezas, H., Argoti, A., Friedler, F., Mizsey, P., Pimentel, J., 2018. Design and engineering of sustainable process systems and supply chains by the P-graph framework. Environ. Prog. Sustain. Energy 37, 624–636. https://doi.org/10.1002/ep.12887

Crini, G., Lichtfouse, E., 2019. Advantages and disadvantages of techniques used for wastewater treatment. Environ. Chem. Lett. 17, 145–155. https://doi.org/10.1007/s10311-018-0785-9

Friedler, F., Tarjan, K., Huang, Y.W., Fan, L.T., 1993. Graph-theoretic approach to process synthesis: Polynomial algorithm for maximal structure generation. Comput. Chem. Eng. 17, 929–942. https://doi.org/10.1016/0098-1354(93)80074-W

Heckl, I., Friedler, F., Fan, L.T., 2010. Solution of separation-network synthesis problems by the P-graph methodology. Comput. Chem. Eng., Selected Paper of Symposium ESCAPE 19, June 14-17, 2009, Krakow, Poland 34, 700–706. https://doi.org/10.1016/j.compchemeng.2010.01.019

Inc, M.& E., Tchobanoglous, G., Burton, F.L., Stensel, H.D., 2002. Wastewater Engineering: Treatment and Reuse, 4th edition. ed. McGraw Hill Higher Education, Boston.

Krotscheck, C., Narodoslawsky, M., 1996. The Sustainable Process Index a new dimension in ecological evaluation. Ecol. Eng. 6, 241–258. https://doi.org/10.1016/0925-8574(95)00060-7

Narodoslawsky, M., 2015. Chapter 3 - Sustainable process index, in: Klemeš, J.J. (Ed.), Assessing and Measuring Environmental Impact and Sustainability. Butterworth-Heinemann, Oxford, pp. 73–86. https://doi.org/10.1016/B978-0-12-799968-5.00003-8

Narodoslawsky, M., Krotscheck, Ch., 2004. What can we learn from ecological valuation of processes with the sustainable process index (SPI) — the case study of energy production systems. J. Clean. Prod., Advances in cleaner production technologies 12, 111–115. https://doi.org/10.1016/S0959-6526(02)00184-1

Saxena, G., Bharagava, R.N., 2016. Organic Pollutants in Tannery Wastewater and Bioremediation Approaches for Environmental Safety 34.

Yenkie, K.M., 2019. Integrating the three E's in wastewater treatment: efficient design, economic viability, and environmental sustainability. Curr. Opin. Chem. Eng. 26, 131–138. https://doi.org/10.1016/j.coche.2019.09.002

Yenkie, K.M., Pimentel, J., Orosz, A., Cabezas, H., Friedler, F., n.d. The P-graph Approach for Systematic Synthesis of Wastewater Treatment Networks. AIChE J. 2021 https://doi.org/10.1002/aic.17253

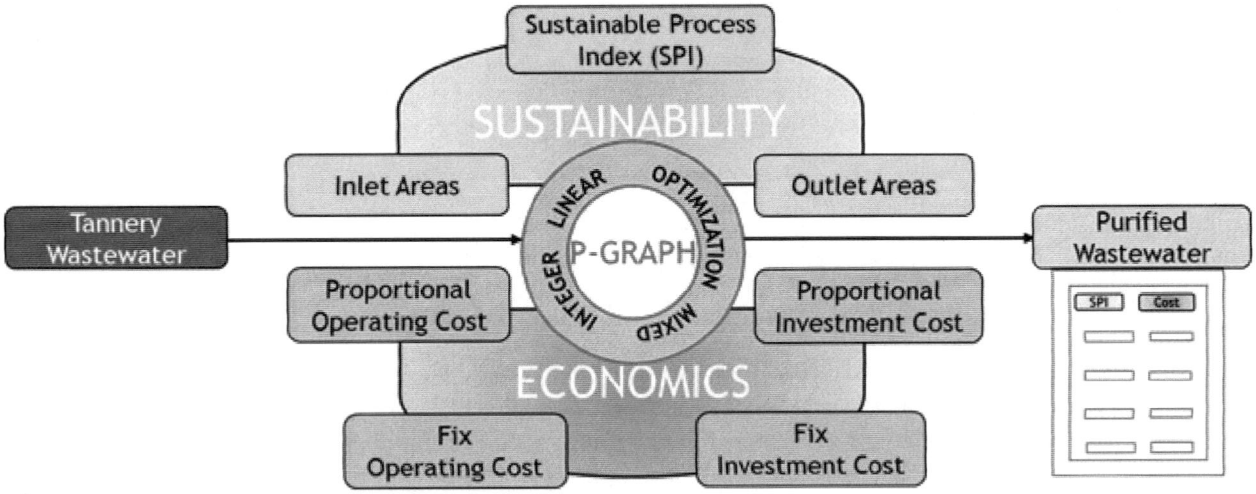

Figure 1. Treatment of tannery wastewater using Excel-based version of P-graph software

Copyright © American Institute of Chemical Engineers. All rights reserved.

624084 Modeling, Simulation and Optimization of a Synergistically Mixed Blowdown Water and Produced Water Wastewater Treatment Process

Abstract:

Water consumption and treatment is an underlying but significant aspect within processes in the energy sector. In particular, producing cost optimum designs of wastewater treatment processes is becoming increasingly important. In thermoelectric power plants, blowdown water streams are a result of cooling tower operations. These streams are purged to remove residual contaminants within the recycled flow, and fresh water is sourced to makeup for this purge and additional evaporative losses. Within the fracking industry, produced water is a high salinity wastewater that results from the extraction process in oil and gas wells. Produced water treatment is notably expensive and thus traditionally this water is disposed of by underground injection. Under analysis of these two separately generated waters, the contaminants of each have chemically complimentary constituents and may be synergistically mixed. A combined treatment process where these two waters are mixed is an identified new approach to alleviate chemical and energy demands of the blowdown water treatment process, maximize reuse of the treated water within the cooling tower, and provide an advantageous and sustainable treatment to a portion of the produced water.

In this work, improvements to the blowdown water treatment process by introduction of produced water are assessed. Specifically, the mixed wastewater treatment design is proposed to maximize water reuse and saleable by-product generation while reducing chemical and energy costs. This treatment train is designed such that usable makeup water and 10-lb Brine are generated with supplementary production of treatment additives including sodium hydroxide, sodium hypochlorite, and chlorine. The proposed design also provides treatment units for direct integration with a thermoelectric power plant. The process flowsheet is modeled in Aspen Plus with inclusion of custom process models for water treatment unit operations not found within the software's internal model library. Results generated from this flowsheet are evaluated in comprehensive techno-economic and sustainability analyses. Established tools in GREENSCOPE[1] and a pollution control unit (PCU) and life cycle inventory (LCI) framework[2] are used to quantify sustainability indicators and holistically evaluate process sustainability. The improved quality of the mixed treatment design is illustrated by the multi-objective sustainability assessment and cost optimization of the modeled process.

References

[1] Ruiz-Mercado, G. J., Smith, R. L., & Gonzalez, M. A. (2012). Sustainability indicators for chemical processes: I. Taxonomy. *Industrial & Engineering Chemistry Research*, *51*(5), 2309-2328.

[2] Li, S., Feliachi, Y., Agbleze, S., Ruiz-Mercado, G. J., Smith, R. L., Meyer, D. E., & Lima, F. V. (2018). A process systems framework for rapid generation of life cycle inventories for pollution control and sustainability evaluation. *Clean Technologies and Environmental Policy*, *20*(7), 1543-1561.

Copyright © American Institute of Chemical Engineers. All rights reserved.

629214 Effluent Reduction in Pulp and Paper Manufacturing: Paper Machine White Water Recycle in Bleaching Section

Abstract:

The pulp and paper industry is one of the largest consumers of water in the US industrial sector. Of the different process areas in a typical integrated bleached Kraft mill, the paper machine and pulp production account for around 75% of the total mill freshwater consumption. Therefore, the most significant freshwater reduction potential exists in these process areas. The overall objective of this study is to develop innovative strategies for water reuse from the paper-making process in the pulping process for reducing overall freshwater consumption.

In previous work, we applied Windows General Energy and Material Balance Systems (WinGEMS) to develop a process simulation model for the bleaching and paper machine sections capturing the material, water, energy, and the Non-Process Elements (NPEs) flows. Steady-state equations for the Elemental Chlorine Free (ECF) bleaching sequence $D_0EopD_1EpD_2$ are included in the simulation. The simulation model captures the effects of carry-over from brown stock washers (BSW) and extraction stage washer on the pulp brightness. It provides a tool to predict changes in the kappa number, brightness, and effluent Chemical Oxygen Demand (COD) values with changes in the operating parameters. We validated the EFC bleaching simulation model outputs using industrial data. Difference between industry operation data and simulation results in terms of key pulp properties such as brightness and kappa number were in the range of 1.8 to 5.5%, respectively.

This presentation focuses on the results of laboratory bleaching experiments for further validating the ECF simulation model results and incorporating Visual-MINTEQ software with the WinGEMS simulation model to determine the distribution of Ca^{2+} ions in the paper machine white water. The design of experiments for the laboratory bleaching study is constructed to replicate industry bleaching operation conditions. The preliminary experimental results revealed that the difference was in the range of 2.9 to 7.9% when comparing the laboratory bleaching work for the first two delignification stages, D_0Eop, with the simulation values for the pulp brightness and the kappa number, respectively. Current experimental work focuses on bleaching experiments for validating the final pulp brightness and the effluent COD values coming out of the bleaching sequence with the simulation values. For determining the distribution of Ca^{2+} ions in the paper machine white water, we will integrate the Visual-MINTEQ software with bleaching and the paper machine section simulations. This comprehensive model will capture the pulp parameters of kappa number and brightness in the bleaching sequence along with the white water Ca^{2+} ions distribution. The WinGEMS and Visual-MINTEQ integrated model will enable us to investigate the scope of white water recycle into the bleaching section. The results of this investigation will be discussed in the presentation.

Copyright © American Institute of Chemical Engineers. All rights reserved.

625944 Evaluation of Dioxins and Their Precursors in Fly Ashes and Wastewater Generated in the WET Scrubbing of Flue Gases from Sugarcane Bagasse Boilers

Abstract:

Sugarcane is one of the most efficient crops in the world for the combined production of sugar, ethanol, and energy. Brazil, India, and China produce together 66% of World's total. Despite the clear benefits to the economy and to the renewable energetic matrix of these countries, the enormous numbers of sugarcane processing also imply important impacts for the environment. The operation of boilers and cogeneration plants based on combustion of sugarcane bagasse releases great amounts of flue gases containing particulate matter (PM), which must be collected before emission to the atmosphere. Wet scrubbers are the most used technology to control PM, presenting the required efficiency to meet the emission limits to the atmosphere. Nevertheless, scrubbers demand the recirculation of large volumes of water that must be treated and represent important costs for the process. Recent studies have alerted that fly ashes conveyed in flue gases from bagasse combustion can contain dioxins, a group of highly toxic and persistent chemical compounds that are harmful to humans and other species. The absorption of dioxins in the scrubbing water could possibly contaminate the whole sugarcane process (due to water recirculation) and the environment (due to wastewater discharge in the field or in aquatic systems). The origin of the precursors of dioxins in the bagasse boilers is still unclear, but the possibility of spreading the contamination to the environment is of special concern in Brazil, where most of the ≈ 400 sugarcane industries in operation are located above the Guarani Aquifer System, one of the largest reservoirs of fresh water in the world. In this work, the collected ashes and the wastewater generated in the scrubbing process of five of the largest sugarcanes industries in Brazil were monitored along the season to detect the presence of dioxins or chlorine and organic carbon compounds, which are dioxin precursors. The metal composition of bagasse ashes was also evaluated to track possible catalysts for formation of dioxins. The results were correlated to the production processes in each industry to identify the sources of contaminants and to propose actions to prevent dioxin formation or to mitigate the environmental impacts.

Copyright © American Institute of Chemical Engineers. All rights reserved.

Assessing ozonation suitability for secondary effluent treatment via virtual piloting with the AMOZONE model

G. Bellandi*, R. Muoio*, M. Daza*, U. Rehman*, I. Nopens, S. Weijers***, P. van Dijk***, R. Schemen***, T. Weijtmans***, S. Hogard****, R. Pearce****, C. Bott****, H. Hauduc*****, I. Takacs*****, W. Audenaert***

* AM-TEAM, Oktrooiplein 1/601, 9000 Ghent, Belgium, Giacomo.Bellandi@AM-TEAM.com, roberta.muoio@am-team.com, miguel.daza@am-team.com, usman.rehman@am-team.com, Wim.Audenaert@AM-TEAM.com

** BIOMATH, Department of Mathematical Modeling, Statistics and Bioinformatics, Faculty of Bioscience Ingmar.Nopens@AM-TEAM.com

*** Waterschap De Dommel, PO Box 10.001, 5280 DA Boxtel, The Netherlands SWeijers@dommel.nl, TWeijtmans@dommel.nl, RSchemen@dommel.nl, PvDijk@dommel.nl

**** Hampton Roads Sanitation District, 1434 Air Rail Ave, Virginia Beach, Virginia, USA, SHogard@hrsd.com, RPearce@hrsd.com, CBOTT@hrsd.com,

*** Dynamita, 7 Lieu-dit Eoupe, La Redoute, 26110 Nyons France, helene@dynamita.com, imre@dynamita.com

Keywords: virtual piloting; ozonation model; water reuse; micropollutants removal; bromate mitigation; advanced oxidation modelling

Introduction

Drinking water treatment and (indirect) potable water reuse (IPR) is nowadays a topic of major attention due to the rising (and interconnected) concerns about water scarcity and micropollutants (MPs) such as pharmaceuticals and pesticides. Advanced tertiary treatments are increasingly adopted at water treatment sites to eliminate compounds with elevated biological resistance and produce safe water for human consumption (e.g. treatment to drinking water standards prior to aquifer discharge). In the US there are multiple examples (e.g. the HRSD SWIFT project (swiftva.com)) aimed at environmental protection from ~~pharmaceuticals and other chemicals~~MPs. Conventional water resource recovery facilities (WRRFs) only partially contribute to these goals, and often in a rather unpredictable manner (Schaar et al., 2010).

Ozonation and (biological) granular activated carbon adsorption is establishing more and more (especially in Europe) as an attractive IPR technology alternative for MP removal, beside the more consolidated membrane filtration (UF/RO) combined with advanced oxidation. However, the extensive piloting efforts needed for a proper assessment of its applicability have hampered the full applicative potential of ozonation.

Ozone has complex reactions with the effluent matrix. Some of those pathways lead to hydroxyl radical (HO*) formation. Both ozone and HO* will oxidise the MPs. However, depending on the natural Br- concentration in the water, the carcinogenic byproduct bromate (BrO_3) can be formed. In addition to this, the specific ozone demand is strongly effluent matrix dependent, leading to significant differences in operational cost, MP removal efficiencies and BrO_3 formation across WRRFs. Next to the very site specific effluent organic matter (measured as chemical oxygen demand (COD), dissolved organic carbon (DOC), …), effluent variability is of key importance. Effluent variability of the upstream biological treatment can impact the downstream ozonation process in various ways (e.g. pH, organics, ammonia, nitrite, alkalinity, bromide). Nitrite is known to consume O_3 (3.43 mg O_3/mg N-NO_2) and HO* significantly affecting the OpEx (Sehlén et al., 2020). In addition to this, ammonia is known to partially inhibit BrO_3 formation (von Gunten, 2003a).

It is this complexity of chemical reactions and site specificity that makes utilities and designers engage in extensive onsite piloting trials prior to full-scale implementation. These piloting efforts easily cost hundreds of thousands of EUR and over a year to several years and need dedicated safety measures and operator training. Extensive analytical measurements of bromate, MPs, ... are an inherent part of the trials, while the whole exercise offers only limited information in proportion to the cost and effort. On top of this, in IPR trains, the additional regulatory compliance with regard to disinfection adds complexity (i.e. O_3 residual needs to be assessed to quantify 'CT').

This paper describes the application of the novel AMOZONE model for virtual piloting: the complete replacement of onsite piloting by computer simulation at (virtual) full-scale using a mechanistic model with the Sumo© platform (Dynamita). The AMOZONE model has similarities with the well-established Activated Sludge Models M~~ASM m~~odels (ASM): it relies on 'grouping' parameters (e.g. Chemical Oxygen Demand (COD) fractionation in ASM) to handle the complexity of real matrices, while being built on mechanistic process understanding.

The Netherlands is currently accelerating implementation of demo and full-scale plants for MPs removal. To be eligible for the national funds, the plant should be operated at removal efficiency of 70% for 7 of 11 guide MPs. For the first time ever, virtual (model-based) pilot testing was performed for two WRRFs in The Netherlands to assess MP removal, CapEx and OpEx and bromate formation.

Methodological Approach

AMOZONE model

AMOZONE is a kinetic model of ozonation and advanced oxidation processes that realistically predicts O_3 and HO* concentrations in real water matrices (Audenaert et al., 2019). AMOZONE was initially developed in Sumo19© to describe batch experiments and pilot data collected by HRSD in the framework of SWIFT. The model includes rapid and slow O_3 consumption, HO* production and scavenging, O_3 and HO* based MP oxidation, bromate formation mechanisms (von Gunten, 2003a), nitrite quenching, NH_2Cl formation and reactions, and several pH equilibria.

Full-scale plants and data collection

The WRRFs of Eindhoven and Soerendonk are conventional activated sludge plants with treatment capacities of respectively 750,000 and 38,000 person equivalents. In the framework of this project, Waterschap De Dommel organized dedicated weekly measurement campaigns to monitor plant effluent characteristics important for ozonation (e.g. Br^-, NO_2, ...). In addition to this, online sensors were installed for a month-long monitoring campaign at the effluent of both WRRFs to measure UV transmittance (UVT), Dissolved Organic Carbon (DOC), NO_3, NO_2, and Turbidity. This data, together with plant flow rates were used as input for the model for the dynamic simulations.

Model calibration and validation

The AMOZONE model needs calibration for the specific WRRF matrix. At each plant, four 24h composite samples were taken during dry weather days and used for batch ozonation tests revealing O_3, HO*, UVT, Br^-, BrO_3, NH and NO_2 profiles as function of ozone dose and time. Two of the batch tests were used for model calibration while the remaining ones were used for validation. More extensive model validation on existing pilot or full-scale plants was performed earlier in the framework of SWIFT and other projects (data not shown).

Results and Discussion

The model was calibrated for both plants with batch ozonation tests performed on three samples of the biological plant effluent. Each sample was collected over 24h with a flow proportional sampler and two specific doses of ozone were tested. Figure 1 shows examples of the calibration and validation results against the batch laboratory tests performed on the secondary effluent composite samples. Additional samples have currently been taken to further quantify model uncertainty under dynamic conditions. Despite the work to be performed, the model already addresses key issues.

Dynamic model simulations were run using the online data collected at the WRRF effluent. The model results show important effects of NH_4-N concentration in the framework of BrO_3 production. In specific, the model addresses significant responsibility to NH_4 in sequestering Br- under the form of mono- and di-bromamine, which makes it unavailable for BrO_3 formation.

Figure 2~~Figure 2~~ shows how upstream WRRF control strategies impact the final water quality for the Soerendonk plant in terms of NH_4-N concentration. O_3, BrO_3 and MP concentrations are affected by the upstream biological plant in various ways in both Eindhoven and Soerendonk WRRFs. Specific MP removals varied significantly as function of effluent quality (Figure 3). Dynamics in BrO_3 (Figure 4) were caused by variations in secondary effluent nitrite, ammonia, and pH directly affecting O_3 and HO* levels. The model allowed to assess the effect of different O_3 dosage control strategies on MP removal (Figure 5, left), BrO_3 formation (not shown), and operational costs (Figure 5, right) based on real influent variability.

Conclusion

Advanced treatment modelling is specifically relevant given the growing importance of MP removal and IPR. AMOZONE is very recent, and the modelling exercise has not been finished yet (e.g. extensive uncertainty analysis under dynamic conditions at many existing plants). However, practitioners and industry leaders already explicitly acknowledged the value, as real-life ozone testing is expensive, complex and characterized by significant uncertainties.
For the first time onsite ozonation pilot testing was completely replaced by computer simulation. This work shows the potential of virtual piloting in accelerating the understanding of ozonation for a specific water matrix. For the first time, a computer model completely replaced onsite ozonation piloting trials. For the Soerendonk WRRF, 70% removal of MPs could be achieved with ozone doses varying between 0.7 and 5 kgO_3/h. At this small WRRF, the operational cost varied between 2 and 3.5 $cEUR/m^3$. The AMOZONE model is expected to play an important role in IPR and MP removal process testing, design and operation. Coupling of the model with conventional WRRF models can be used for overall cost saving and performance maximisation.

References

Audenaert, W., Bellandi, G., Pearce, R., Takács, I., Buehlmann, P., Hogard, S., Salazar-Benitez, G., Rehman, U., Nopens, I., Wilson, C., Bott, C.B., 2019. A novel kinetic ozonation model for

prediction of bromate formation, bromate mitigation and trace organic contaminant removal, in: 24th World Congress of the International Ozone Association.
Sehlén, R., Nilsson, J., Stapf, M., Schütz, J., Bester, K., Kharel, S., Lukas, M., Bogusz, A., Putna-Nimane, I., 2020. Evaluation and experiences of full-scale ozonation followed by MBBR post-treatment and comparison with previous pilot tests.
von Gunten, U., 2003a. Ozonation of drinking water: Part II. Disinfection and by-product formation in presence of bromide, iodide or chlorine. Water Res. 37, 1469–1487.
von Gunten, U., 2003b. Ozonation of drinking water: Part I. Oxidation kinetics and product formation. Water Res. 37, 1443–1467.

Figures and Tables

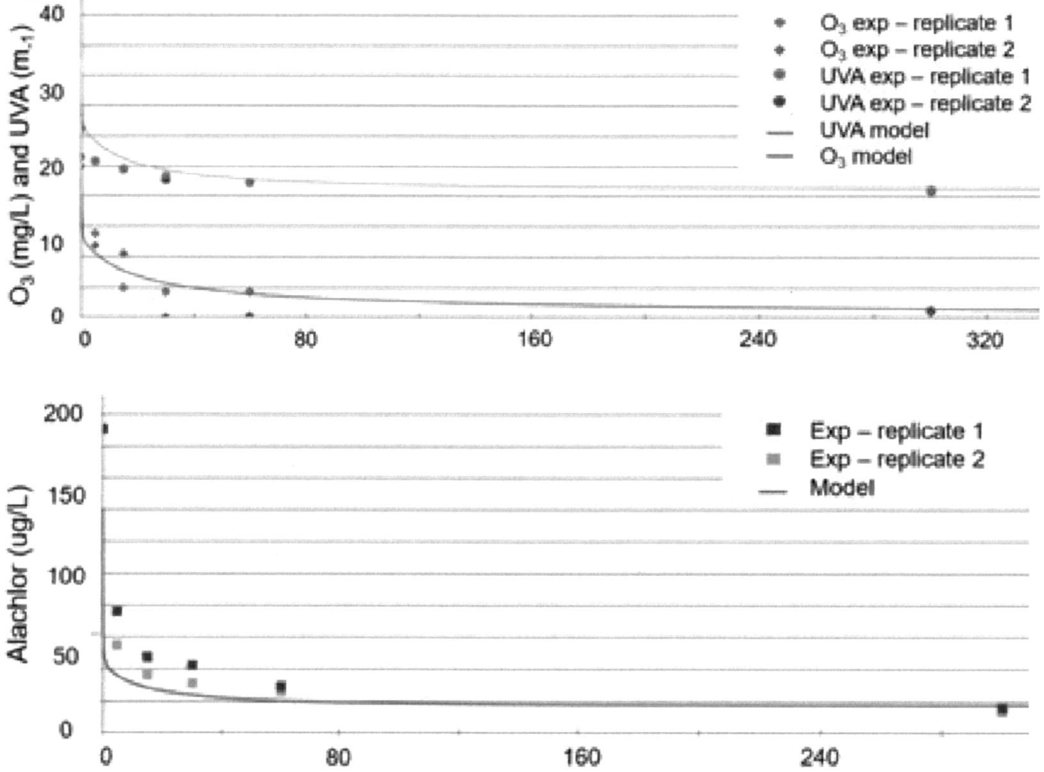

Figure 1 – Examples of lab batch model calibration on O_3 and UVA profiles (top), and validation on the chemical Alachlor (bottom).

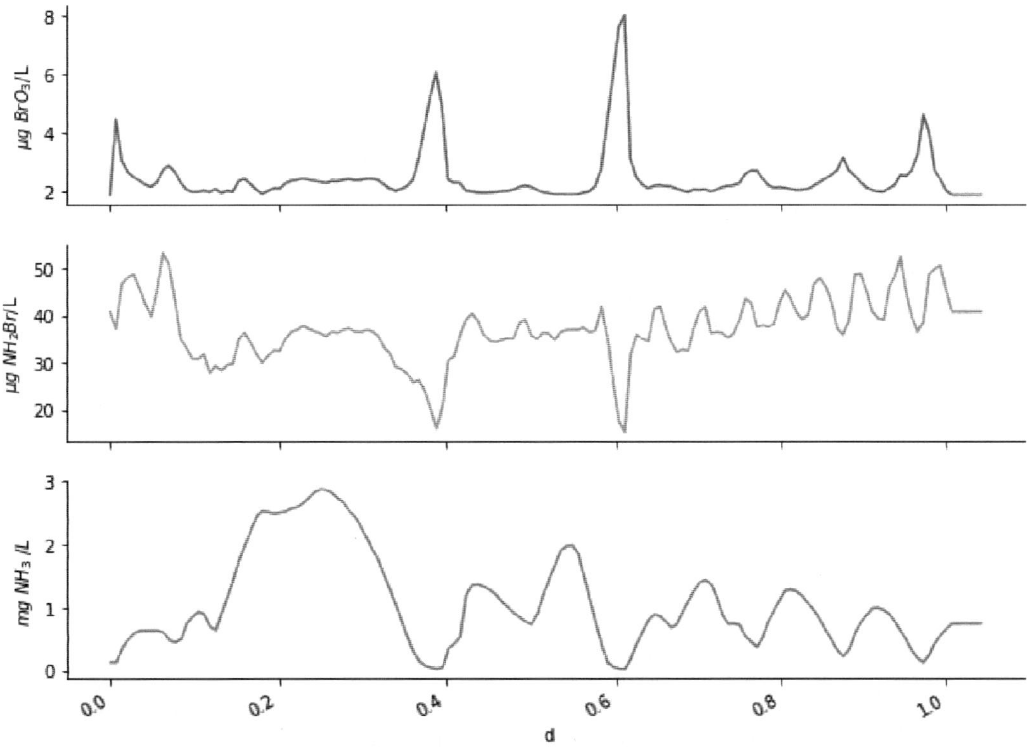

Figure 2 - A typical measured ammonia profile in a dry day at Soerendonk WRRF (bottom) and the modelled mono-bromamine and bromate (middle and top respectively) have been combined for the first time.

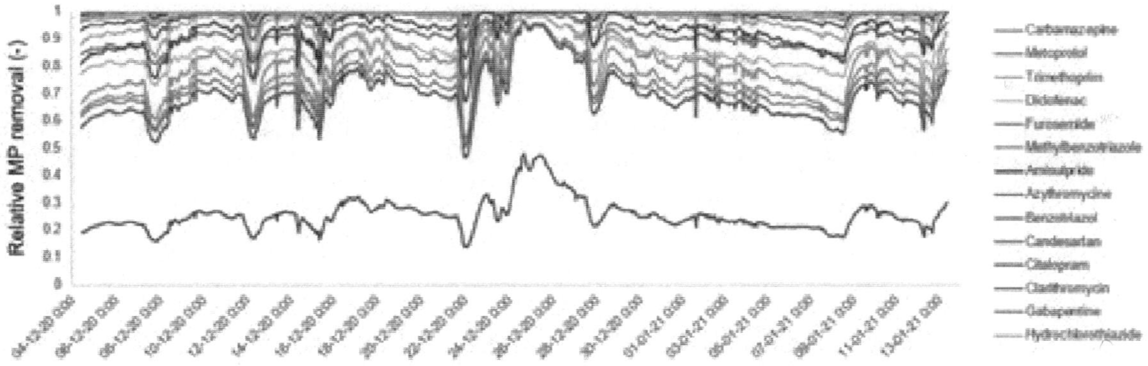

Figure 3 – Time series of specific MPs removal (%) for the WRRF of Eindhoven simulating the flow proportional controller with 8 mg O_3/L setpoint along 1 month of dynamic data.

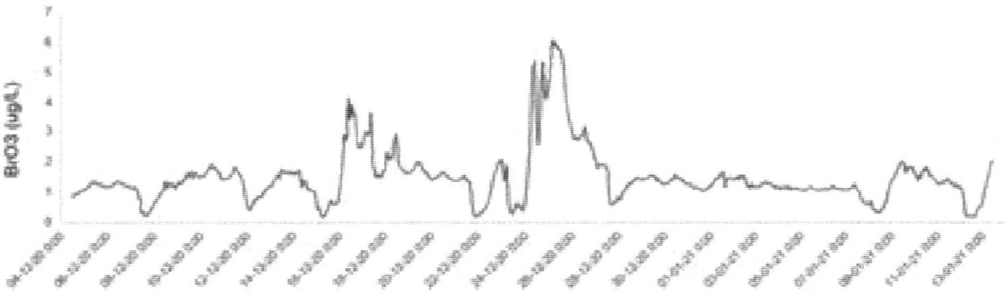

Figure 4 - Time series of BrO_3 for the WRRF of Eindhoven simulating the flow proportional controller with 8 mg O_3/L setpoint along 1 month of dynamic data.

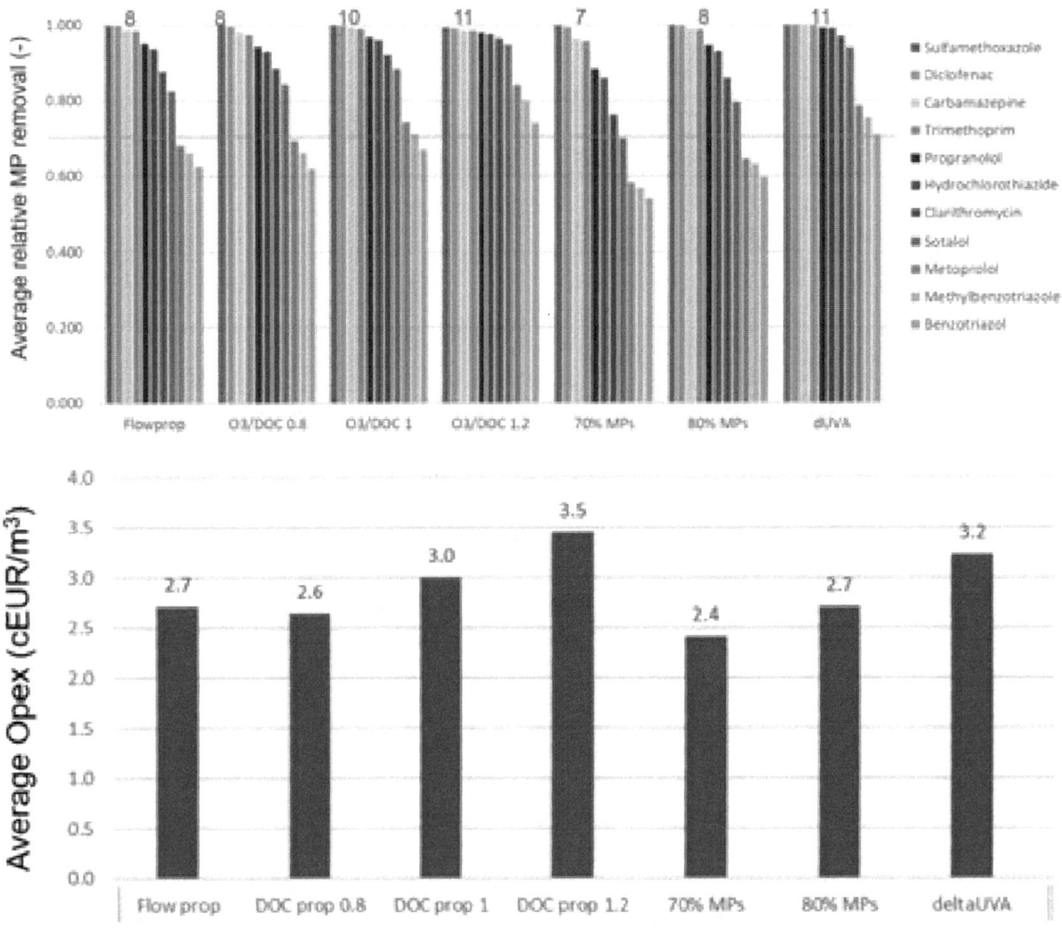

Figure 5 – Comparison of the average MP removal (top) and OpEx at Soerendonk WRRF (bottom) of different proportional controllers based on flow, DOC, MP removal, and deltaUVA.

620651 Abiotic Dehalogenation of 1,4-Dichlorobenenze with Reactive Iron Mineral Coatings in a Contaminated Sediment Core

Abstract:

Subsurface systems contaminated by dichlorobenzene (DCB) is a significant issue due to its past use as a solvent, a chemical intermediate, and a degreasing agent. As a result of its widespread release into the environment, understanding natural attenuation processes is important. However, with a stable aromatic ring, (a)biotic degradation is a challenge. This study focuses on 1,4-DCB as it is one of the constituents of concerns at a site with historical contamination. An 18.3 m anoxic core was collected at the site with the sediment redox condition preserved. After subsampling the core into a total of 225 5.08 cm thick samples, five redox transition zones (RTZs) were identified through screening analyses, which included sediment Fe and S concentrations, sediment pH, sediment oxidation-reduction potential, total volatile organic compounds in the sediment headspace, and abundant bacteria. Reactive Fe mineral coatings in the sediments from five RTZs have been characterized using field emission scanning electron microscopy with energy dispersive X-ray analysis, X-ray diffraction, and sequential extraction. The RTZs abundant with reduced iron mineral coatings including Fe sulfide minerals were targeted for this investigation: mackinawite (FeS), pyrite (FeS_2), and siderite ($FeCO_3$). Batch studies conducted in the glovebox revealed pseudo-first-order rate expressions where 90% degradation of 1,4-DCB was achieved between 52 d and 202 d for the three RTZs studied. The zone most rich in the poorly crystalline Fe sulfides revealed the greatest rates of dehalogenation.

Copyright © American Institute of Chemical Engineers. All rights reserved.

630450 Spatio-Temporal Features of NO_x and Hydrocarbons Trapping and Conversion in a PNA+HCT+OC Sequential Monolith Configuration

Abstract:

Introduction

In the past 10 years there has been considerable progress in addressing NO_x (x = 1, 2) and CO/hydrocarbon (HC) emissions from diesel vehicles using urea-SCR[1] technology and DOC technology[2], respectively. However, these catalysts only achieve high NO and CO/HC conversions (>90%) once the exhaust temperature has exceeded 200 °C. During the "cold start" period a large fraction of the exhaust HCs and NO_x remain unreacted and are emitted to the environment. Precious metal dispersed zeolites (Pd/SSZ-13, Pt+Pd/BEA) have been shown to be effective for trapping NO and HCs at low temperature, followed by release at high temperature and subsequent conversion. To date there have been few studies of the combined trap. In this study we investigate the combination of PNA, HCT (HC trap) and OC (oxidation catalyst) to address a number of questions about material selection, device configuration, and operation strategy to achieve emission performance targets. Using spatially-resolved mass spectrometry, this is the first study to provide insight into the spatiotemporal features of transient behavior of sequential PNA, HCT and OC. Conventional measurements of the inlet and outlet values provide only a partial picture of the reactor behavior. The instantaneous profiles provide insight into the storage and reduction dynamics and the mass coupling between the three catalysts.

Materials and Methods

For the spatial-temporal measurements, PNA monolith containing 1 wt. % Pd/CHA, HCT monolith containing 1 wt% Pt/BEA and OC monolith containing (0.5 wt. % Pd + 0.5 wt. % Pt)/CeO_2 (each 2cm in length with 8x8 Channels in a 400 CPSI monolith) were installed in sequence in a quartz tube reactor. All three catalysts were provided by Johnson Matthey Inc. The catalysts were degreened at 600 °C for 4 h in 12% O_2 and 6% H_2O before usage. The feed gas mixture contained 400 ppm NO, 12% O_2, 6% H_2O and balance Ar. NO uptake was conducted for 30 min over 100°C followed by a temperature ramp up to 600 °C for 25 min. Outlet Species concentrations were analyzed by an online Thermo Scientific 6700 Nicolet gas analyzer while the transient spatial profiles were obtained via a mass spectrometer (Hiden Analytic Ltd., HPR-20) with multiple capillary sampling system (Hiden Analytic Ltd[3]). The spatial system comprises of a rail mounted inlet assembly and HPR-20 QIC benchtop gas analysis system. The system is capable of acquiring data from 8 capillaries and 8 thermocouples. Three capillaries probes (OD: 363 and ID 220) and three thermocouples (OD: 250, type K) were employed for this study with three monoliths. Each capillary and thermocouple was positioned 2 cm apart starting at the entrance of PNA catalyst.

Results and Discussion

Figure 1. a - c shows typical performance of each catalyst using a simulated diesel exhaust feed. Common exhaust components C_2H_4/CO have an incremental benefit on NO uptake. Significant formation of NO_2 in a sequential PNA+HCT configuration is beneficial for downstream SCR. The Pt-Pd based OC shows low temperature oxidation performances for CO and hydrocarbons. The Pt-based oxidation catalyst used as a conventional DOC is more prone to CO poisoning, while adding Pd to Pt in the catalytic system induces slower particle sintering and moderating CO inhibition compared to monometallic Pt-based catalyst[4]. Figure 1d shows transient NO concentration at three different positions in the PNA along with the transient effluent NO (measured by FTIR). The spatial profiles provide detail on NO uptake in the PNA. The spatially resolved measurements can help to unravel the complex dynamical behavior that is obviously inaccessible without the spatial probes.

Significance

Profiles spanning the entire PNA+HCT+OC device will be will be further explored for application in the optimum design of multi-functional catalytic system. Such measurements help to optimize the multi-functional device design and operating strategy for achieving high adsorption and conversion of NOx and HCs species during the cold start period. The Spaci Mass Spectrometer method provides a more accurate and comprehensive picture of structured catalysts under operating conditions which will help us in accurately model and scale up the system.

References

1. Zheng, D. Luss, M.P. Harold, SAE Int. J. Engines. 7 (2014).
2. Mukai, K., Kanesaka, H., Akama, H. and Ikeda, T., SAE, 0148, (2004).
3. Sá, D.L.A. Fernandes, F. Aiouache, A. Goguet, C. Hardacre, D. Lundie, et al.,Analyst 135 (2010) 2260–2272.
4. Kang, S.B., Hazlett, M., Balakotaiah, V., Kalamaras, C., Applied Catalysis B: Environmental, 223, 67-75 (2018)

Acknowledgements

We acknowledge the financial support of this research by the Department of Energy (DE- EE0008233). We also thank Johnson Matthey, Inc. for supplying the monolith samples.

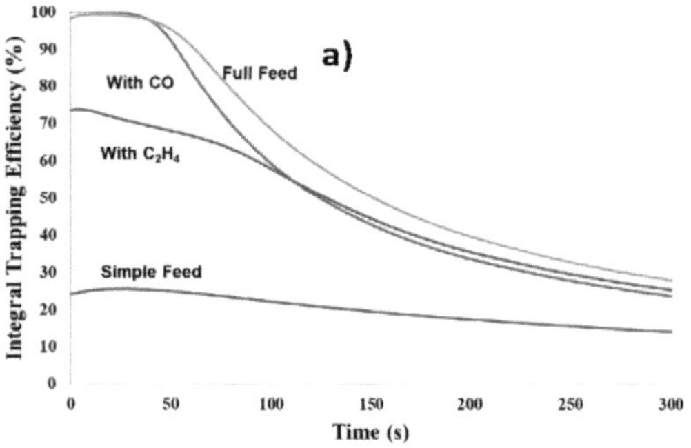

PNA: Integrated trapping efficiency of NO by itself, in the presence of CO or Ethylene and the full feed mixture in PNA Catalyst. **Feed:** 200 ppm NO, 200 ppm C_2H_4, 500 ppm CO, 100 ppm H_2, 42 ppm C_7H_8, 58 ppm $C_{12}H_{26}$, 6% H_2O, 6% CO_2, 12% O_2, balance N_2.

- ***Nearly complete uptake of NO is achieved at 100 °C for about 50s***

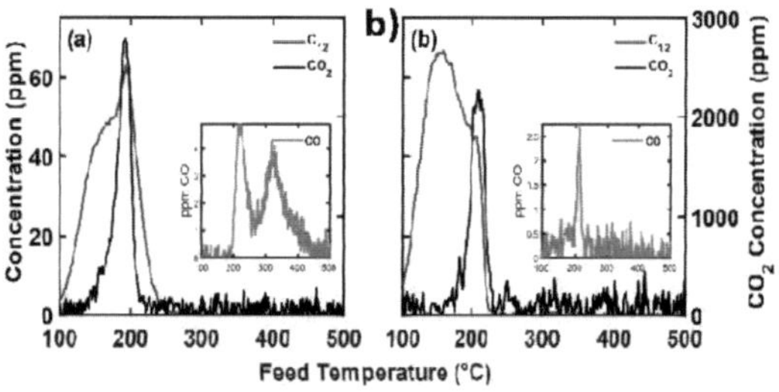

HCT. $C_{12}H_{26}$, CO_2, and CO (inset) TPD profiles for (a) Sequential HCT → PNA and (b) Sequential PNA → HCT. **Feed:** 400ppm NO, 200 ppm $C_{12}H_{26}$, 2% O_2 and 3% H_2O.

- ***Pd/SSZ-13 + Pt/Pd/BEA dual layer configuration meets performance targets with 50% volume reduction.***

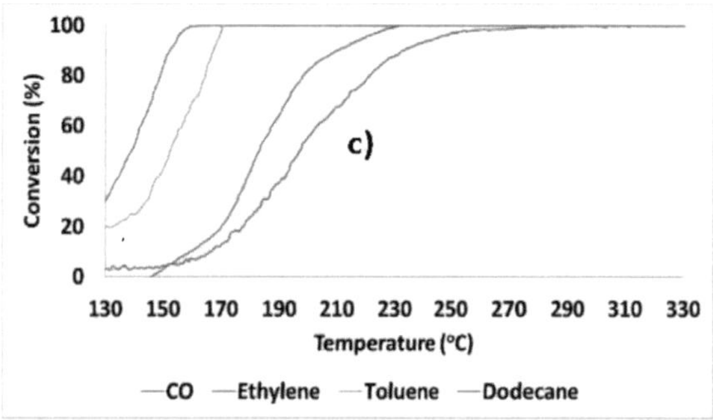

OC: CO and HC light in Oxidation Catalyst for the stimulated gas feed **Feed:** 200 ppm NO, 200 ppm C_2H_4, 500 ppm CO, 100 ppm H_2, 42 ppm C_7H_8, 58 ppm $C_{12}H_{26}$, 6% H_2O, 6% CO_2, 12% O_2, balance N_2.

- ***Pd-Pt based catalyst with Ceria/Alumina base we were able to achieve more than 80% conversion of the HCs below 200°C for oxidation catalyst.***

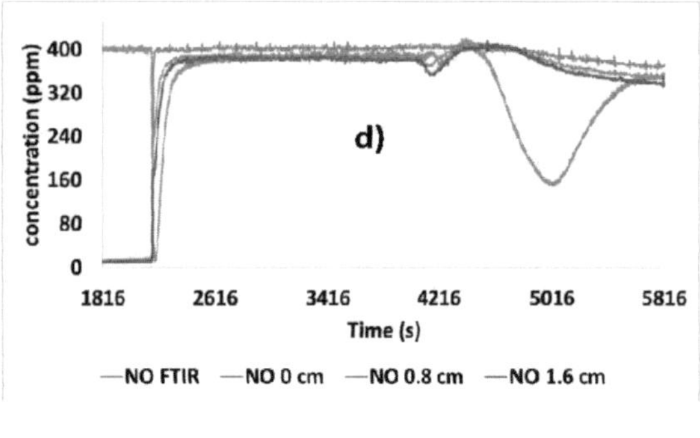

PNA+HCT+OC: NO profile for three different positions in a 2cm long PNA monolith in a sequential monolith configuration along with NO outlet profile from the reactor **Feed:** 200 ppm NO, 6% H_2O, 12% O_2, balance Ar

- ***Incremental NO uptake can be observed for the three positions along with the variation in its profile when compared to FTIR***

Copyright © American Institute of Chemical Engineers. All rights reserved.

631974 Energy-Efficient Nanoparticle Contamination Control Using Functionalized Fiber Filters

Abstract:

Nanoparticles have always been an important part of environment. Recent advances in nanotechnology have enabled us to utilize engineered nanoparticles (ENPs) in a plethora of applications to improve the performance of products and also as process enhancers. Although the advantages of ENPs are undeniable, there is a growing uncertainty concerning the impact of excessive release of ENPs into environmental systems. In recent years, nanoparticle (NP) contamination has come to the forefront as a critical challenge for environment and human health. The hazards of nanoparticles mostly stem from their small size resulting in highly reactive materials that can penetrate through surfaces effectively. Nano-contaminants released into atmosphere can span a wide variety of materials that make up nanoparticles (NNPs), incidental nanoparticles (INPs), and engineered nanoparticles (ENPs) that includes viruses, nano-plastics, and colloidal metals to name a few. Due to the interrelation of earth and water systems, drinking water treatment plants act as critical nodes for controlling the exposure of humans to NPs as they exist at the interface of natural and man-built systems. Despite ongoing efforts, the understanding about the biological impact and quantification methods available for ENPs are still at primitive stage. In addition, the concentration of ENPs in drinking water systems is much lower compared to natural and incidental nanoparticles (NNPs and INPs) redirecting the focus of scientific community to address the endless list of emerging contaminants in our water streams.

The lack of focus towards developing novel techniques for nanoparticle (NP) removal from drinking water which is also exacerbated by the fact that no clear regulatory (for example US EPA) requirement exists for incidental and engineered NP removal, financial burden to replace legacy equipment, and the multi-barrier approach commonly used in water treatment that apportions the removal of NNPs such as clay particles and enteric viruses to either coagulation and sedimentation or disinfection technologies. However, simple methods to remove NPs from aqueous matrices are energy intensive and challenging to implement in traditional centralized drinking water plants and also point of use applications (household and disaster relief type scenarios). Filtration is actually the most common unit operation used in water purification and adaptable at different scales, scenarios, and resource availabilities but most common media filtration methods do not lead to removal of NPs, necessitating the use of more energy intensive and expensive mode of membrane filtration. In this study, we show the feasibility of using a sustainable filtration technique fabricated from easily-accessible materials that can be deployed with minimal cost to achieve highly efficient removal of both NNPs and ENPs from water.

Human enteric viruses are high-risk NNPs that needs immediate attention as infectious disease spread is a major and continuing health burden faced by humanity. The advent of antibiotics has placated the impact of bacterial infections but viral disease outbreaks continue to cause global epidemics. The recent COVID-19 pandemic underlines the importance of technologies that can limit the sustained transmission of viruses. Due to their small size (~5 to 300 nm), viruses pose unique challenges requiring chemical and energy intensive technologies for their effective removal and inactivation from air and water to mitigate disease transmission. This is particularly evident in the context of water purification, due to the reliance on size-exclusion based virus filtration techniques, which gives rise to a trade-off between the productivity and the level of pathogen removal achieved. This trade-off presents a unique challenge, which is a striking example of the interrelation between clean water production and associated energy consumption (the energy-water nexus). Recent studies in the literature have used chemical functionalization of low-pressure membranes or specialized membrane fabrication techniques such as electrospinning and use of nanofibrous materials to overcome the limitation imposed by the trade-off curve. However, due to the use of advanced fabrication and modification strategies the widespread use of these solutions is limited and energy intensive. In this work, we focused on developing a simple, low-cost, and low (embedded and operational) energy alternative that is applicable for NP removal in a wide range of public health scenarios using viruses as a test surrogate. This strategy of utilizing viruses for devising NP treatment techniques can be useful moving forward as the analytical methods for quantifying virus concentration are robust and advanced compared to ENPs.

Broad application of currently available virus removal technologies across the world is often limited by their cost, complexity, and accessibility. Due to the low efficiency of conventional filtration, disinfection has become the most commonly practiced method to meet drinking water standards imposed by US Environmental Protection Agency (EPA) which mandates $4\log_{10}$ (99.99%) of virus removal. However, advanced disinfection technologies such as ozone and UV are expensive, require high energy inputs, and chemical disinfection (using chlorine) produces toxic disinfection by-products. In addition, all disinfection strategies include materials or chemicals with high embedded energy which incurs a total energy requirement on par with the filtration techniques and are most commonly used in conjunction with and downstream of filtration technologies. Additionally, widespread disease outbreaks still occur in centralized distribution systems practicing disinfection due to inadequate or interrupted disinfection. Due to the above challenges with conventional filtration and disinfection technologies, the focus has shifted towards application of membrane filtration technology which rely on size-based virus removal. Presently, membrane filtration is the most widely accepted alternative water treatment technology and Reverse Osmosis, Ultrafiltration and Nanofiltration membranes with sub-micron size pores were shown to be effective against viruses. The fundamental challenge with membrane filtration arises from reliance on small pore sizes as a physical barrier for viruses. First, the inherent small pore sizes require high pressure differences to drive flow which increases operation costs limiting their widespread use. In addition, membrane filtration needs near continuous monitoring of their integrity as any microscopic defects can cause the passage of viruses.

In this study, we show that the commonly available natural fibers, upon simple functionalization with water extract from *Moringa oleifera* (MO) seeds can achieve highly efficient NP removal under practically relevant conditions. The MO tree is prevalent throughout tropic and sub-tropic regions, and its seeds have historically been used as a natural coagulant. The seeds of this tree contain cationic proteins *MO* coagulant protein (MO2.1) and *MO* chitin binding protein (MoCBP) with established antifungal and coagulant activities. Natural fibers functionalized with these antimicrobial proteins from the seeds of the MO tree acted as an effective filter media for capture and removal of an array of NPs including MS2 and T7 bacteriophages ($\sim 7\log_{10}$) along with silver and polystyrene latex particles ($>4\log_{10}$). We chose to use unprocessed cotton, wool, and flax for this study to show that our functionalization procedure can be implemented effectively worldwide even in resource-limited areas. Our experiments further show that the proposed filters can effectively remove NPs at superficial velocities of 2 m/hr, which is an order of magnitude faster than slow sand filtration and close to the lower range of rapid sand filtration, two most commonly used filtration techniques in water treatment. Using fundamental characterization techniques, we present an overall description of the filters proposed and compare their permeability and removal efficiency with conventional membrane filtration techniques to show that MO coated cotton is a highly energy-efficient sustainable alternative for effective virus removal.

Caption

Figure 1: **Permeability and total energy trade-off curves of traditional virus treatment technologies.** The literature review of various filtration and disinfection techniques that are typically used for virus removal from water show that there exists a trade-off curve between the permeability (productivity of a filter water for unit trans-filter pressure) and the efficiency which has been observed in various membrane separations. Here we also show that a new productivity based on the total energy requirement of the given techniques also exhibits a trade-off. In this study the proposed filter was designed to overcome these two trade-off curves in order to achieve a wholistically sustainable filter with high productivity.

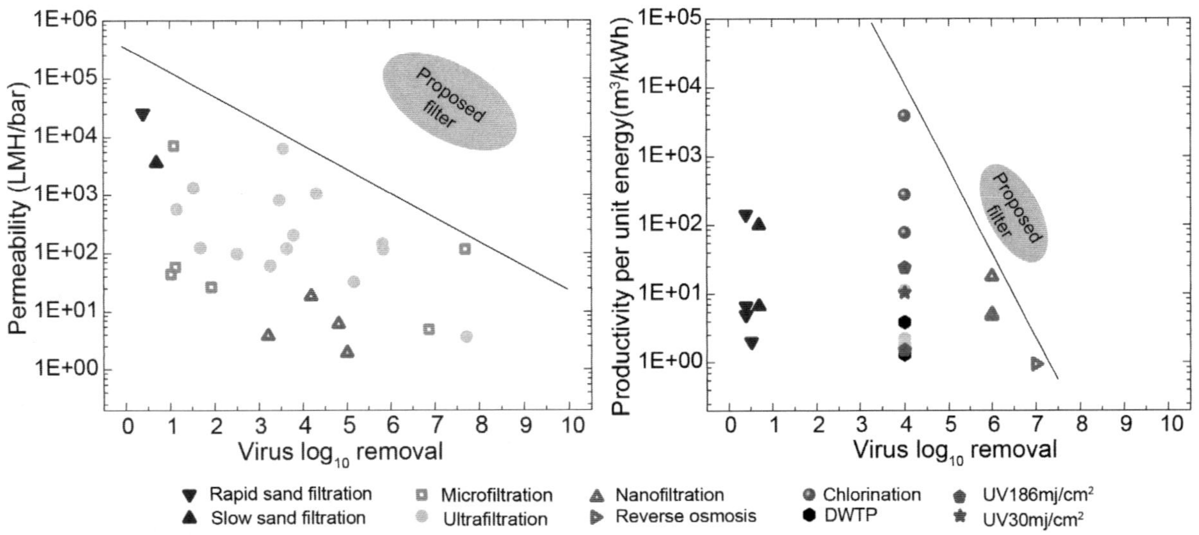

Copyright © American Institute of Chemical Engineers. All rights reserved.

632036 Recent Advances in the Development of Fly Ash Based Geopolymers for the Adsorption of Aqueous-Phase Heavy Metal Contaminants

Abstract:

Geopolymers are novel inorganic polymeric materials synthesized from aluminosilicate sources via alkaline activation at relatively low temperatures. Although geopolymerisation is not a new concept and has been used as barriers in immobilizing toxic and radioactive metals via the combination of physical encapsulation and chemical stabilization, the application of this technology to produce effective adsorptive materials for mitigation of liquid-phase contaminants is relatively recent. Due to their inorganic structure and stability properties, heavy metals do not decompose in nature and pose risk of accumulation in living species as highly toxic and carcinogenic elements, some even at very low concentrations. The removal of trace toxic heavy metals (TTHMs) from solutions has been commonly carried out by several processes such as adsorption, ion exchange, chemical precipitation, reverse osmosis and solvent extraction. However, recently, adsorption is the one driving most attention among these methods because of simpler equipment and smaller corrosiveness, lower energy consumption, high removal efficiency, reversible nature allowing regeneration of adsorbents and flexibility in design. Adsorption has been regarded as the main interaction mechanism for geopolymers in heavy metal removal investigations. A number of studies on the use of geopolymers derived from fly ash for the sequestration of aqueous and gaseous pollutants has been recently reported. This paper will provide a state-of-the-art overview of the recent scientific advances on geopolymers widely synthesized from abundantly available coal fly ash (CFA) as raw material; and sustainable environmental applications. It focuses on the adsorption of TTHM contaminants in water; evaluating the effectiveness against other adsorption technologies utilizing conventional materials such as activated carbon and zeolites.

Copyright © American Institute of Chemical Engineers. All rights reserved.

625168 Recovery of High-Value Elements from Produced Water Using Ligand Modified Silica Nanoparticle

Abstract:

Recent advances in crude oil production have unlocked significant petroleum reserves, and the amount of produced water (PW) generated with it has increased dramatically. Oil and gas production in Oklahoma alone generates over 100,000 acre-feet per year of PW. Data from the Oklahoma Corporation Commission (OCC) indicate that nearly 3 billion barrels of PW in Oklahoma were injected into the earth in 2017. Deep injection has led to increased seismicity and other issues in several regions of the world. PW has various valuable elements including lithium (Li), cobalt (Co), iodine (I) and rare earth elements (REE) including lanthanum (La), neodymium (Nd), and europium (Eu). These elements are critical materials for numerous technologies including consumer electronics, automotive, energy, and defense. PW is an attractive potential source for REE and lithium production. This work seeks to develop an adsorbent that is capable of removing these valuable elements from aqueous systems. Hydrophilic silica nanoparticles were modified using N-[(3-trimethoxysilyl) propyl] ethylenediamine triacetic acid (TMS-EDTA). This modified silica was used for the adsorption and separation of these valuable elements from aqueous solutions. The presence of the EDTA on the surface of the nanoparticles was confirmed using FTIR. Batch studies were conducted to determine the adsorption capacity, effect of pH, temperature, dosage and kinetics of adsorption. The mechanism of adsorption would be verified using existing models along with the kinetics of adsorption and thermodynamic parameters.

Copyright © American Institute of Chemical Engineers. All rights reserved.

626820 Development of a Compact, Rapid, and Zero Liquid Discharge Wastewater Treatment System

Abstract:

Wastewater production is an unavoidable consequence of human activities and industrial operations. The quantity and quality of wastewater is determined by many factors such as source and processing method. In the U.S., more than 32 billion gallons per day of wastewater is generated from domestic, commercial, and industrial facilities.

Using apt chemical engineering principles, we aim to develop a compact and novel wastewater treatment system. The design simplifies the conventional multistep sewage sludge separation process and accomplishes a high level of solid dissolution by polymer conditioning in a single coagulating/flocculating mixing tank. The solids then undergo enhanced dewatering and thermal drying, yielding a highly dehydrated solid cake. The liquids are unified and treated by chlorination, de-chlorination, and filtering. This produces clean drinking-quality water, free from dangerous pathogens and chlorinated organics.

These simple features in combination translate to a highly compact land footprint, creating the possibility for a portable treatment plant. Speed of water recovery in this design is a vital aspect as it can allow for zero-liquid discharge operation. This means that a small quantity of freshwater can be recycled in a closed-loop treatment system, dramatically mitigating water consumption. The treatment system will therefore be capable of operating in areas where fresh water supply is diminished or unreliable, further enhancing its portability and value.

Copyright © American Institute of Chemical Engineers. All rights reserved.

634325 Development of a Low-Cost Non-Amine Sorbent-Based Technology for Direct Air Capture of CO_2

Abstract:

Worldwide emissions of CO_2 now exceed 40 Gt/yr annually, accelerating the greenhouse gas effect that has already increased global median temperature by 1°C. Negative emissions technologies (NET) must be rapidly developed and deployed in order to remove CO_2 from the atmosphere and limit the global temperature rise to <2°C. Direct air capture (DAC) of CO_2 is a promising NET, but the cost of capturing CO_2 via DAC remains too high ($300–1,500/tonne CO_2) to be a viable commercial solution. For sorbent-based DAC systems, the major cost contributor is the energy required to release CO_2 and regenerate the sorbent.

Susteon Inc. has developed a platform technology for DAC based on low-cost supported alkali-based sorbents with high CO_2 capture rate and capacity. Compared to conventional DAC using amine sorbents supported on high surface area solids (e.g., metal-organic frameworks), the alkali-based DAC sorbents developed by Susteon are capable of removing >70% of CO_2 from ambient air (~415 ppm) at an order-of-magnitude greater rate and releasing CO_2 under mild regeneration conditions. This enables significantly lower energy requirement for regeneration and allowing for the use of renewable electricity to provide low-temperature heat for desorption. Moreover, the presence of water (from the moisture in air) favorably impacts the DAC sorbent performance, representing another advantage over amine sorbents. Extensive experimental work demonstrates excellent sorbent regenerability and long-term, multi-cycle performance.

Susteon has explored multiple potential avenues for the DAC platform technology, including (i) the development of a novel reactor system to selectively heat the alkali adsorbent layer to facilitate CO_2 desorption while minimizing energy costs; and (ii) the development of a dual-functional material (DFM) to sequentially perform DAC and utilize the captured CO_2 *in situ* to produce renewable natural gas (RNG). Altogether, the technologies developed by Susteon will significantly lower the overall cost of DAC.

Copyright © American Institute of Chemical Engineers. All rights reserved.

625507 Experimental Characterization of Sorbents for Direct Air Capture of Carbon Dioxide

Abstract:

Direct air capture (DAC) technology is an unprecedented and important solution to climate change. Even if we reduced CO_2 emissions to zero in a relatively short period of time, we will still need to remove 1,000 gigatons of CO_2 from the atmosphere by the end of the century to prevent the planet from warming above 1.5 °C[1]. Temperature, pressure, and moisture swing DAC processes are amenable to a range of advanced sorbents which promote efficient CO_2 sorption and regeneration[2,3]. The development, selection, and application of appropriate sorbents require a full understanding of essential characteristics during both sorption and desorption: 1) reaction kinetics; 2) mass transfer; 3) heat transfer; 4) equilibrium capacity; 5) cyclability (lifetime). The definition of these characteristics enables the development, design and scale-up of system processes and equipment to attain technical and economic feasibility[4]. There are numerous candidate materials with potential for DAC, each can be used in a variety of process cycles or regional climates. We have developed a systematic approach to evaluate sorbents from the milligram to tonne scale considering the important parameters mentioned above. Our systems comprise 1) a closed system (150 mL total volume) for fast screening of sorbents in the 1-10 mg scale; 2) an open flow system to test 0.1-1 g scale; 3) a wind tunnel system to test 1-100 g scale under representative air flow velocities and with appropriate form factor; 4) a bench scale reactor to test 0.1-2 kg scale; 5) and finally, our mechanical tree prototype which will be able to test at the 0.1-1 tonne scale. We will describe those systems and provided relevant results for representative sorbents which demonstrate favorable rate and capacity characteristics.

References:

1. Intergovernmental Panel on Climate Change (IPCC). Special Report Global Warming of 1.5 C°, Available at: <www.ipcc.ch>
2. Tao Wang; Klaus S. Lackner; and Allen B. Wright. Moisture-swing sorption for carbon dioxide capture from ambient air: a thermodynamic analysis. *Phys. Chem. Chem. Phys*., **2013**, 15, 504.
3. Xiaoyang Shi; Hang Xiao; Habib Azarabadi; Juzheng Song; Xiaolong Wu; Xi Chen; and Klaus S. Lackner. Sorbents for the Direct Capture of CO_2 from Ambient Air. *Chem. Int. Ed.* **2020**, 59,6984 –7006
4. Habib Azarabadi; Klaus S. Lackner. A sorbent-focused techno-economic analysis of direct air capture. *Applied Energy* 250 **(2019)** 959–975.

Copyright © American Institute of Chemical Engineers. All rights reserved.

623022 Coupling Accelerated Mineral Carbonation and Direct Air Capture Processes

Abstract:

Global warming and climate change concern sparked significant interest in gigaton scale direct air capture (DAC) of carbon dioxide. Alkali hydroxide solutions such as sodium hydroxide solution are the benchmark solvents for CO_2 reactive absorption from thin air[1]. However, recycling sodium hydroxide from aqueous sodium carbonate solution, which is the product of CO_2 absorption, is critical for the success of the process. Technologies based on lime-soda ash were previously proposed, which however requires regeneration of lime by high-temperature calcination[2]. Also, the generated pure CO_2 from calcination needs to be sequestered through additional chemical production or geological storage processes. Instead, coupling of accelerated mineral carbonation and sodium hydroxide-based DAC process would remove the need for lime and energy-intensive calcination. Mineral carbonation works by converting Ca-rich aluminosilicate minerals or industry-derived residues into calcium carbonate in an aqueous solution. For successful coupling with DAC, we proposed to accelerate the slow carbonation of silicates in alkaline media (pH>12). So far, researchers mainly preferred acidic media for enhancement of its rate and alkaline pH received little attention. Based on our new observations in the silicate dissolution-carbonation mechanism, the solution chemistry of carbonate-bearing solutions may be tailored to achieve high carbonation efficiency. Using coal fly ash as an example, we present the ash carbonation kinetics and its dependence on solution chemistry and mineral phase heterogeneity. We will discuss the challenges with other feedstocks and present far-from-equilibrium thermodynamics as a new tool for rapid feedstock screening and process design.

References:

1. Lackner, K. S. *et al.* The urgency of the development of CO_2 capture from ambient air. *Proc. Natl. Acad. Sci. U. S. A.* **109**, 13156–13162 (2012).
2. Sanz-Pérez, E. S., Murdock, C. R., Didas, S. A. & Jones, C. W. Direct Capture of CO2 from Ambient Air. *Chem. Rev.* **116**, 11840–11876 (2016).

Copyright © American Institute of Chemical Engineers. All rights reserved.

634336 Flexible Low Temperature CO_2 Capture System, E-Cachys™

Abstract:

This project focuses on the design, integration, and optimization of a flexible natural gas combined cycle plant with carbon capture, capable of operating in a highly variable renewable energy environment. The objective of the project is to design and size a low temperature, hybrid sorbent-based CO_2 capture system (E-CACHYS^{TM}) that balances the degree of capture and flexibly schedules sorbent regeneration such that capital costs of the capture system are optimized to achieve the lowest levelized cost of capture. Innovative integration of the proposed system reduces capital expenditures for CO_2 capture by more than 30 percent compared to state-of-the-art technologies.

The technical approach is based on an innovative but proven multi-phase sorbent technology for post combustion capture of CO_2 from flue gas. The hybrid sorbent technology, which consists of a regenerative sorbent and a novel heat exchange system for optimal energy recovery, seamlessly integrates into the natural gas combined cycle architecture. The sorbent, which has high capacity for CO_2 (10-15 g CO_2/ 100 g sorbent) has fast reaction kinetics, is insensitive to oxygen degradation and non-toxic, and has low material and manufacturing costs. The sorbent has also shown to be applicable for capture from low CO_2 concentration streams (i.e., direct air capture).

In this study, we will report initial data on the sorbent regenerator design and laboratory-scale results of sorbent performance regarding working capacity and energy requirements as a function of key regenerator operating variables such as temperature, pressure, and steam/CO_2 ratio. We will also report on component process model algorithms to be used to design the overall process.

Copyright © American Institute of Chemical Engineers. All rights reserved.

625800 Cyclic Adsorption and Persulfate Oxidation on Iron-Activated Activated Biochar for Removal of Microcystin-LR in Water and Wastewater

Abstract:

In this study, $FeCl_3$-activated biochar (FA-BC) was prepared and applied for the removal of microcystin-LR (MC-LR) in lake water and DI water by adsorption with persulfate (PS) oxidation-driven regeneration. The increase in temperature resulted in a faster increase in the MC-LR adsorption rate on FA-BC due to the improvement in the desorption of MC-LR from FA-BC and the diffusion of MC-LR in water. The lower pH and higher PS concentration led to the higher regeneration efficiency of MC-LR-spent FA-BC by PS oxidation. In addition, the MC-LR oxidation and PS reaction rates were remarkably improved by factors of 3.4 and 3.5 with increasing temperature from 20 to 50 °C. Both diffusion and desorption of MC-LR from FA-BC controlled the MC-LR oxidation rate during the PS oxidation-driven regeneration of MC-LR-spent FA-BC. Sulfate radical was proven to be the main oxidant participating in the PS oxidation process. Under the conditions of 200 mg/L PS, pH 6, and 50 °C, 83 - 99% regeneration efficiencies were achieved during four cycles of adsorption-PS regeneration in DI water and lake water. Moreover, XRD analysis showed that the crystal structures of α-Fe_2O_3, Fe_3O_4, and Fe^0 in FA-BC were stable during PS oxidation. Therefore, this study indicates effective regeneration of MC-LR-spent FA-BC by PS oxidation, which would allow cost-effective treatment of MC-LR in water.

Copyright © American Institute of Chemical Engineers. All rights reserved.

627563 Electrochemical Recycling N from Wastewater Towards Fossil-Free NH_3 Production Using Dispersed Ru Atom in Cu Metal Catalyst: An Ab-Initio Study.

Abstract:

The use of fertilizes has led to nitrate buildup in water systems, and now routinely exceed health and safety levels. In this work, we use density functional theory calculations to examine electrolytic nitrate reduction pathways on Ru in Cu single atom alloy (SAA) catalysts as compared to the parent Cu and Ru bulk material. We examine two potential pathways, reduction to N_2, which is useful if the water is to be consumed, or NH_3, which could be used in concentrated water to cyclically produce fertilizes, closing the nitrate cycle with lower energy input. Specifically, we map the minimum reaction pathways for nitrate reduction to N_2, NH_3 and other reduction products on the Cu[111], Ru single atom in Cu[111] and Ru[001] surfaces. The Ru SAA and Ru bulk enhance the electrocatalytic performance because they bind NO_3 more strongly, thus increasing the surface concentration of the desired reactants. Similarly, the Cu catalyst preferentially directs reduction along the N_2 pathway because the N-surface bonds are weaker. Thus the formation of N-N bonds is favored over N-H bonds by 1.0 eV. The Cu-Ru has 0.3 eV smaller reaction barriers for NH bond association as compared to N_2 association. With higher reaction barriers for N_2 and nitrous oxide reaction pathways over Cu-Ru catalyst, it synergistically reduces by-product formation, with selectivity towards NH_3 production. Lastly, we use microkinetic models to build correlations between applied potential and Nitrate concentration to expected reactor performance, showing the criticality of moderate, but not overly high potentials, in order to maximize production of NH_3 or N_2 rather H_2.

Copyright © American Institute of Chemical Engineers. All rights reserved.

630918 Methodology Development for the Identification of Environmental Microplastic Samples Using Py-GC/MS

Abstract:

Microplastic pollution has gained traction over the past few years for its increasing threat to the environment and while several researchers have conducted studies to understand the impact, there remains significant research gaps in the characterization and identification of the plastic waste. Apart from being complex in size, shape, color, etc., environmental microplastics undergo weathering which can significantly affect the original properties of the material. Additionally, field samples are likely a mixture of organics and inorganics that are difficult to separate. These drawbacks make traditional characterization methods, such as microscopy or spectrometry, challenging. In their place, this research proposes a pyrolysis-gas chromatography/mass spectrometry (Py-GC/MS) method that can accurately be used to identify environmental microplastics by their characteristic pyrolyzates and additives. Microplastics used in this study were sampled from the Concord River and treated with either a sodium hypochlorite mixture (household bleach) or potassium hydroxide (KOH) to understand the significance of digesting environmental samples and its effect on analysis results. The data obtained from this study suggests high concentrations of rubbers, styrenes, and acrylics in local waterways though in different proportions depending on the pretreatment method. This work provides foundational knowledge on the prevalence of microplastics in the environment and is intended to be a foundation for future microplastic studies.

Copyright © American Institute of Chemical Engineers. All rights reserved.

631199 Mesoporous Fe- or Cu-Doped MgO Nanoparticles for Photo-Fenton-like Degradation of Salicylic Acid

Abstract:

Water scarcity combined with its pollution by toxic contaminants is one of the key challenges facing humanity. Among emerging water contaminants, pharmaceuticals and personal care products (PPCPs) have been identified as a rising threat due to their presence in wastewater, surface water, and groundwater. Advanced water treatment techniques, such as microfiltration, reverse osmosis, and advanced oxidation processes (AOPs) have been demonstrated to be highly active toward removing such recalcitrant organic matter. AOPs are of particular interest since they allow for the degradation of organic contaminants that are otherwise recalcitrant and ubiquitous in municipal, industrial, and agricultural wastewater. While the homogeneous Fenton reaction is highly effective, it intrinsically faces two challenges that prevent successful large-scale implementation including (a) the solution pH of ~3 and (b) the waste iron sludge generated. Therefore, research has focused on heterogeneous Fenton catalysts due to their easier separation and reusability, as well as a solution pH closer to the circumneutral which can be discharged into the watershed. This presentation will focus on mesoporous MgO based nanoparticles as photo-Fenton-like catalysts that possesses dual functionality. They both degrade target organic molecules *via* ROS generation and control the pH near the catalysts thus preventing dissolution and, potentially, loss of iron into the solution. In particular, mesoporous Cu- and Fe-doped MgO nanoparticles were synthesized using a facile sol-gel method and utilized for photo-Fenton-like degradation of salicylic acid (SA). The MgO surface dissolution facilitated an increase in the pH under the reaction conditions that allowed the metal doped MgO catalyst to be active without detectable metal ion leaching. Under simulated solar radiation, salicylic acid, a model compound since it is a common ingredient in nonsteroidal anti-inflammatory drugs (NSAIDs), was completely degraded with an initial rate constant of 0.048 min^{-1} at the optimal reaction conditions of 500 ppm loading of 5% Fe-MgO, 20 mM H_2O_2 concentration, and 50 ppm SA concentration. The catalyst was stable over 5 reaction cycles. The Fe-MgO catalyst was shown to have a surface area up to 171 m^2/g and contain hematite (Fe_2O_3) nanoparticles with octahedrally coordinated iron catalytic centers, as inferred from diffuse reflectance UV-vis measurements. Post-reaction catalyst characterization showed that some Fe^{2+} was present in the catalyst due to the redox cycle during the chain initiation and propagation steps of the reaction.

Copyright © American Institute of Chemical Engineers. All rights reserved.

631660 Bisphenol a Degradation and Biofilm Formation By Engineering *Shewanella Oneidensis*

Abstract:

Bisphenol A (BPA) has been widely used as a plasticizer in the production of synthetic polymers such as food storage containers and feeding/non-returnable bottles. BPA has drawn global attention due to its ability to interfere with endocrine systems, causing carcinogenicity, immunotoxicity, and embryotoxicity. Wastewater treatment bioprocess barely removes BPA due to the poor BPA degradation ability and efficiency of applied microorganisms. *Shewanella oneidensis* has been studied and a in the biodegradation process of wastewater treatment because of its excellent extracellular electron transfer properties. In this work, we engineered *S. oneidensis* MR1 to enable BPA degradation. We integrated *bisdA* and *bisdB* encoding a BPA degradation pathway from *Sphingomonas bisphenolicum* AO1 into *S. oneidensis* MR1. The engineered *S. oneidensis* MR1 exhibited over 95% of 10 μg/mL BPA degradation within 10 min, while the engineered *Escherichia coli* with the same BPA degradation pathway showed 80%. Similar BPA degradation percentages were obtained when the BPA concentration was increased to 100 μg/mL. However, high concentrations of BPA was toxic to cells: we observed that 100 μg/mL BPA killed and inhibited the cells after initial BPA degradation. In addition, we improved the biofilm formation of *S. oneidensis* by enhancing curli biogenesis and c-di-GMP production. Since biofilms are used to degrade toxic organic compounds during wastewater treatment, we study BPA degradation in functional *S. oneidensis* biofilms. We will discuss the relation between BPA degradation, enhanced electron transfer, and biofilm formation of *S. oneidensis*.

Copyright © American Institute of Chemical Engineers. All rights reserved.

632590 Sorption of Heavy Metal Ions By Composite Cryogels Based on Bentonite Clay with Ionic and Nonionic Polymer

Abstract:

Composite cryosorbent based on bentonite clay (BC) mineral from domestic origin Manyrak field and ionic polymer – polyacrylic acid (PAA), nonionic polyacrylamide (PAAm) was synthesized and characterized physical and chemical properties, sorption capacity in relation to cadmium and lead metals using individual and model aqueous solution with the mixture of given two metal ions. The synthesis of PAA, PAAm cryogel was carried out by radical polymerization methods at -20°C for 8 hours. The composite BC/PAA cryogel was obtained with the ratio 1:1, 1:5 and 1:10. The structure and morphology of composite cryogels were investigated using FTIR, SEM and AFM. The content of bentonite clay was determined and found that the clay consist of mineral montmorillonite with high dispersion. The SEM and AFM images show the high porosity of polymer/clay composite cryogels compare to pure cryogels. The sorption capacity of PAA/BC for [Pb^{2+}] was 97% and for [Cd^{2+}] was 98%, and as for PAAm/BC for [Pb^{2+}] was 88% and for [Cd^{2+}] was 89%. Overall, the sorption capacity of PAA/BC was higher than PAAm/BC composite cryogels. Moreover, during the evaluation of sorption capacity of cryogels using the model solution, the value of both PAA/BC and PAAm/BC composites for [Pb^{2+}] was higher than that of [Cd^{2+}]. This is predicted by their differences in coordination number, atomic mass and radius.

Copyright © American Institute of Chemical Engineers. All rights reserved.

631612 Life Cycle Assessment of Resource Recovery Technologies on Large-Scale Dairy Farms

Abstract:

Large-scale production of livestock for dairy products has a significant impact on the environment through greenhouse gas emissions, water pollution, and resource consumption. In particular, manure management at large-scale concentrated animal feeding operations (CAFOs) poses risks to air, water, and soil quality.

However, manure is also a rich source of organic matter, nutrients (nitrogen and phosphorus), and water. Resource recovery technologies offer a way to recycle valuable compounds in manure into beneficial products while simultaneously reducing harmful emissions. In this research, a novel Nutrients, Energy, and Water Integrated Recovery (NEWIR) system is compared to a typical California dairy CAFO with life cycle assessment (LCA) methodology. NEWIR consists of three innovative processes for integrated resource recovery from dairy manure: 1) hydrothermal carbonization (HTC) for energy, 2) algal raceway ponds for nutrients, and 3) membrane distillation (MD) for high-quality water. HTC creates a carbon-rich hydrochar and a nutrient-rich aqueous stream. Hydrochar can be burned in conventional coal power plants, replacing a fossil fuel with a renewable form of bio-energy. The aqueous HTC effluent serves as a water and nutrient source for microalgae that are used as a protein-rich cattle feed supplement to offset soybean production. Finally, MD uses waste heat from HTC to recover high-quality water from the algae effluent that is reused on-farm for cleaning and sterilization, replacing conventional reverse osmosis. LCA results are used to evaluate how resource recovery technologies affect relevant environmental impact categories such as carbon footprint, water footprint, and eutrophication. The life cycle inventory is developed from experimental data and modeling for a full-scale NEWIR system and analyzed using the ReCiPe impact assessment method. Results will be presented for a full-scale model of the NEWIR system, including avoided products due to resource recovery. By creating useful products from a biowaste stream, NEWIR paves the way towards sustainable food production on a large scale.

Copyright © American Institute of Chemical Engineers. All rights reserved.

632629 Consumption-Based Accounting for Tracing Nutrient Footprints of Beef Supply Chains in the United States

Abstract:

Food supply chains span multiple spatial and temporal scales and have evolved to become increasingly complex. Food products go through many different processing steps before being ready for consumption. While highly connected food chains provide numerous benefits, they lack traceability. As such, understanding their resiliency, security, and associated environmental burdens of consumption become difficult. One specific class of impacts is the nutrient footprint of food consumption. Application of synthetic fertilizers and release of reactive nitrogen and phosphorus has been linked to degradation in water quality and loss of biodiversity across the globe. Beef production involves consumption of large quantities of animal feed, the production of which is heavily dependent on fertilizer application. Although there is a rich body of work on nutrient footprint estimation for beef production, there is a gap in the literature regarding the spatial distribution of the nutrient releases throughout the supply chain. Additionally, bulk of the studies focus on production related impacts making it difficult to trace environmental impacts of beef consumption that typically occur far from the point of consumption.

We develop an optimization-based framework to model supply chain networks of beef products at the county level. Using publicly available data, we construct a weighted network of nutrient flows based on nutrient mass balance, including synthetic fertilizers, manure production, and crop uptake and residues. Preliminary results show that the beef supply chain is highly concentrated in the central region of the United States, with a majority of slaughtering facilities owned by a small number of companies. Additionally, there is a large inflow of cattle to the central region for feeding and slaughtering. Feed is sourced from several different counties and states depending on the specific crop type. Analysis reveals that most states consume far more meat than they produce within their boundaries, and highly populated areas are reliant on the inflow of products from the Midwest. Our results also highlight the disparity between production and consumption-based impacts and spatial heterogeneity in nutrient footprint of beef. Implications of these findings for reducing the environmental impacts of beef supply chain, consumer food choice, and policymaking will be discussed.

Copyright © American Institute of Chemical Engineers. All rights reserved.

623757 Assessment of Global Water Stress Using the Generalized Global Sustainability Model (GSM)

Abstract:

Water is essential for human survival and critical for other spheres of human civilization, namely, agriculture, industrial production, and energy production. Many regions in the world are expected to face water scarcity issues in the near future. However, the water sector cannot be studied in isolation. Agriculture, responsible for food and feed production, is the largest consumer of water as compared to other anthropocentric consumption routes such as municipal, industrial, livestock, and energy [1]. Energy is an important contributor to human development and well-being, and it is also a key input to agriculture. With electricity-powered pumps, water can be drawn from deeper wells or transferred for long distances, highlighting the role of energy in improving water accessibility. However, energy production itself requires water, and some energy production technologies are highly water-intensive. Biopower and biofuels have been shown to have high water footprints [2]. Additionally, excessive water withdrawal due to cheap energy may further deplete freshwater resources. Thus, ensuring adequate water availability for everyone is a problem that cannot be addressed in isolation. Instead, it is a part of the food-energy-water (FEW) nexus.

The inter-dependencies of these three sectors are complex. Scanlon et al. [4] highlighted the criticality of understanding the FEW nexus. Several researchers have studied the FEW nexus with different objectives. For example, D'Odorico et al. [5] closely examined the individual links between food-energy, energy-water, and food-water and identified the critical areas for action. However, this study did not capture the integrated nature of the FEW nexus. The scale of the FEW nexus is also an essential factor. However, Nie et al. [6] studied the optimization problem involving the FEW nexus on a small scale. Franz et al. [7] studied the FEW nexus at the regional scale, which again represents a limited spatial scale. However, as global systems are large, inter-dimensional influences may take time to manifest. Hence, integrated cross-disciplinary models exploring the interactions between these various components has become increasingly important. Such models and their applications can improve our understanding of the complex relationships between these sectors and provide prescriptive recommendations using scenario analysis. Development of such global models began with World3 by Meadows et al. [8]. The World3 model included subsystems such as food production, industry, population, pollution, and non-renewable resources. Some of the other notable examples of global models are GUMBO [9], Earth3 [10], and HANDY [11]. More recently, our group [12] developed the generalized Global Sustainability Model (GSM). This is conceptually similar to the earlier version developed by Kotecha et al. [13]. However, we revised the model by selecting parameters as well as variable ranges based on global historical data. As compared to other global models, its ecological dimension is represented with features such as multiple trophic levels and intra-trophic level diversity. Moreover, the explicit integration between the economic and ecological dimensions is another strong point.

One key limitation of the revised model was that it did not explicitly capture water as a resource. It was assumed to be a part of the generalized resource pool. Water, though, is a crucial resource and is a part of a nexus, as highlighted previously. The critical novel contribution of this work is the incorporation of a global water network in GSM. The revised model with the water network is then used to study scenarios and policy alternatives. The analysis work tackles the following questions using the GSM: with business-as-usual conditions, can we have a sustainable water supply for all? How would the situation change if there is a population explosion? Can the improvements under Sustainable Development Goal 6 help in preventing undesirable situations?

Figure 1 shows GSM with the water network system in place. The GSM consists of three ecological trophic levels and human society. There are eight ecological compartments, human households, and their inter-dependencies give rise to a complex food web. Lotka-Volterra type equations are used to represent these relationships. The economic dimension is represented in the form of four sectors: agriculture, livestock, industry, and energy. Typically, economic state variables, prices, and production targets for the four sectors are governed by their products' inventory and the wage rate. As a price-setting mechanism is used in this model in lieu of the Computable General Equilibrium Approach, the prices are used to compute human society's demands for each industry. The supply is constrained by the raw material availability limited by the ecological dimension. This is where the interaction of ecological and economic dimensions comes into the picture.

The water system in the GSM is developed in two steps: first, the global water reservoirs and flows are established, and then, the demands by different sectors are characterized. The water reservoir represents the shared pool of global exploitable water. Inaccessible water whose utility cannot be harvested owing to technological or economic limitations or pollution is shown by the non-exploitable water reservoir. The water recycling compartment represents the mass of water undergoing water recycling. Agriculture, industrial, energy and municipal consumption of water is shown by flows to P1, IS, EE and HH compartments. In order to model the water demands, we define sectoral water intensity as the quantity of water withdrawn per unit of the desired variable. Agriculture and livestock are part of the ecological system, and, hence, would require water for their survival. Considering this constraint, P1 and H1 masses are used as shadow variables. On the other hand, the industrial and energy sectors do not need continuous replenishment of water like the ecological sectors. These sectors withdraw water during the production process only. Hence, IS and EE production terms are used as shadow variables. For municipal water withdrawal, the human population is used as a shadow variable. To summarize, P1 and H1 mass for the agricultural and livestock sector, IS and EE production for the industrial and energy sector and population for the municipal withdrawal are used as shadow variables. A sectoral intensity trend is a relationship between the historical data and corresponding shadow variable. These trends are used for projecting the future global water demands for those sectors.

The region greatly influences the demand-supply dynamics of water under consideration. To study this aspect, countries are categorized into six groups based on the continent where they are located: Africa, Asia, Europe, North America, South America, and Oceania. As exact details of water demand and supply distributions for various sectors are not available, the shadow variables approach is used to obtain the regional distribution of the water variables for each sector. The availability of exploitable water in a region is relatively stable and can be considered proportional to its area. Thus, the area of the country is considered as a shadow variable here. Again, shadow variables represent the water demand for various sectors: municipal water demand - population; agricultural water demand - area under agriculture; industrial and energy sectors - GDP (measured in international $ 2011 and adjusted for inflation). This historical country-level data is aggregated, and the contribution of each of the six regions towards the total value is computed. The approach of the piece-wise linear fit [14] is used to obtain the linear regression fits. Global and regional demand-availability dynamics, the potential adverse effect of the population explosion, and the positive influence of water resource management are analyzed.

The model was parameterized using historical data until 2013, and simulations were performed until the year 2120. Simulation results indicate that overall global water resources are adequate to meet the demands of human society. Here, the metric used to quantify the relation between water demand and availability is net water stress. It is the ratio of total water demand and exploitable water resources. This metric is more realistic as it considers the techno-socioeconomic feasibility of water utilization. Accordingly, global net water stress is close to 60%. However, deeper investigation considering distribution across various regions in the world reveals a disparity in water availability. Asia is the most water-stressed region in the world. Its water demand is expected to exceed availability by 2036. The population explosion scenario also affects water stress in the long run. For Asia, the peak water stress for the population explosion scenario increases from 109% to 124%. Water use efficiency improvement offers a way to prevent or at least delay this impending crisis. Scenario analysis indicated that at least a 30% reduction in water demand in the agricultural sector is necessary to avoid the water crisis in Asia altogether.

To summarize, the global FEW nexus's sustainability analysis identifies Asia as a flashpoint of the impending water crisis and advocates aggressive reforms in water use agricultural patterns resulting in substantial efficiency improvement to prevent the same. The model will be used to identify specific technology improvement targets and policy alternatives to ensure global water sustainability.

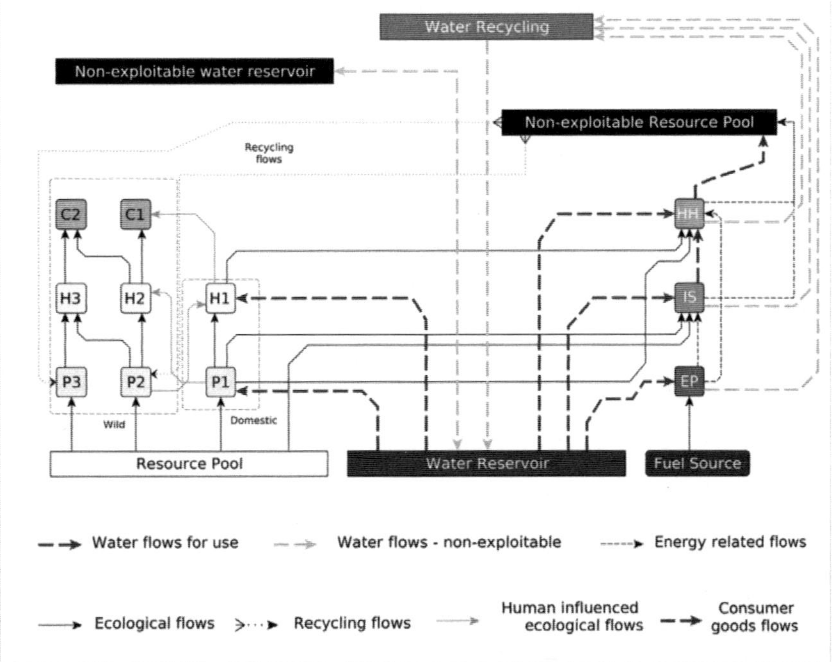

Figure 1: The Generalized Global Sustainability Model (GSM) is shown here. It has a trophic level structure. In the ecological dimension shown on the left side, higher trophic level depends on the lower trophic level. These are three ecological trophic levels are plants, herbivores and carnivores. These are classified into domesticated/human controlled and wild. Domesticated ecological compartments, that is, P1 and H1 represent the agricultural sector and the livestock respectively. P2 which is a part of the uncontrolled part of the ecosystem and represents grasslands is consumed by H1. Such multiple dependencies among the compartments give rise a complex food web. On the right side, HH represents human households, and IS and EP represent industrial sector and energy production sector respectively. The water dimension related components are shown in the center. Different types of flows are summarized below main figure.

Copyright © American Institute of Chemical Engineers. All rights reserved.

624215 Bacteria Containment in Filtration

Abstract:

Drinking water systems contain a range of planktonic bacteria and bacterial communities that are most stable in the filtration system. Filters can have a negative impact if potential pathogens escape into distribution systems and contribute to infrastructure decay due to corrosion. Previous studies have determined that drinking water biofilms, especially those in biologically active sand filers, are stable communities that are able to persist the fluctuations in the source water community, and to develop resistance to disinfectants commonly used in drinking water. Focusing on the understanding of how bacteria biofilms can be contained further in filters by varying the filter material, and mechanical aspects of water. The effects of varying filter media and varying flow rates on bacteria biofilms by running trials in a plug flow reactor were measured.

Copyright © American Institute of Chemical Engineers. All rights reserved.

627378 Use of the Aqueous Phase and Char from Hydrothermal Liquefaction of Biomass in Compost and Heavy Metal Adsorption Applications

Abstract:

Hydrothermal liquefaction (HTL) has shown much promise as a technique for converting waste biomass into bio-crude oil without the need for drying. Implementation of HTL at the large scale has been hindered by the limited options for value-added use of the aqueous phase (HTL-AP) and char (HTL-char) co-products.

This study investigated the applications of HTL-char as a mesoporous material for adsorption of heavy metals from water and of HTL-AP as an additive for compost. Food waste HTL-char, and a magnetic modified HTL-char prepared through co-HTL of food waste and red mud, were characterized and tested for removal of Pb and Cu from wastewater. Batch experiments were used to study the effects of magnetization, initial wastewater pH, contact time, and initial metal concentration on adsorption performance. Preliminary results from isotherm analysis showed that HTL-char derived from food waste has a higher adsorption capacity for Pb than for Cu. The maximum adsorption capacity of Pb (32.9 mg Pb g^{-1} dry char) was obtained at pH = 3.5, T = 23.1°C, and initial PB concentration of 80 mg/L. The feasibility of utilizing HTL-AP to add moisture (and some carbon/nutrients) to compost was assessed in terms of the influence on microbial activity, nutrient availability (C/N ratio), humidity of the compost pile over the course of production, and quality of the finished compost.

Copyright © American Institute of Chemical Engineers. All rights reserved.

628007 Analysis of the Water-Energy-Food Nexus in Agriculture Dependent Rural Communities By Using Mathematical Programming

Abstract:

The food-energy-water (FEW) nexus has become a useful tool to address the complex and interrelated nature of global resource systems, on which society depends to achieve different social, economical and environmental goals. A rapidly growing world population is causing an unsustainable pressure on resources. The demand for water, food and energy is expected to increase by 30-50% in the next two decades. An effective connection between the elements of the nexus can be achieved through the use of new technologies or forms of energy production. For example, the use of agricultural and food biomass wastes for the production of biomaterials and bioenergy. The application of optimization tools in the analysis of the FEW nexus represents a useful alternative for decision-making as well as for the implementation of public policies when managing the elements of the nexus.

This work presents a mixed integer linear programming model to approach the food-energy-water nexus of agriculture based rural communities. The model considers the optimal management of water, food production involving different irrigation and cultivation schemes, and the use of biomass for the production of biofuels. The formulation includes the estimation of a socioeconomic development indicator and the potential implementation of public policies, such as government financing and incentive programs to agricultural activities. The effect of climate spatial variations on physical and biological resources and phenomena are also considered by using a multi-period approach. Three objectives are included in the model: maximization of annual net profit, minimization of water consumption and minimization of unsatisfied demand for food. The model was applied to a case study of an agricultural population, which has been solved through the GAMS modeling environment. The model applied to such case study consists of 23810 equations, 22054 continuous variables and 698 binary variables. The results show that the proposed methodology is capable of obtaining the optimal configuration of the food and biofuel production system with desirable results in terms of the sustainable management of resources and the application of financing programs and government incentives. The following are some of the conclusions obtained: *i)* it is shown that the application of government incentives has a positive impact on the modernization of agricultural activities, *ii)* the most important program for socio-economic development is government financing, *iii)* the modernization of irrigation allows a significant reduction in water consumption, *iv)* the installation of a biorefinery brings economic benefits and favors the production of some foods, *v)* it is possible to distribute food production, and therefore economic resources, in such a way that the same per capita income is obtained for each of the rural regions under analysis.

Copyright © American Institute of Chemical Engineers. All rights reserved.

627509 Flue Gas (CO_2) Extraction and Bio-Sequestration (FGXB) for Increased Crop Yields and Water Utilization Efficiency (WUE)

Abstract:

There are no existing technologies and markets that can use the billions of tons or gigatons per year (GT/Y) of CO_2 produced by stationary industrial and power plant, and mobile transportation sources, until now. The Black sWan Cycle (BWC) proposed in this presentation uses stationary and mobile sources of CO_2 bearing exhaust to make food (agricultural products), increase biomass, and make water/reduce water consumption with Flue Gas eXtraction and Biosequestration (FGXB.) The increased biomass is used for fuel. BWC with Membrane Air Enrichment (MAE) performs Direct Air Capture (DAC) of CO_2, water, and oxygen and uses Oxy-Combustion to decrease fuel consumption, increase capacity, increase the heating value of low-quality biofuels, such as solid biomass and landfill/sanitation anaerobic digester biogas, and concentrate the CO_2 in the flue gas for more economic use and capture.

The first step of BWC utilizes the conventional science of Crop Carbon (dioxide) Enrichment (CCE) that has been practiced in greenhouses for almost a hundred years, and researched in crop Open Top Chambers (OTC) and Free Air Carbon (dioxide) Enrichment (FACE- meaning without any enclosures) systems for decades by the Brookhaven National Laboratory and USDA. CCE is the agricultural/ horticultural practice of raising a plant's biosphere from ambient 400 ppmv to under 1000 ppmv CO_2 for the purpose of increasing biomass, yield, growth rate, and water utilization efficiency of plants by minimum 40%. Dr. Bruce Kimball et al of USDA has produced dozens of studies for wheat, cotton, citrus, grapes, and soybeans validating this science and its performance, including a 17 year study showing increase of citrus yield by average 70% by using 550 ppm CO2. Conventionally the CO_2 is provided at a high operating cost ($300/Ton) by fit-for-purpose fuel fired flue gas (at 5 to 10% CO_2) generators, or supplied from trucked-in/ locally stored near pure (90+%) CO_2 liquid or dry ice, and then diluted down to 1000 ppmv CO_2.

The Kolodji Corporation, an energy carbon management engineering company, utilizes patented and patent pending BWC technologies to commercialize and monetize the use of CO_2 in emissions with BWC FGXB by making food and water, saving water, and producing biomass for fuel. Another crucial aspect is the low energy/cost BWC Wig Membrane which uses Direct Air Capture (DAC) to recycle the fugitive CO_2 in a carbon enriched orchard by producing an enriched oxygen and water, and further CO_2 enriched permeate stream for substitution of combustion air in a conventional fired system by using the BWC Membrane Air Enrichment Oxy-Combustion (MAE) process. The benefits of MAE are reduced firing fuel consumption, increased firing efficiency, temperature, and heating value, as well as water recovery, Direct Air Capture of Water and CO_2, and increased concentration of the CO_2 in the flue gas for ease and more economic use and capture from the flue gas.

Four pilot scale applications of under 1000 Ton CO_2 per year BWC FGXB and MAE have operated in orchards in Kern County, California since 2018. The largest pilot facility is in a 50-acre almond orchard in the midst of 100s of acres of crops, across from a refinery emitting 200,000 tons per year

CO_2, and is funded by the California Department of Food and Agriculture (CDFA.) A demonstration scale MAE application that increases the sustainability and CO_2 capture potential of a biogas/natural gas fired 2.5 MW power plant, and a second smaller (under 1 MW) in two California sanitation districts are planned. These projects emphasize the key benefit of MAE to eliminate the need for supplemental firing of the 100% sustainable renewable biogas fuel with un-sustainable non-renewable natural gas by raising the heating value of the biogas using oxy-combustion. Further, the flue gas is concentrated from 5 to 15% CO_2 making it easier to capture or use the flue gas. Funding potential is for $4.4 Million in grants pending award by the California Energy Commission in April/June 2021.

Shared in this presentation are pilot plant results, and demonstration unit plans including a Bryan Research and Engineering (BR&E) PROMAX(TM) process simulation models, heat and material balances and process flow diagrams, as well as utility consumption benefits and a capital/ operating cost-benefit analysis to profit (return on investment) from fuel savings, increased agricultural yields, and water make/ savings, and as well, highlighted carbon (dioxide) balance. With wide acceptance of this BWC technology, Kolodji Corp will justify targets of carbon neutrality in the state of California (negative 300 Million Tons per year CO_2 production) by 2025, twenty years earlier than the 2045 mandated statutory requirement. As well, there is the goal of US wide CO_2 deceleration (-0.05 Tons/year/year or negative 5 GT/Y) of CO_2 by 2030, and global neutrality by 2035.

Copyright © American Institute of Chemical Engineers. All rights reserved.

629119 Cryogenic Separation of CO_2, SO_2, and NO_x from Flue Gas

Abstract:

Energy use is the dominant source of U.S. CO_2 emissions, with the share estimated to be about 75%. Addressing climate change requires deep decarbonization of our energy, chemical, and manufacturing sectors. One approach involves carbon dioxide capture and storage (CCS) where CO_2 is captured at point emission sources and then transported to sequestration sites for long-term storage in geologic formations. A key CCS challenge is the high cost of CO_2 capture, which varies significantly depending on source concentrations and volumes [1]. For instance, power plant capture costs exceed $55/tonne, but capture costs can be below $33/tonne in the cement, iron, and steel industries. Although significant advances are underway in developing solvent, sorbent, and membrane capture technologies, their costs remain high.

We have developed a novel cryogenic separation process that can remove CO_2, SO_2, and NO_x from power plant flue gas [2]. Cryogenic separation has been substantially ignored by carbon-capture researchers. However, it has great potential because it requires only conventional equipment (e.g., compressors, expanders, heat exchangers, pumps). Furthermore, compared to competing technologies, it does not require expensive materials and chemicals that typically have limited availability (e.g., ionic liquids, metal organic frameworks, graphene).

In the cryogenic process, raw flue gas is compressed and cooled, which causes water vapor to condense. Subsequently, residual water is removed using a drier. Then, the dry flue gas is countercurrently cooled via heat exchange with exhaust gas. The chilled flue gas is refrigerated to condense CO_2 (>90%), SO_2 (>99.99%), and NO_x (>99.999%). The non-condensable nitrogen is warmed by passing countercurrently through the heat exchanger and then exhausted through the expander to capture energy. Then, liquid CO_2, SO_2, and NO_x are pumped to a high pressure (e.g., 150 bar) and are countercurrently warmed and captured. Hence, our cryogenic technology mitigates not only CO_2 emissions, but also SO_2, and NO_x pollutants.

Compared to other processes, the cryogenic process consumes less energy because significant energy integration is utilized. Moreover, appropriate use of waste heat also improves energy efficiency. The flowsheet is further optimized by using our in-house process synthesizer SPICE [3], which further reduces energy consumption and capture costs. Sensitivity analysis on several important decision variables (e.g., rotating equipment efficiency, utility cost) is also performed. Preliminary results indicate that even with the most conservative assumptions, the cryogenic process is very attractive compared to alternative process. Without credits for removing SO_2 and NO_x, current estimated capture and pressurization cost is less than $25/tonne CO_2.

References:

[1] Hasan, M. M. F.; First, E. L.; Boukouvala, F.; Floudas, C. A. A multi-scale framework for CO_2 capture, utilization, and sequestration: CCUS and CCU. *Computers & Chemical Engineering* **2015**, *81*, 2–21.

[2] Holtzapple, M.; Monjur, M. S.; Hasan, M. M. F. Cryogenic Separation of Carbon Dioxide and Sulfur Dioxide from Flue Gas. *U.S. Patent Application No. 63/132,148, filed Dec. 30* **2020**.

[3] Demirel, S. E.; Li, J.; Hasan, M. M. F. Systematic Process Intensification using Building Blocks. *Computers & Chemical Engineering*, **2017**, 150, 2–38.

Copyright © American Institute of Chemical Engineers. All rights reserved.

624246 Optimization and Analysis of Carbon Capture in Ethanol Biorefineries

Abstract:

Bioenergy with carbon capture and sequestration (BECCS) is a promising method for global warming mitigation that combines fuel production, to replace current fossil fuels, with permanent carbon dioxide removal (CDR). While typical industrial power plants that use natural gas as fuel have only a single point of carbon dioxide emissions, an ethanol biorefinery has three point sources of carbon dioxide emissions: from fermentation (~99 wt%), in biogas (~73 wt% CO_2), and in flue gas from solid residue combustion (~20 wt%). The relative flows of these streams are dependent on both the feedstock and pretreatment method used. However, previous studies on carbon capture in biorefineries have limited their analyses to biorefineries with a fixed feedstock and pretreatment technology, and predetermine from which of the three point sources of carbon dioxide.

Previous studies also often assume a fixed capacity of the biorefinery, typically of 2,000 tons of feedstock per day. However, increasing capacity has two important effects: it decreases fixed costs due to favorable sublinear scaling of equipment costs, but also increases the feedstock cost as larger transportation distances are required. Different methods of transportation such as by truck or rail, and the total distance of transport also affect the overall emissions balance and energy consumption of the entire biorefinery and supply chain system.

To address the limitations of previous studies, we use a mixed-integer nonlinear programming (MINLP) model of an ethanol biorefinery and supply chain system. By solving thousands of optimization problems with varying parameter values, we analyze the impact of feedstock selection, pretreatment method, biorefinery capacity, and carbon sequestration credit on the economic, energetic, and environmental performance of an ethanol biorefinery with carbon capture.

We find that biorefineries using feedstocks and pretreatment methods that lead to higher emissions from fermentation and anaerobic digestion tend to have lower average costs of capturing carbon dioxide. Biorefineries with high biomass to ethanol yields, such as those using dilute acid pretreatment, can capture high percentages of the carbon in the feedstock if energy is purchased, but less excess energy available from residue combustion limits carbon capture at an energetically self-sufficient biorefinery. Biorefineries processing feedstocks with high lignin content or that have low energy requirements create the most excess energy, which enables higher capture rates. We also find the inclusion of carbon capture technologies can significantly increase the cost-optimal biorefinery capacity.

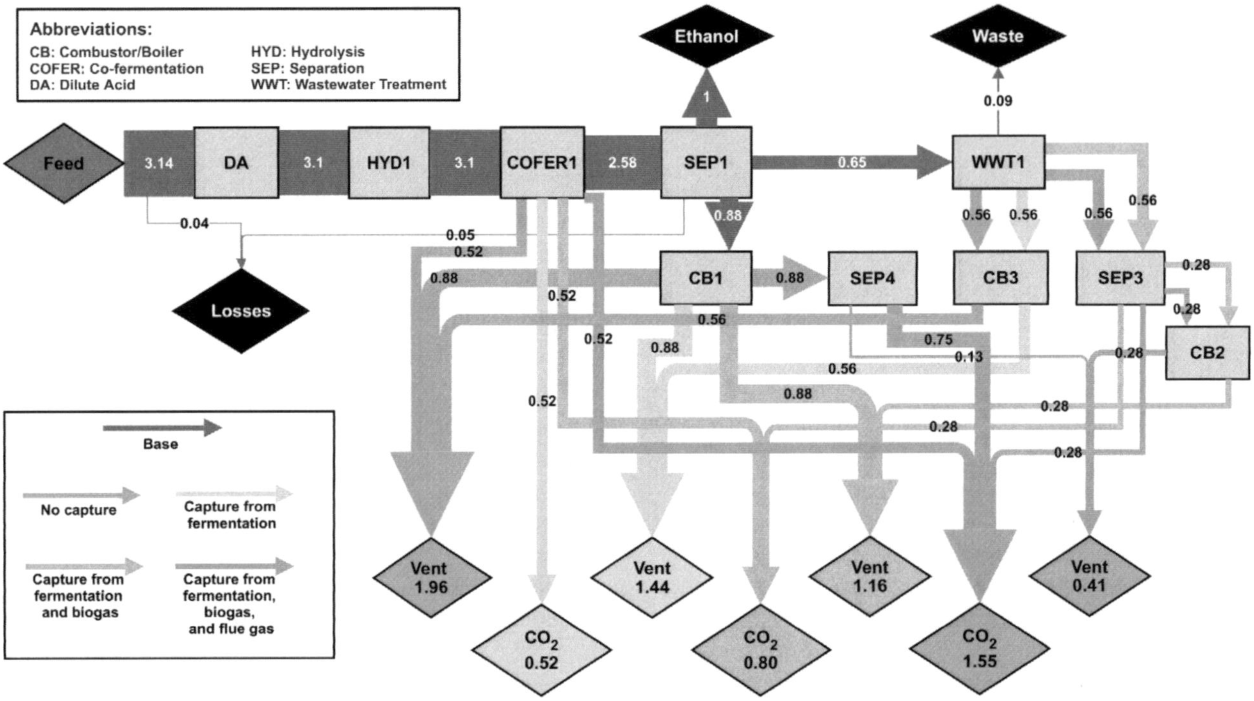

Copyright © American Institute of Chemical Engineers. All rights reserved.

Modeling of a flexible carbon capture-equipped power plant integrated with lime-based direct air capture

Moataz Sheha[a], Edward Graham[a], Dharik Mallapragada[a], Emre Gençer[a], Phillip Cross[b], James Custer[b], Adam Goff[b], Ian Cormier[b], Howard Herzog[a]

[a] MIT Energy Initiative, Massachusetts Institute of Technology, Cambridge, MA, United States

[b] 8 Rivers Capital LLC, Durham, NC, United States

The cost-effective integration of variable renewable energy (VRE) generation is critical for power sector decarbonization but is contingent on designing power systems to be more flexible. Deployment of carbon capture and storage (CCS) equipped fossil fuel power plants on the supply-side and direct air capture (DAC) technologies on the demand side can address the dual challenge of lower carbon emissions while providing grid flexibility. Here, we evaluate the operation of a novel low-carbon power plant concept that combines flue gas CO_2 capture with a lime-based DAC process in a way that minimizes impact on power generation equipment flexibility. The process involves three main elements: 1) flue gas from a conventional natural gas combined cycle (NGCC) power plant, is fed to a calcium looping system that includes a calciner and a carbonator, where limestone is thermally decomposed to lime and CO_2. 2) the CO_2 rich gas from the calcium looping system goes to an electrically-driven separation system based on well-established membranes and cryogenic distillation units, to recover high purity liquid carbon dioxide, which can be sent for sequestration. 3) the lime produced from the calciner would be used to capture additional CO_2 from air via passive contacting DAC process that is operated in a batch manner. The coupling between the operation of the CO_2 capture process (calcium looping + separation system) and the power plant is minimized because the calcium looping system operation is also dependent on additional CO_2 supplied from the feed limestone.

In this presentation we outline the steady state model of the process and discuss some special issues that needed to be addressed in its development. The process model was developed within Aspen Plus and integrates a custom membrane model developed in Aspen Custom Modeler. We analyze key performance metrics as a function of loading levels of the NGCC plant (net power output). The impacts of changing the power plant loading level on the power balance and equipment sizing (which is related to plant capital cost) will be highlighted. In the next phase of the project, the developed model will be used for developing surrogate models for individual unit operations and for estimating the capital and operating costs of the individual units along with input data from technology vendors. The surrogate models will be developed using ALAMO (Automated Learning of Algebraic Models). The surrogate models and cost information will be incorporated into an integrated design and operations optimization model to identify cost-effective system design and operation under various CO_2 prices and time-varying electricity price profiles. The optimization problem will be solved using a global optimization solver such as BARON.

Why global cooperation will be needed to achieve carbon dioxide removal at the Paris Agreement's scale

Solene Chiquier[ab], Mathilde Fajardy[c], Niall Mac Dowell[ab*]

[a]*Centre for Process Systems Engineering, Imperial College London, UK*
[b]*Centre for Environmental Policy, Imperial College London, UK*
[b]*Cambridge Judge Business School, UK*

Abstract

Through the 2015 Paris Agreement, most nations have committed to limiting global warming to well below 2°C, and ideally to 1.5°C above pre-industrial levels [1]. To achieve this, the strong reliance on carbon dioxide removal (CDR) — the direct removal of CO_2 from the atmosphere — as part of climate change mitigation pathways in integrated assessment models (IAMs) has been evidenced in the latest *IPCC Special Report on Global Warming of 1.5°C* [2]. Among almost all 1.5°C-consistent scenarios, IAMs have estimated that a cumulative removal of 189–1,186 $GtCO_2$ by 2100 would be necessary, relying predominantly on Bioenergy with Carbon Capture and Storage (BECCS) and Afforestation/Reforestation (AR). [2–4].

Other CDR solutions exist — including Direct Air Capture and Carbon Storage (DACCS), ocean fertilisation, enhanced weathering, biochar or soil carbon sequestration — but most of them remain largely theoretical [5]. With the exception of AR which has been long well established, only BECCS and DACCS have been conducted at the demonstration scale, yet nowhere near the scales required to deliver the Paris Agreement [6–8]. Consequently, only DACCS has been recently also considered in IAMs, but other alternative CDR solutions are still generally excluded [9–12].

Further, most of the IAMs scenarios have been developed with the objective of finding the least-cost mitigation pathway for limiting global warming by 2100, without considering realistic scale-up rates of CDR solutions, or seldom considering trade-offs with other objectives — such as the sustainable development goals for instance [13,14]. They have consequently raised some concerns about the feasibility and sustainability of the large-scale deployment of CDR solutions in the near future [5,15–17].

Yet, CDR solutions present different trade-offs between bio-geophysical resources (*e.g.,* availability of land and bioenergy, geological sinks for CO_2) and socio-economic factors (*e.g.,* low-carbon energy), themselves heterogeneously distributed across the globe. To name but a few, some CDR options, such as BECCS or AR, are land-intensive, and could therefore compete with the food and the biofuel sectors. Some, such as biochar or AR, may present significant co-benefits (*e.g.,* protecting biodiversity, increasing yield, soil carbon and/or other nutrients content), and might be rather inexpensive, but are highly uncertain in term of CDR permanence. Others, lastly, are to this day capital- and energy-intensive with unknown co-benefits, such as DACCS.

Modelling such challenges and opportunities in IAMs is overall all the more difficult, given the current speculative knowledge on alternative CDR solutions, if deployed at scale, and the low granularity (*i.e.,* not spatially-disaggregated enough) of IAMs. Yet it is vital to understand and anticipate the drivers, barriers and potentials of the deployment of CDR options at the national, and even regional level, as it is the level at which climate policy is implemented.

In this study, we investigate the potential of a portfolio of CDR options — AR, BECCS and DACCS — in meeting the Paris Agreement's 1.5°C-consistent CDR objectives between 2030–2100. We use the Modelling and Optimization of Negative Emissions Technologies (MONET) framework [18–20] to analyse alternative international cooperation scenarios: international cooperation with shared CDR burden and shared bio-geophysical resources use, (current) competition with domestic CDR burden and domestic bio-geophysical resources use, and national isolation with domestic CDR burden and domestic bio-geophysical resources use.

There are two distinct aspects to the MONET framework — modelling and optimisation. The MONET modelling framework firstly provides a regionally-disaggregated and time dependent whole system analysis of AR, BECCS and DACCS by evaluating each CDR option' resources use (*i.e.,* land and water), energy use or generation, and cost per unit of CO_2 removal. Five nations are considered — Brazil, China, the European Union (EU-27 and the UK), India and the United States — themselves disaggregated into 169 cells.

Secondly, the MONET optimisation framework uses these results to determine the optimal co-deployment of these three CDR options to limit global warming to 1.5°C, on a decadal time-step, by minimising the total net system cost, subject to region-specific constraints including: CDR targets; geological CO_2 storage availability for BECCS and DACCS; respectively marginal and potential reforested land availabilities for BECCS's biomass production and AR; respectively afforestation expansion and built rates for AR, and BECCS and DACCS; and water stress constraints. Following the IPCC P3 — also called "middle of the road" — scenario' global cumulative CDR target of 407 Gt of CO_2 removal by 2100, we apply the responsibility burden-sharing method based on historical CO_2 emissions to determine the domestic shares of the five MONET nations, resulting in cumulative CDR targets of 7–101 Gt CO_2 (254 Gt CO_2 together).

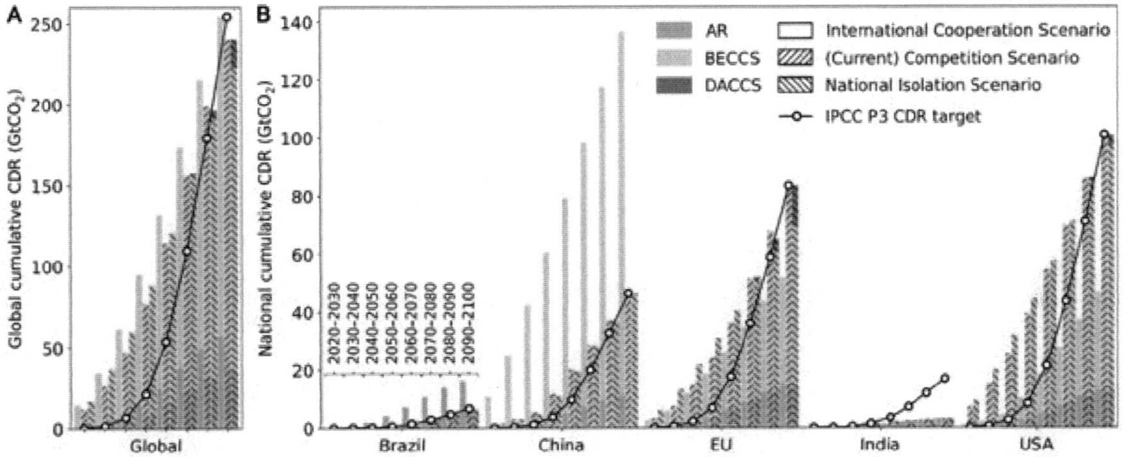

Figure 1. Cost-optimal CDR deployed between 2030–2100 to deliver the Paris Agreement: (A) globally, and (B) nationally. The International Cooperation Scenario (no fill) meets the 2100 global CDR target, and relies significantly on BECCS deployed in China. Conversely, neither the (Current) Competition Scenario (stripped fill) nor the National Isolation Scenario (squared fill) meet the 2100 global CDR target, owing to India which fails at meeting its share of the global CDR burden.

Cost-optimal cumulative CDR, at the global level (A) and at the national level (B), to deliver the Paris Agreement by 2100, for all scenarios are shown in Fig. 1.

Our results suggest that the Paris Agreement can be successfully and most cost-efficiently delivered in the total international cooperation, with BECCS predominantly (78%), deployed in regions with low-carbon electricity mix, with low-cost biomass supply (local or imported from low-cost producing countries such as Brazil or India) and low-cost CCS infrastructures, and AR (22%), deployed as early as possible to anticipate for its inherent CDR saturation. International cooperation allows for most capable and cheapest nations, such as China and Brazil, to provide CDR surplus on behalf of least capable ones, such as India, or more expensive ones, such as the European Union and the United States, at an overall CDR cost of \$68/tCO_2 by 2100.

Delivering domestically the Paris Agreement, based on national responsibility for climate change, leads to a more challenging — less accessible bioenergy resources, less available low-cost energy or geological CO2 storage — and more expensive portfolio of CDR solutions, increasing the overall CDR cost by 2100 by 51–69%. In the current competition scenario, this is due to the exhaustion of local biomass for BECCS in the European Union or the United States, and the resultingly import of biomass from Brazil or India, whereas in the national isolation scenario, this is because of the deployment of DACCS as a backstop CDR solution. Crucially, nations lacking from identified well-distributed and high geological CO_2 sinks, such as India, are found to fail at delivering their share of the CDR burden.

Lastly, our results show that the creation of an international negative emissions market, similar to already existing carbon emissions markets, such as the EU Emissions Trading Scheme, could provide an incentive for delivering the Paris Agreement in the most cost-efficient manner, whilst allowing for a fair distribution of the CDR burden, as it would allow for Negative Emissions Credits (NECs) trading.

Overall, we show that no CDR option is a silver bullet. Delivering the Paris Agreement requires the deployment of a portfolio of CDR options, best adapted to the location and time of deployment. We also emphasize the need for international cooperation to deliver the Paris Agreement at least-cost in a sustainable manner. We introduce the concept of an international market for negative emissions trading, similar to already existing carbon emissions markets, such as the EU Emissions Trading Scheme and we show that it could provide an incentive for delivering the Paris Agreement in the most cost-efficient manner, whilst allowing for a fair distribution of the CDR burden. Finally, we highlight that the more we delay the development of such market, the more expensive the CDR pathway deployed will be to deliver the Paris Agreement at scale. Such market will have to be associated with robust institutions that can ensure the monitoring, verification and accreditation of negative emissions trading, in order to ensure the effective and successful deployment of CDR at the Paris Agreement's scale.

Keywords: Carbon Dioxide Removal, BECCS, DACCS, Afforestation/Reforestation, Cooperation

---

1 UNFCC, *The Paris Agreement*, 2015.

2 Intergovernmental Panel on Climate Change, *Global Warming of 1.5°C. An IPCC Special Report on the impacts of global warming of 1.5°C above pre-industrial levels and related global greenhouse gas emission pathways, in the context of strengthening the global response to the threat of climate change,* World Meteorological Organization, Geneva, Switzerland, 2018.

3 J. Rogelj, A. Popp, K. V. Calvin, G. Luderer, J. Emmerling, D. Gernaat, S. Fujimori, J. Strefler, T. Hasegawa, G. Marangoni, V. Krey, E. Kriegler, K. Riahi, D. P. van Vuuren, J. Doelman, L. Drouet, J. Edmonds, O. Fricko, M. Harmsen, P. Havlík, F. Humpenöder, E. Stehfest and M. Tavoni, *Nat. Clim. Chang.*, 2018, **8**, 325–332.

4 D. Huppmann, E. Kriegler, V. Krey, K. Riahi, J. Rogelj, S. K. Rose, J. Weyant, N. Bauer, C. Bertram, V. Bosetti, K. Calvin, J. Doelman, L. Drouet, J. Emmerling, S. Frank, S. Fujim and R. Zhang., *Integr. Assess. Model. Consort. Int. Inst. Appl. Syst. Anal.*, , DOI:10.22022/SR15/08-2018.15429.

5 S. Fuss, W. F. Lamb, M. W. Callaghan, J. Hilaire, F. Creutzig, T. Amann, T. Beringer, W. De Oliveira Garcia, J. Hartmann, T. Khanna, G. Luderer, G. F. Nemet, J. Rogelj, P. Smith, J. V. Vicente, J. Wilcox, M. Del Mar Zamora Dominguez and J. C. Minx, *Environ. Res. Lett.*, , DOI:10.1088/1748-9326/aabf9f.

6 Drax Group Plc, https://www.drax.com/about-us/our-projects/bioenergy-carbon-capture-use-and-storage-beccs/, (accessed 9 December 2020).

7 Climeworks, https://www.climeworks.com/co2-removal, (accessed 9 December 2020).

8 Carbon Engineering Ltd, https://carbonengineering.com/our-technology/, (accessed 9 December 2020).

9 A. Marcucci, S. Kypreos and E. Panos, *Clim. Change*, 2017, **144**, 181–193.

10 J. Strefler, N. Bauer, E. Kriegler, A. Popp, A. Giannousakis and O. Edenhofer, *Environ. Res. Lett.*, , DOI:10.1088/1748-9326/aab2ba.

11 G. Realmonte, L. Drouet, A. Gambhir, J. Glynn, A. Hawkes, A. C. Köberle and M. Tavoni, *Nat. Commun.*, 2019, **10**, 1–12.

12 J. Fuhrman, H. McJeon, P. Patel, S. C. Doney, W. M. Shobe and A. F. Clarens, *Nat. Clim. Chang.*, 2020, **10**, 920–927.

13 J. Fuhrman, H. McJeon, S. C. Doney, W. Shobe and A. F. Clarens, *Front. Clim.*, 2019, **1**, 11.

14 M. Honegger, A. Michaelowa and J. Roy, *Clim. Policy*, , DOI:10.1080/14693062.2020.1843388.

15 P. Smith, S. J. Davis, F. Creutzig, S. Fuss, J. Minx, B. Gabrielle, E. Kato, R. B. Jackson, A. Cowie, E. Kriegler, D. P. van Vuuren, J. Rogelj, P. Ciais, J. Milne, J. G. Canadell, D. McCollum, G. Peters, R. Andrew, V. Krey, G. Shrestha, P. Friedlingstein, T. Gasser, A. Grübler, W. K. Heidug, M. Jonas, C. D. Jones, F. Kraxner, E. Littleton, J. Lowe, J. R. Moreira, N. Nakicenovic, M. Obersteiner, A. Patwardhan, M. Rogner, E. Rubin, A. Sharifi, A. Torvanger, Y. Yamagata, J. Edmonds and C. Yongsung, *Nat. Clim. Chang.*, 2016, **6**, 42–50.

16 V. Heck, D. Gerten, W. Lucht and A. Popp, *Nat. Clim. Chang.*, 2018, **8**, 151–155.

17 G. F. Nemet, M. W. Callaghan, F. Creutzig, S. Fuss, J. Hartmann, J. Hilaire, W. F. Lamb, J. C. Minx, S. Rogers and P. Smith, *Environ. Res. Lett.*, 2018, **13**, 063003.

18 M. Fajardy and N. Mac Dowell, *Energy Environ. Sci.*, 2017, **10**, 1389–1426.

19 M. Fajardy, S. Chiquier and N. Mac Dowell, *Energy Environ. Sci.*, 2018, **11**, 3408–3430.

20 M. Fajardy and N. Mac Dowell, *One Earth*, 2020, **3**, 214–225.

625080 Determining the Role of Negative Emission Technologies (NETs) in the Electricity System in the UK

Abstract:

In line with the 2015 Paris Climate Accord[1], the Inter-governmental panel for Climate Change (IPCC) highlighted the crucial role of negative emission technologies (NETs). In particular, bioenergy with carbon capture and storage (BECCS) is inevitable to meet the 1.5°C target above pre-industrial levels, but presents technical, sustainable and political challenges[2] which need to be assessed. Also, only a few studies have been looking into NETs synergies as part of a broader NETs portfolio. Moreover, it is essential to model regional granularity to integrate NET's key deployment parameters, such as biomass supply chain for BECCS[3] or climate ecological zones for Afforestation (AF). Meanwhile, the UK global economy, including the power sector, is required to reach net-zero emissions in 2050, with an estimated 90 $MtCO_2$/year of Greenhouse Gas Removal (GGR) in 2050 (Committee on Climate Change, 2020). However, no studies mentioned the eventuality of a national net-negative electricity sector, implying substantially more BECCS deployment and a profound impact on the design and operation of the electricity system. Whereas there are works modelling and optimising the UK electricity system[4], also including BECCS and DACCS, none of them studied the impact of nature-based NETs deployment on the electricity capacity mix.

To fill these gaps, our contribution is double. First the implementation of the Modelling and Optimisation of Negative Emission Technology (MONET-UK) that integrates three NETs (BECCS, AF and Direct Air Capture (DACCS)) and accounts for national bio-geophysical resources at a regional level. Second, the combination of the Electricity System Optimisation (ESO) -a pre-existing model[4]- and MONET-UK in a joint system. These two models allow the system to find the global minimum of the total system cost under technological and national CDR constraints. The joint system is optimised by successively optimising MONET and ESO, until a global minimum is reached. The interaction terminates if a convergence criterion verifying the stability of total system capacity is satisfied. (see Fig. 1)

Our results show a typical UK NET optimal portfolio for delivering 90 $MtCO_2$ of GGR in 2050 (see Fig. 2). AF, the cheapest solution (105 \$/tCO_2) is fully deployed and is limited by national forest expansion rate. BECCS, slightly more expensive (155 \$/tCO_2), is deployed until available national biomass and imported pellet are out of stock. As a last resort DACCS is deployed as it is the most expensive technology (529 \$/tCO_2).

Then, in figure 3, we compare the path to follow for a net-zero and a carbon negative electricity system in 2050. The coloured dots show the associated optimal NET portfolio responsible for other sector's emissions removal estimated to 90 $MtCO_2$ by 2050. The net-zero electricity system presents a high share of renewables but also requires firm thermal generation (natural gas, biomass), as renewables are not flexible and require additional storage that is not available. We notice a progressive build of renewable capacity (solar, wind), due to the need of low-carbon electricity generation, as well as a system demand increase (+15%) caused by significant DACCS deployment. In contrast, a carbon negative electricity system is achieved through substantially more BECCS capacity, 17.5GW in 2050 (versus 2.5GW in the net-zero scenario) displacing renewable capacity and firm low-carbon generation. These results confirm that the use of intermittent renewable technologies is not sufficient, although important, to deliver required CDR targets. Moreover, they also show that, in order to meet these targets, having a net-zero electricity system and relying on other NETs (AF and DACCS) is drastically more expensive (+33%) than aiming at a net-negative electricity sector, which requires less external removal (AF and DACCS).

Finally, we studied how nature-based solutions affect the power system and total system cost. For example, if we increase biomass availability, average cost of carbon removal decreases due to DACCS share replaced by BECCS. Simultaneously, it increases BECCS share in the electricity sector, which proves cheaper than further IRES growth which requires further energy storage and grid flexibility. In figure 4, a cost analysis shows that increasing biomass availability from 100 to 200TWh decreases total investment of 12% per year, however further increase doesn't change much. We recognize that heavy dependence on imported pellet poses global security concerns, making the power sector vulnerable to the instability of global supply chains. In our next steps, we will focus on other natural based solutions -other NETs e.g. biochar, afforestation potential increase- to avoid such a dependence.

References

1. IPCC, 2018: Global Warming of 1.5°C. An IPCC Special Report on the impacts of global warming of 1.5°C above pre-industrial levels. In Press.

2. Galik, C. S. (2020, January 1). A continuing need to revisit BECCS and its potential. *Nature Climate Change*. Nature Research.
3. Fajardy, M., Chiquier, S., & Mac Dowell, N. (2018). Investigating the BECCS resource nexus: Delivering sustainable negative emissions. *Energy and Environmental Science*, *11*(12), 3408–3430.
4. Heuberger, C. F., Rubin, E. S., Staffell, I., Shah, N., & Mac Dowell, N. (2017). Power capacity expansion planning considering endogenous technology cost learning. *Applied Energy*, *204*, 831–845.

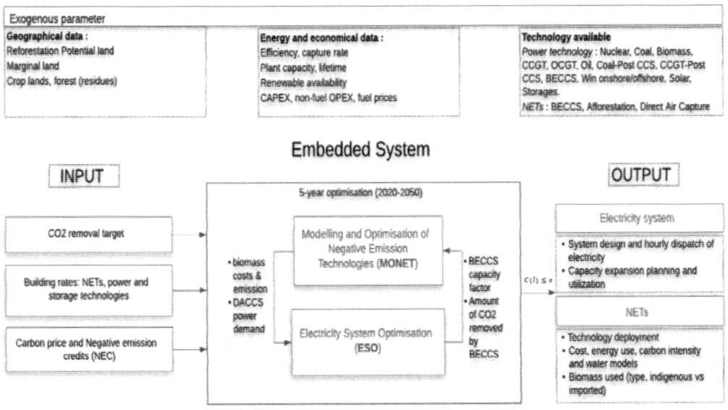

Figure 1: Embedded System, MONET-ESO framework

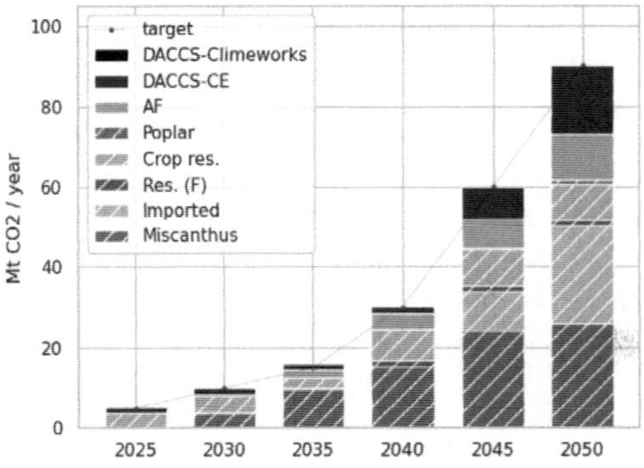

Figure 2: Annual CO2 removed by NETs in the UK under CCC's 2050 target. Imported pellet limited.

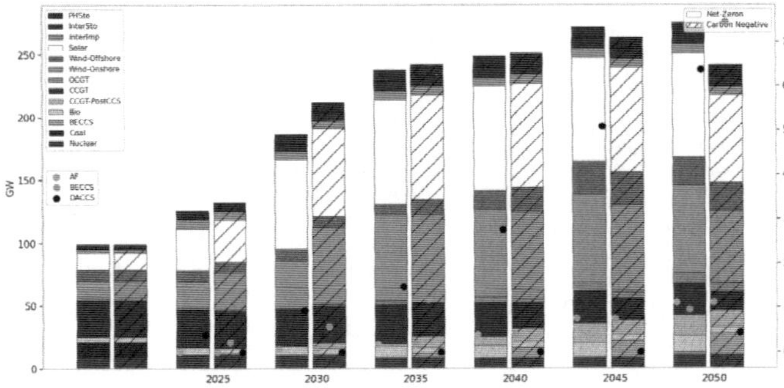

Figure 3: Capacity Expansion of power generating technologies (bars, left axis), and NETs removal (dots, right axis) from embedded system optimisation for 2 scenarios.

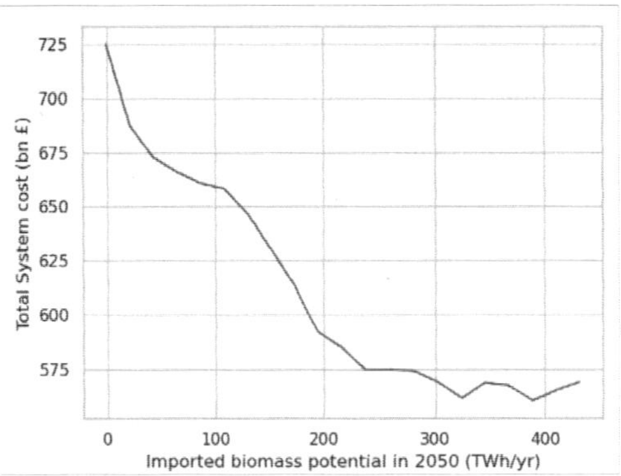

Figure 4: Effect of biomass availability increase (woody imported pellet) on embedded system cost.

Copyright © American Institute of Chemical Engineers. All rights reserved.

623851 Thermally Enhanced Advanced Oxidation-Driven Regeneration on Iron-Activated Biochar for Removal of Microcystin-LR in Water and Wastewater

Abstract:

In this study, a cyclic process of adsorption and persulfate (PS) oxidation-driven regeneration using $FeCl_3$-activated biochar (FA-BC) was suggested as a cost effective process for removal of microcystin-LR (MC-LR) in lake water. For enhancing overall treatment efficiency, the effects of temperature on adsorption and PS regeneration were investigated. While the increase in temperature resulted in a drastic increase in the MC-LR adsorption rate on FA-BC, the MC-LR oxidation and PS reaction rates were remarkably improved by factors of 3.4 and 3.5 with increasing temperature from 20 °C to 50 °C. In addition, the lower pH and higher PS concentration led to the higher regeneration efficiency of MC-LR-spent FA-BC by PS oxidation. Both diffusion and desorption of MC-LR from FA-BC were thought to control the MC-LR oxidation rate during the PS oxidation-driven regeneration of MC-LR-spent FA-BC. The sulfate radical was proven to be the main oxidant participating in the PS oxidation process, however, the crystal structures of α-Fe_2O_3, Fe_3O_4, and Fe^0 in FA-BC were stable during PS oxidation. Under the conditions of 200 mg/L PS, pH 6, and 50 °C, 83 - 99% regeneration efficiencies were achieved during four cycles of adsorption-PS regeneration in DI water and lake water. Therefore, this study indicates high regeneration of MC-LR-spent FA-BC by PS oxidation, which would allow cost-effective treatment of MC-LR in water.

Copyright © American Institute of Chemical Engineers. All rights reserved.

631509 Reversible Adsorption and Desorption of Pfas Via Switching the Electric Field on Carbonaceous Adsorbents

Abstract:

Per- and polyfluoroalkyl substances (PFAS) are a group of organofluorine compounds that possess a carboxylic or sulfonic acid attached at one end while a fluoroalkyl chain (CF_2 or CF_3) attached to the other end. The fluoroalkyl chain can provide excellent thermal stability and chemical durability that enable PFAS to be used in a wide range of practical applications.

Recent studies showed that PFAS was detected in groundwater and surface water near a PFAS manufacturing facility and military bases. More recently, PFAS was also detected in drinking water systems which have raised serious concerns. This is because PFAS is persistent and accumulative in both environment and the human body. Thus, the U.S. administrations established regulations on the manufacturing, use, and disposal of PFAS. These regulations were mainly targeted to PFAS with a long fluoroalkyl chain (i.e., $(CF_2)_6$ or longer). However, there is a sharp increase in the production and use of PFAS with a short or ultrashort (i.e., $(CF_2)_3$ or shorter) fluoroalkyl chain to replace those with a long fluoroalkyl chain. Thus, the total amount of PFAS discharged into the environment remains the same or even increased.

Conventional separation technologies including coagulation and flocculation showed the limited capability to remove PFAS from water. This can be attributed to extremely low concentration (i.e., parts per trillion) or the chemical durability of PFAS. While physio-chemical processes such as chemical oxidation and reduction, and plasma-based technology can decompose PFAS, this often results in fragmented PFAS which can cause secondary pollution.

Adsorption is perhaps the most prevailing technology to remove PFAS from the water. Carbonaceous materials (e.g., activated carbon, graphite, graphene) based adsorbents are common because of their excellent chemical durability and thermal stability. Also, they can provide a large specific surface area. One of the primary challenges is that these adsorbents often suffer from a relatively low adsorption capacity (i.e., the amount of PFAS taken up by the adsorbent per unit mass) for PFAS with a short fluoroalkyl chain. Also, they exhibit a decrease in the PFAS removal efficiency upon repeated adsorption-desorption processes.

To improve the adsorption capacity and reusability of the adsorbents, applying an electric field can be a viable option. Electric-field aided sorption (i.e., electrosorption) has demonstrated that it can remove the ionized contaminants from water. When an external electric field is applied, ionized substances can get attracted and adsorb to an electrode surface. Given that PFAS is dissociated in water, anionic PFAS and proton will be attracted and adsorbed to an anode and a cathode, respectively. Also, PFAS with a higher polarity (i.e., short fluoroalkyl chain) can be subjected to a stronger coulombic force resulting in a higher adsorption capacity. Furthermore, the electrostatic repulsive force produced by reversing the voltage across the electrodes will enable the desorption of anionic PFAS from the electrode surface.

Herein, we demonstrate reversible adsorption and desorption of PFAS with varied fluoroalkyl chains by applying an electric field across inexpensive graphite electrodes. The results show that it can exhibit very high adsorption capacity for both long and short fluoroalkyl chain PFAS (perfluorooctanoic acid and perfluoropentanoic acid, respectively). We also demonstrate that adsorbed

PFAS can be released to water upon reversing the voltage which makes the adsorbent reusable. Finally, we develop a mathematical relation that can describe the kinetics of adsorption and desorption of PFAS by utilizing a pseudo-first-order kinetic model.

Copyright © American Institute of Chemical Engineers. All rights reserved.

633154 Investigating Lead Biosorption Mechanisms at Trace Concentrations

Abstract:

Heavy metals are highly toxic and acutely hazardous for living organisms and human health, even at very low concentrations. Lead, in particular, is one of the most abundant and toxic heavy metals. As a result of various incidents of lead-contaminated drinking water, relevant water regulations are being revised, while according to the US Environmental Protection Agency *no level of lead* in drinking water is considered safe.

Conventional water/wastewater treatment processes either fail to completely remove dissolved lead ions, or result in significant energy consumption, increased operational and maintenance costs, and generation of undesirable by-products. Biosorption, *i.e.* adsorption of heavy metal ions by inactive biomass, can offer a promising, low-cost and environmentally friendly alternative to conventional treatment processes. Although biosorption of heavy metals using a wide variety of biomaterials, including microorganisms and plants, has been extensively studied at the parts per million (ppm) contaminants scale, limited research has been conducted, so far, at the challenging trace concentrations of parts per billion (ppb) and below.

In this presentation the lead adsorption mechanisms at the ppb scale will be discussed. *Saccharomyces cerevisiae* has been selected as the most suitable biosorbent, as it is inexpensive and easy to cultivate at large scales, while it has been proven a good lead biosorbent at the ppm scale. Kinetic and equilibrium experiments have been conducted, using synthetic lead-containing aquatic solutions, to measure and characterize the stoichiometry, equilibrium, and selectivity of biomass under different pH values, initial lead concentrations (at the range of 30 – 1000 ppb) and biosorbent dosages. Residual lead concentrations have been measured using inductively coupled plasma - mass spectrometry. In addition, advanced characterization techniques, including X-ray photoelectron spectroscopy, transmission electron microscopy and Fourier-transform infrared spectroscopy, have been used to examine the biomass, in an effort to identify the functional groups and biomass sites responsible for lead uptake. Through these experiments, the adsorption equilibrium has been quantified and the underlying mechanisms have been identified, paving the way for the development of novel, highly effective and sustainable treatment processes.

Copyright © American Institute of Chemical Engineers. All rights reserved.

624188 Micelle-Laden Hydrogel Microparticles for the Removal of Hydrophobic Micropollutants

Abstract:

Micropollutants, including various emerging contaminants which occur at low concentration in water, are ubiquitous and hazardous to the ecosystem over long periods of time. Their chemical diversity and low concentrations make it difficult to remove them from water using current methods like activated carbon adsorption. This talk introduces hydrogel-based polymeric absorbents containing immobilized micelles for the removal of a broad class of hydrophobic micropollutants from water. The micelles have hydrophobic cores, into which hydrophobic micropollutants can easily partition. To speed uptake by increasing surface area, an off-the-shelf microfluidic device is used to prepare monomer microdroplets, which are polymerized to form absorbent microparticles. NMR spectroscopy is used to study cross-linked hydrogel formation, and coarse-grained molecular dynamics simulations help optimize hydrogel composition. Hydrophobic micropollutants are rapidly removed by our hydrogels, with mass transfer coefficients greater than those determined for a commercially available activated carbon used for water purification. Finally, this talk will discuss a sustainable and easy method of regenerating spent absorbent, which has so far not been possible with commercial absorbents.

Copyright © American Institute of Chemical Engineers. All rights reserved.

631718 Adsorption and on-Demand Desorption of Perfluoroalkyl Substances in Water By Thermo-Responsive Poly(N-isopropylmethacrylamide)

Abstract:

Per- and polyfluoroalkyl substances (PFAS) have been utilized in a wide range of applications including surfactants, aqueous film forming foams (AFFF), and oleophobic coatings due to their thermal stability, low surface free energy, and chemical inertness. Recent reports have revealed that soil and groundwater at which firefighting activities have been undertaken such as military base sites are contaminated. Also, industrial effluents and disposal of PFAS-containing products have contributed to the contamination. Paradoxically, PFASs' excellent chemical and thermal durability makes them persistent in both environment and the human body. While it still remains an active debate field, it is generally agreed that PFAS can cause a variety of human health disorders and pose adverse effects to both aquatic and terrestrial ecosystems. Thus, the U.S. Environmental Protection Agency (EPA) included perfluorooctanoic acid (PFOA) and perfluorosulfonic acid (PFOS) in the Contaminant Candidate List in its 2016 annual report.

A typical PFAS concentration detected in water is in the order of parts per trillion (10^{-12}, ppt) which can cause a grand challenge not only for removing them but also for their detection. To measure the concentration of PFAS by using commercial instruments such as high-performance liquid chromatography (HPLC), the concentration should be in the range of parts per billion (10^{-9}, ppb). Various methodologies including activated carbon, ion exchange, and reverse osmosis membrane have been employed to remove PFAS from water. While they have proven effective in removing and isolating PFAS, their regeneration after releasing them has been rarely reported.

Herein, a poly(isopropylmethacrylamide) (P-NIPMAM)-based adsorbent that can adsorb and retain PFAS from water at a temperature higher than its lower critical solution temperature (LCST) while releasing them at a lower temperature is reported. P-NIPMAM was prepared by a free radical polymerization of N-isopropylmethacrylamide (NIPMAM) with potassium persulfate (KPS) and N,N'-methylenebisacrylamide (MBAA) as an initiator and a cross-linker, respectively. The resulting P-NIPMAM exhibits a LCST behavior. At a temperature below the LCST, hydrogen bonding between hydrophilic groups (i.e., $-NH_2$) of P-NIPMAM and water dominates. When a temperature becomes above the LCST, the solvation entropy for hydrophobic moieties (i.e., isopropyl and methyl groups) of P-NIPMAM becomes dominant resulting in a globule-like collapsed structure. These hydrophobic moieties exhibit a low affinity to water and can partition with the compounds bearing non-polar hydrophobes such as PFAS as adsorbate.

The adsorption and desorption of P-NIPMAM were systematically studied by using PFAS possessing varied fluoroalkyl chain lengths. The adsorption capacity was measured at $T = 50$ °C (> LCST). The results show that P-NIPMAM exhibits a higher adsorption capacity for PFAS with longer fluoroalkyl chain length. This is a direct consequence of a stronger hydrophobic interaction between the fluoroalkyl chain and the hydrophobic moieties of P-NIPMAM. We also demonstrated that P-NIPMAM can release the adsorbed PFAS to water at T (= 23 °C) < LCST. The results indicate that PFAS with a shorter fluoroalkyl chain is released faster than that with a longer chain. This can be attributed to different hydrophobic interaction which governs the adsorption and desorption. Finally, we fabricated a prototype sampler by utilizing our P-NIPMAM that can adsorb and retain PFAS at an elevated temperature (T > LCST) while releasing them into the water at a lower temperature (T < LCST).

Copyright © American Institute of Chemical Engineers. All rights reserved.

628259 Could Benzyl Hydroxyl Shielding Promote the Radical Induced Pyrolysis of Lignin?

Abstract:

Introduction

Due to the inherent complexity and changing chemical structure of lignin, a strategy, which has been gradually accepted in liquid depolymerization, is to shield the benzyl hydroxyl to inhibit the self-condensation reaction of lignin. This information on whether benzyl hydroxyl shielded in lignin structure affected the pyrolysis behaviors of lignin, especially explained by the radical theory, however, is scarce in the literature. Accordingly, the influence of the shielded benzyl hydroxyl in lignin structure on thermal behavior and radical evolution was investigated in this study. The benzyl hydroxyl in the lignin structure was first stabilized by propionaldehyde and prepoxidized by 2,3-dichloro-5,6-dicyano-1,4-benzoquinone (DDQ) respectively, compared with the raw lignin as a control (ball-milling wood lignin, MWL) for subsequent investigation. Then, the raw lignin and pretreated lignin were pyrolyzed at different temperatures, and the corresponding products were determined to find the best condition for obtaining organic liquids. In addition, through the electron paramagnetic resonance (EPR) spectroscopy, the evolution of pyrolytic radicals had been comprehensively studied to reveal the connection of the products formation with the radical induced reaction by different method of pretreatment.

Materials and Methods

The maple wood was first prepared for subsequently extraction of different types of lignin. Milled wood lignin (MWL) were obtained from the maple wood via the ball-milling method. Propionaldehyde stabilized lignin (PAL) was obtained by extraction in dioxane: water solutions with HCl and the propionaldehyde. To prepare the DDQ oxidized lignin (DOL), the maple wood was firstly oxidized by the DDQ oxidation with O_2(g) as the co-oxidants. Then the DOL was obtained by the ball-milling method extracted from the oxidized wood. TGA, 2D HSQC NMR and Ultimate analysis of raw and pretreated lignin samples were conducted to investigate the characterization and thermal stability.

Analytical pyrolysis of MWL, DOL and PAL was conducted with pyrolysis gas chromatography/mass spectrometry (Py-GC/MS) at 350-650 °C. Lab-scale pyrolysis was performed on a fixed bed to harvest chars from lignin. The radical concentration of sample was detected on a Bruker EMX-10/12 spectrometer.

Results and Discussion

The results showed that the process of inhibiting the coking of lignin cannot be achieved by shielding the α-OH. The maximum total yields (11.10×10^9 area/mg) of MWL lignin was obtained at 450 °C compared with the oxidized lignin (3.71×10^9 area/mg of DOL, 5.02×10^9 area/mg of PAL), which indicated that the yields of bio-oil was inhibited by oxidized α-OH of the lignin.

To understand the pyrolysis radical induction, the radical evolution of MWL was first detected by EPR from 350 to 650 °C. The pyrolysis radical concentration increased rapidly from 26.82×10^{17} spin/g to 245.04×10^{17} spin/g at 550 °C, which means the lignin undergo an explosive fission reaction. For PAL radical detected at 550 °C, the lower radical concentration (169.32×10^{17} spin/g) and higher char yields (37.10%)was obtained compared to the MWL. As for DOL radical, the liquid yield of the DOL was lowered than the MWL with the higher spin concentration (259.32×10^{17} spin/g for DOL). Therefore, the possible pathway of radical evolution leading to various products distribution from raw and pretreated lignin was given a new sight to understand the lignin pyrolysis reaction.

Conclusions

The shielded α-OH in lignin structure showed significant effect on the thermal stability and yield distribution during fast pyrolysis. It was proved that the radical reactions play an important role and process of inhibiting the coking of lignin cannot be achieved by shielding the benzyl hydroxyl. The bio-oil yield was decreased after shielding the benzyl hydroxyl, while the char yield had been increased. Facts have proved that in the pyrolysis process, due to the violent radical-induced reaction, the caped hydroxyl groups might be quickly converted into oxygen radicals, which further increases the probability of formation of benzoquinone and caused carbonization. Thus, the effect of capping pretreatment on the distribution of lignin pyrolysis products is clarified, and the failure of benzyl hydroxyl shielding pretreatment and the possible inducement of promoting the condensation of lignin to carbon were proved from the perspective of radical induction.

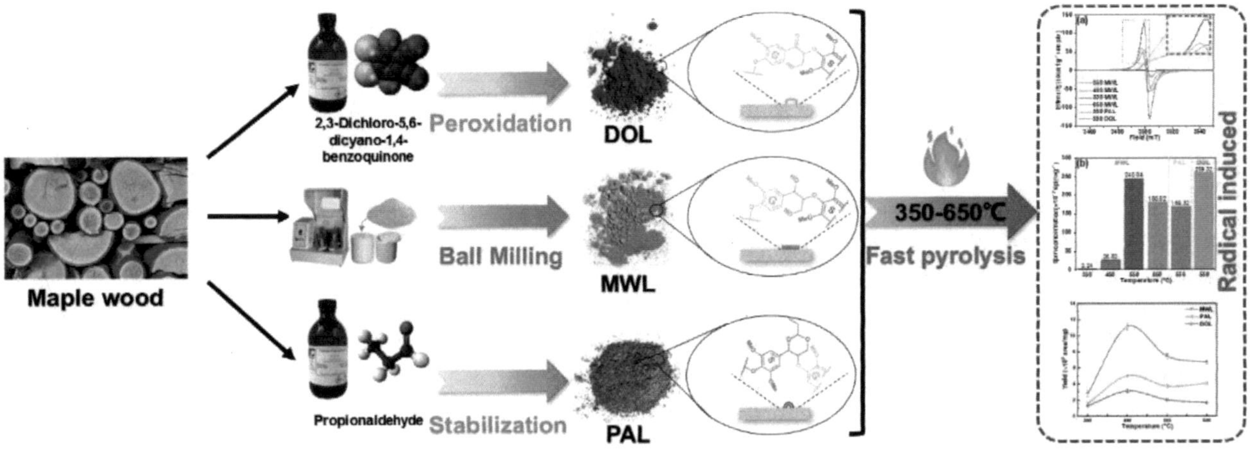

Copyright © American Institute of Chemical Engineers. All rights reserved.

622248 Decarbonising the UK Heating Sector: A Whole-Energy System Analysis

Abstract:

The decarbonisation of heating is crucial in reaching the ambitious emission targets adopted by many countries, including the UK, which aims to reach net-zero emissions by 2050. While the electricity sector has received much attention from policymakers, private companies and researchers and considerable progress has been made on the decarbonisation of power generation, the heating sector has seen only little change. The majority of heat in the UK is still generated from conventional, unsustainable sources, with more than 85 % of households using natural gas boilers as main heating technology [1]. On top of that, 57 % of the commercial heat demand and 51 % of the industrial heat demand are provided from natural gas; with electricity, oil, solid fuel (coal) and biomass & waste also being significant contributors [2]. Consequently, emissions associated with heating are high. Residential combustion of fuel, mainly for heating purposes, alone accounted for 18 % of the UK's carbon emissions in 2019, and commercial and industrial heating add to that total [3]. It was estimated that in 2009, 38 % of the UK's carbon emissions could be attributed to heating [4].

Decarbonisation of heating is challenging due to the systems typically being distributed. Millions of devices in people's homes, commercial buildings and industrial facilities need to be replaced by low-emission heating technologies. These technologies typically rely on carbon-neutral fuels since carbon-capture at the end-user is impractical. Heat pumps and resistive electric heaters can utilise low-carbon electricity for heating, and hydrogen boilers can combust low-carbon hydrogen generated from e.g., methane-reforming with carbon-capture and storage (CCS) or water electrolysis without carbon emissions at the end-user. Solar-thermal collectors directly use a renewable energy source. Another potentially attractive option, especially for urban areas, are low-emission district heating networks that can benefit from centralised low-carbon generation and economies of scale. Additionally, improved building insulation and other energy conservation measures can be valuable as they decrease the demand and therefore reduce the required generation capacities. Various studies presented in literature show that no single technology is clearly superior to others, and most likely a combination of different technologies is most suitable for heating decarbonisation.

We present a newly developed heating sector optimisation model to identify optimal decarbonisation pathways. The model is a deterministic generation-expansion and unit-commitment optimisation model that simultaneously optimises technology portfolios and their operation. It is implemented as mixed-integer linear-programming (MILP) model in GAMS, using Python for data pre- and post-processing. A graphical user interface is also available to make the model accessible and easy to use. The model uses a similar formulation as the Electricity Systems Optimisation (ESO) model [5,6]. Starting from today's technology portfolio, the model optimises the evolution of the heating sector, considering investment decisions every five years. Build-rate constraints limit the deployment of technologies to technically feasible rates. For each of the decision years, the operation schedule of the technologies is optimised. Regarding emissions, the model can consider end-point constraints (e.g., net-zero in 2050), interior-path constraints and cumulative carbon budgets. So far, the model considers residential heating demand in form of representative households, commercial and industrial demand is to be added later. Modelled technologies are vapour-compression heat pumps, electric heaters, gas and hydrogen boilers and solar-thermal systems, and it is planned to include additional technologies such as thermal heat pumps, co-generation plants and district heating in future.

The emissions and fuel costs of electric and hydrogen heating technologies depend strongly on the respective generation mixes. A representation of the hydrogen industry is included in the heating sector model. Considered are generation from natural gas using methane reforming with or without CCS and from electricity via water electrolysis. Large-scale hydrogen storage is also included in the model. To determine the costs and emissions associated with electricity use, it is planned to connect the heating sector model with the existing ESO model. The result will be a comprehensive model of the UK electricity and heating energy systems that allows a simultaneous optimisation of technology capacities and dispatch in both sectors. The integrated approach allows to reduce the model externalities and necessary assumptions. The model can be a powerful tool to e.g., determine the optimal degree of heating electrification, prioritise the use of hydrogen for heating or electricity generation and to consider potential synergies such as co-generation or electricity load-shifting with heat pumps and thermal storage tanks. It can also be used to investigate the effectiveness of different policy measures. Guidelines on how to best support and facilitate the transition towards low-carbon heating and electricity provision can be developed. We present an analysis which robustly quantifies and qualifies the role and value of a range of technologies for providing low carbon heat.

References

[1] Committee on Climate Change, "Heat in UK buildings today" 2016. [Online]. Available: https://www.theccc.org.uk/wp-content/uploads/2017/01/Annex-2-Heat-in-UK-Buildings-Today-Committee-on-Climate-Change-October-2016.pdf.

[2] Department for Business Energy & Industrial Strategy, "Energy consumption in the UK (ECUK) - End use tables." 2019, [Online]. Available: https://www.gov.uk/government/statistics/energy-consumption-in-the-uk.

[3] Department for Business Energy & Industrial Strategy, "Final UK greenhouse gas emissions national statistics: 1990 to 2019." 2021, [Online]. Available: https://www.gov.uk/government/statistics/final-uk-greenhouse-gas-emissions-national-statistics-1990-to-2019.

[4] Department of Energy & Climate Change, "Overview of UK's greenhouse gas emissions from heat related activities" 2012. [Online]. Available: https://www.gov.uk/government/statistics/uk-emissions-from-heat.

[5] C. F. Heuberger, I. Staffell, N. Shah, and N. Mac Dowell, "A systems approach to quantifying the value of power generation and energy storage technologies in future electricity networks" Comput. Chem. Eng., vol. 107, pp. 247–256, Dec. 2017, doi: 10.1016/j.compchemeng.2017.05.012.

[6] C. F. Heuberger, E. S. Rubin, I. Staffell, N. Shah, and N. Mac Dowell, "Power capacity expansion planning considering endogenous technology cost learning" Appl. Energy, vol. 204, no. August, pp. 831–845, 2017, doi: 10.1016/j.apenergy.2017.07.075.

Copyright © American Institute of Chemical Engineers. All rights reserved.

632205 Techno-Economic Assessment of Three Potential Pathways for Biomass Liquefaction

Abstract:

Biomass liquefaction is of interest within the biorefinery as it is a potential answer to the bottleneck formed from front-end materials handling. If the biomass could be transformed into a flowable slurry at the process entry point, then the material could more readily flow into the biorefinery pretreatment. Increased flowability will reduce downtime from any materials plugs that may form in the current design. Currently, the heterogeneous mixture of pelleted biomass and water can impede machine functionality and operability, which results in costly maintenance shutdowns. Liquefaction of biomass can potentially mitigate process bottlenecks. If a liquefaction process were to be installed at the front-end of an existing biorefinery, it would likely reduce or eliminate bottlenecks due to solids agglomeration. However, installation cannot be more costly than downtime losses. Through Techno-economic assessment (TEA), it is possible to quantify the reduction in losses via implementation of a liquefaction process. Three liquefaction pathways were considered: enzyme, maleic acid, and a combination of maleic acid+enzyme. The trade-offs in viscosity and equipment requirements were assessed using TEA to effectively compare the pathways to one another. All three pathways were simulated in Aspen Plus® Software.

Copyright © American Institute of Chemical Engineers. All rights reserved.

623756 Altering Physico-Chemical Properties of Pine-Derived Carbon Quantum Dots By Changing Hydrothermal Treatment Conditions

Abstract:

Carbon quantum dots (CQD) are a relatively new and exciting material that are finding their place in a vast range of industries. These dots are aptly named as they exist within a diameter range of 0-10 nanometers. They are being utilized in everything from material coatings that self-clean or self-repair, to medicines that deliver cancer treatments to targeted areas in the body using UV light, to use in home entertainment systems to produce high quality screen visuals. Efficient methods and sources for producing these quantum dots are still being studied. It is found that the properties of the dots largely depend upon the surface functional groups and the size of the dots. Typically depending upon the method used to develop the dots (top-down or bottom-up) and the feed source the products are derived from will determine the surface groups and sizing.

A large amount of attention has recently been given to producing carbon quantum dots in the environmentally friendly manner of hydrothermal treatment. Many feedstocks have been studied and almost everyone has produced a variant dot. The goal of this research is to determine the effects of varying the hydrothermal treatment temperature and residence time on the characteristics of CQDs. Loblolly pine was chosen is chosen for this study due to abundancy and low cost. The experimental procedure focuses on three different reaction temperature conditions of 180, 220, and 260 °C. The time also varies in increments of 30 min up to 3 hours. After hydrothermal treatment the process fluid will undergo vacuum filtration followed by dialysis in order to separate the dots for characterization. Next the dots will be freeze dried for characterization that requires dry samples. The process fluid and the hydrochar itself will be analyzed for carbon quantum dots and these will be characterized using Fourier Transform Infra-Red Spectroscopy, Ultraviolet-Visible Spectroscopy, Fluorometry, Tunneling Electron Microscopy, and X-Ray Diffraction to determine trends. Altering the treatment time and temperature/pressure the quantity, surface functionalization, and the size can be controlled due to altering the formation of the passivation layer and condensation mechanism that forms the dots. As a result the quantum yield, absorbance, emission, and applicability of the dots are altered and can be fine tuned per application in industry.

Copyright © American Institute of Chemical Engineers. All rights reserved.

623863 Understanding the Effects of Mixed Municipal Solid Waste Streams on HTL Product Characteristics

Abstract:

In 2018, each person in the United States was responsible for the production of nearly five pounds of municipal solid waste (MSW) per day.[1] Co-HTL of food and green waste components of MSW can help reduce waste and achieve economies of scale and the interactions between these feeds must be understood.

This work aims to understand the effect of mixing food and green waste on oil yield and product composition. Real food waste obtained from a veteran's affairs hospital resulted in 35.9% oil yield non-catalytically, in contrast to 12.13% for green waste. Current studies have revealed sustained oil yield (34.1%) with addition of up to 25% lignocellulosic waste (**Figure 1**). Additionally, results show increasing char yields (21 – 33.5%) with increased green waste content, due in part to its high (up to 13%) ash content.

Despite a drastic difference in carbon content in the feeds, 52% for food waste and 19% for green waste, FTIR indicates minimal differences between oil phase functional groups and vibrations. The primary difference occurs at 3400 cm^{-1} where the addition of green waste increases the OH stretching band. Solid phase FTIR mirrors the observed similarities, however, pure green waste solid phase which contains a significant OH stretch band. The similarities observed via FTIR are not sustained with GC/MS, as food waste oil contains large fatty acid peaks that decrease with the addition of green waste, only to be replaced by the presence of ketones and aromatics.

Qualitative observations also show improved homogeneity of food waste oil, noting that it is less sticky. Green waste oil, on the other hand appears to contain two distinct phases, a free-flowing liquid, and a sticky residue. This works provides useful insight into the feasibility of mixed MSW feeds for use in hydrothermal liquefaction.

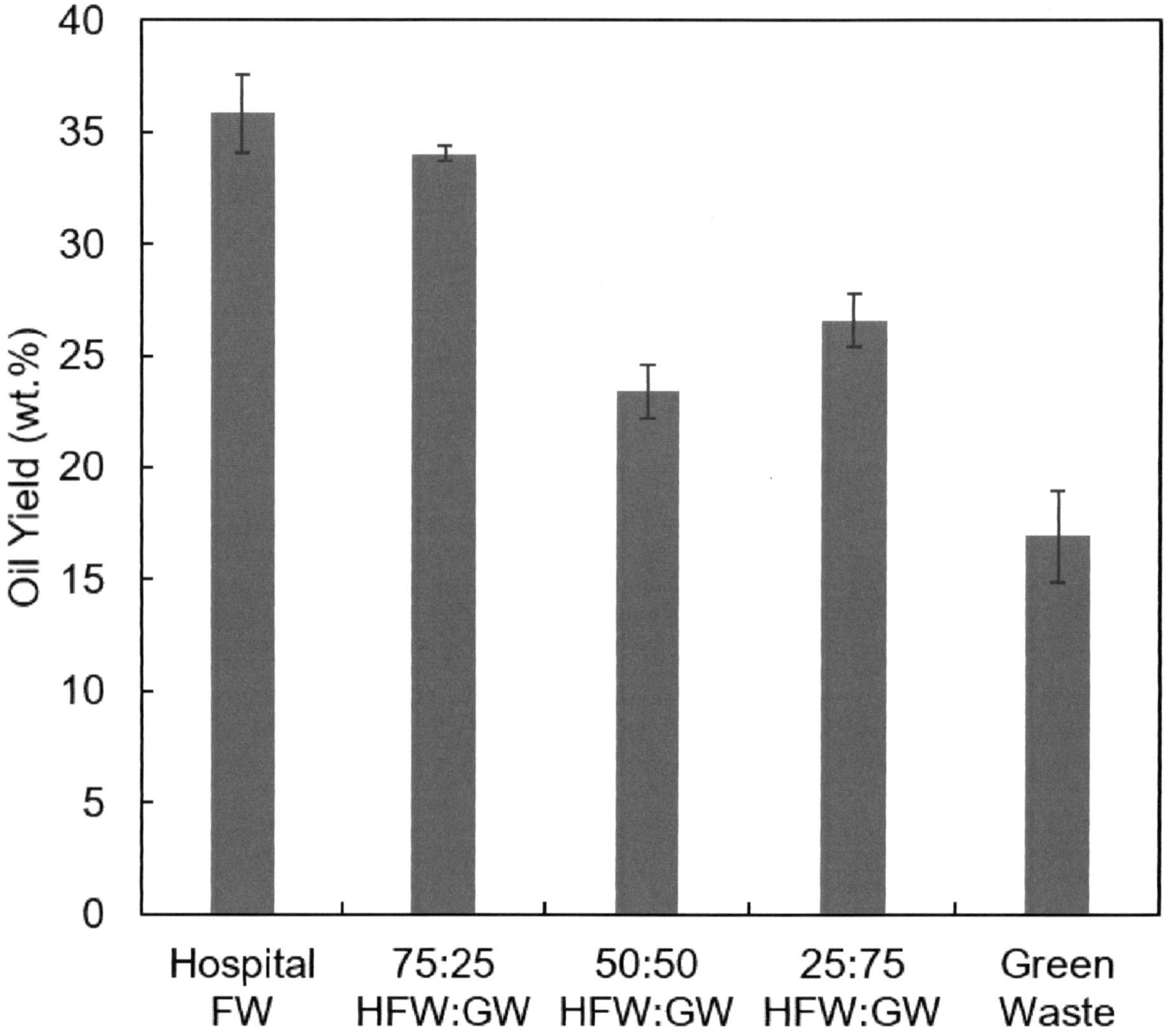

Copyright © American Institute of Chemical Engineers. All rights reserved.

632957 Adsorption Kinetics Studies of Catalytic HTL Derived Biochar

Abstract:

Ultra-porous carbonaceous powdered material derived from lignin rich biomass can be useful for water purification, energy storage, soil amendment, cement additive, wastewater treatment, gas purification etc.. In this study, we used corn stover agricultural residue as a cheaper feedstock and treated with hydrothermal liquefaction (HTL) at 250-300°C and 800-1600 psi to derive a solid residue, which was thermally and chemically post-treated to obtain mesoporous carbon. This carbon material was thoroughly characterized using the SEM/EDX, TEM and FTIR. Nitrogen adsorption isotherms such as Langmuir, Freundlich and Brunauer-Emmett-Teller (BET) were analyzed to determine the kinetic parameters. Barrett, Joyner, and Halenda (BJH) method was used to determine the pore-size distribution. Results obtained with specific surface area, pore size distribution, adsorption rate constant, monolayer/multilayer adsorption characteristics and mesoporosity will be presented.

Copyright © American Institute of Chemical Engineers. All rights reserved.

630345 An Optimization-Based Decision Framework for Integrating Energy Storage with Fossil Power Plants

Abstract:

Rapidly declining costs of clean renewable energy and increased policy efforts have driven the integration of renewables in electricity grids. However, the intermittency in availability of renewable energy requires electricity grids to be more flexible, resilient and reliable. To incorporate flexibility in electricity grids, the traditionally considered techniques are the increased cycling operation of conventional power plants and the integration of grid-level energy storage. However, the conventional plants are typically designed to operate at base-load and witness increased failure rates when forced to rapidly cycle their output, or while undergoing startup and shutdown operations. Such cycling operation also reduces the power generation efficiency of the plant [1,2]. On the other hand, to effectively manage energy supply-demand fluctuations, grid-scale energy storage integration requires large capital investment with only a limited number of candidate storage technologies [3-5]. To address these power generation and storage challenges, we propose the integration of localized energy storage with individual fossil power plants. Such an energy storage scheme requires less capital investment, reduces the cycling associated with fossil-fueled power plants to improve grid flexibility, and increases the number of candidate storage technologies. Moreover, the operational synergies that exist between the power plant and the localized storage systems provide additional routes for cost savings and system flexibility.

There exists a variety of candidate energy storage technologies which include electrochemical, thermal and/or mechanical storage. However, these technologies have several associated trade-offs in terms of efficiency, lifespan, availability, cost, environmental footprint and safety. To handle these trade-offs and the complex dynamics between energy generation, storage and dispatch systems, we develop an optimization-based framework for the downselection of viable storage alternatives. The overall problem is formulated as a mixed-integer nonlinear programming (MINLP) model considering two objectives: (i) maximizing the total profit obtained from an integrated power generation and storage system, and (ii) minimizing the power plant cycling over a finite time-horizon of operation. To extensively compute integration profitability, the problem considers spot market revenues, power plant operational costs, energy storage investment and operating costs, and power plant cycling operation costs. The overall problem is especially challenging due to the presence of numerous decision variables, dynamic operations of the power plant and storage, and the complex interactions between the various systems components. To address this, we employ surrogate models for both power plant and storage, which are developed based on high-fidelity storage models. We demonstrate the utility of the framework using a case study on storage technology selection for natural gas combined cycle (NGCC) power plants. To ensure that the system meets a time-variant grid demand, detailed dynamic models of the NGCC plant and sodium sulphur (NaS) battery-based electrochemical storage are used where this candidate technology is synergistically integrated with the NGCC plant [6,7]. A sensitivity analysis is performed to determine the combination of cost parameters which facilitate the storage selection. The results indicate that for the objective of profit maximization, the selection of energy storage is economically viable for low storage investment cost as well as for high NGCC variable operating costs, which includes the fuel costs and CO_2 emission cost. In addition, the integration of energy storage of capacity 23% of the power plant nominal output is optimal to minimize power plant cycling while meeting grid demand.

References:

[1] Lefton SA, Besuner PM, Grimsrud GP. Managing utility power plant assets to economically optimize power plant cycling costs, life, and reliability. In: Proceedings of International Conference on Control Applications; 1995:195-208.

[2] Mechleri E, Fennell PS, Mac Dowell N. Optimisation and evaluation of flexible operation strategies for coal-and gas-CCS power stations with a multi-period design approach. Int. J. Greenhouse Gas Control. 2017;59:24-39.

[3] Peng X, Root TW, Maravelias CT. Storing solar energy with chemistry: the role of thermochemical storage in concentrating solar power. Green Chem. 2017;19(10):2427–38.

[4] Palys MJ, Daoutidis P. Using hydrogen and ammonia for renewable energy storage: A geographically comprehensive techno-economic study. Comput Chem Eng. 2020;136:106785.

[5] Heuberger CF, Staffell I, Shah N, Mac Dowell N. A systems approach to quantifying the value of power generation and energy storage technologies in future electricity networks. Comput Chem Eng. 2017;107:247-256.

[6] Wang Y, Bhattacharyya D, Turton R. Evaluation of novel configurations of natural gas combined cycle (NGCC) power plants for load-following operation using dynamic modeling and optimization. Energy & Fuels. 2019;34(1):1053–70.

[7] Vudata SP, Bhattacharyya D, Turton R. Optimal thermal management of a high-temperature sodium sulphur battery. In: 2018 AIChE Annual Meeting.

Copyright © American Institute of Chemical Engineers. All rights reserved.

624452 Optimization of a Hybrid Biomass and Gas to Liquids Process Via Fischer–Tropsch Synthesis with Integrated Product Refining

Abstract:

The production of liquid transportation fuels still relies heavily on nonrenewable resources like petroleum. To address environmental challenges and ensure sustainable development, it is necessary to investigate alternative processes for liquid fuels production. Biomass as a renewable energy source is an attractive alternative feedstock since it could reduce greenhouse gas (GHG) emissions during photosynthesis. In order to avoid potential influence on the food market, the usage of nonfood-based biomass is necessary. However, biomass route requires large amount of capital investment because of the gasification and cleanup processes. Incorporation of additional carbon-based feedstocks like natural/shale gas as raw material could contribute to the reduction of capital cost.

A process synthesis framework for the production of liquid fuels from biomass and natural gas is introduced. The main challenges associated with BGTL processes are the high capital investment and uncertainties in product prices. The hydrocarbon products consist of gasoline, diesel and kerosene produced from syngas obtained from biomass gasification and natural gas reforming via Fischer–Tropsch synthesis and further upgrading process. The conceptual design of the process and its feasibility is tested in AspenPlus, along with the validation of thermodynamic models. The mathematical model and global optimization are implemented in PYOMO.

We consider simultaneous heat, power, and water integration. Heat is extracted through combustion of light gases obtained from the process. Electricity is produced via gas turbine engine. Excess water from the process is electrolyzed to produce H_2 and O_2. An optimization model is implemented to determine optimal process configuration at minimal total cost. The number and existence of some units, like separators for Claus process, air separation unit for O_2 production are determined and environmental considerations for GHG emission are met by CO_2 recycling and sequestration.

Copyright © American Institute of Chemical Engineers. All rights reserved.

624475 An Optimization Based Framework for the Design of Biofuels/Fossil Fuels Blendstocks

Abstract:

Biofuels have been identified as an important component in the development of a sustainable renewable energy pathway. With this goal in mind, significant efforts have been made to develop biofuels for the replacement of gasoline, diesel, and jet fuel. While in some cases biofuels can be used directly in engines, the most likely use of them is in blends with fossil fuels (e.g. ethanol is typically blended with gasoline at 10%, and biodiesel at 5% with diesel). It is likely that in the middle term, blending will remain common, especially if we consider that satisfying the whole fossil fuel demand using only bio-renewable sources is difficult. However, little attention has been paid to the opportunities that stem from the nature of the blending process. In particular, we note that it is possible to tailor the biofuel fraction such that, when blended with a lower quality fossil fuel, it produces a blend (biofuel plus fossil fuel) that satisfies all quality constraints and specifications. Having said that, designing biorefineries to produce such fuels is a challenging task. One alternative that has been gaining momentum is based on the use of ethanol upgrading [1]. This alternative is attractive because there are several chemistries associated with ethanol upgrading, providing access to a diverse palette of chemical components that can be used to tailor the properties of a biofuel [2].

In this work, we develop an optimization-based framework for the design of ethanol upgrading biorefineries for the production of fuels that are going to be used in blends with fossil fuels. This framework explicitly accounts for the properties of the fuels produced considering the blending with a fossil fuels stock, and it allows estimating the capital and operating costs associated with the chemistries used to upgrade ethanol. The framework uses a superstructure representation in which the selection of the chemistries and catalysts to be used in an upgrading strategy is explicitly modeled. Importantly, we include 29 alternative chemistries and 150 different catalysts, representative of the state-of-the-art for ethanol upgrading. The explicit treatment of the blending problem, coupled with an adequate representation of the process synthesis problem allows us to explore the different relations and trade-offs among the fossil fuel properties, the biodiesel properties, the chemistries, and catalysts selected, and the economics of the process. The model that we present is used to construct a complete technology map, identifying optimal biorefining technologies as a function of different parameters of interests like feedstock cost, and diesel fuel cetane number. We also discuss how changes in these parameters affect the fuel yield, composition, and properties.

1. Eagan, N. M., Kumbhalkar, M. D., Buchanan, J. S., Dumesic, J. A. & Huber, G. W. Chemistries and processes for the conversion of ethanol into middle-distillate fuels. *Nat. Rev. Chem.* **3**, 223–249 (2019).
2. Sun, J. & Wang, Y. Recent advances in catalytic conversion of ethanol to chemicals. *ACS Catal.* **4**, 1078–1090 (2014).

Copyright © American Institute of Chemical Engineers. All rights reserved.

628160 Assessment and Optimization of the Economic and Carbon Sequestration Potential of Renewable Energy and Negative Emission Technologies

Abstract:

Due to the delayed action in climate change mitigation, we need to multiply our efforts in the next decade to avoid the growing threat of abrupt and irreversible climate change. The practical implementation requires us to know where the resources locate and how to plan the utilization of the resources. Renewable energies (RE) are critical alternatives to fossil fuels for emission mitigation and negative emission technologies (NET) are promising candidates to mitigate emissions by removing CO2 from the atmosphere. As the intersection of both, bioenergy with carbon capture and storage (BECCS) and biochar (BC) are two key NETs based on the thermal conversion of biomass. However, the amount of renewable resources is usually constrained by geographical boundaries. There is a lack of geographical survey on the carbon mitigation and reduction potential that could be synthetically achieved by RE, BECCS, and BC in different regions around the world. This study sheds light on the carbon mitigation and reduction potential achievable by such combination of technologies globally and provides an overview of the best options for technology implementation to meet the Paris Agreement from a global perspective.

In this work, we assess the potential application of NETs and HERES at regional and global scales. At the regional scale, a stochastic multi-objective decision-support framework is developed to identify the optimal design of the energy mix under various weather scenarios was developed to address the tradeoff between economic and environmental criteria and the uncertainties caused by changing meteorology conditions. The economic and environmental performances of the system were evaluated using the net present value (NPV) and the greenhouse gas emissions (GHGe), respectively. The maximum NPV and the minimum GHGe were determined by finding the optimal capacities of the solar, wind, combustion, gasification, pyrolysis, and energy storage components in the system using mixed-integer nonlinear optimization programming (MINLP).

At the global scale, a snapshot of the current waste, emission, and policy information in different countries around the world is obtained according to real-world data sources. General Circulation Model (GCM)-based simulations of future climate and the Integrated Model to Assess the Global Environment (IMAGE) are used to generate possible future scenarios. Based on these inputs, mixed-integer linear programming (MILP) optimization is used to calculate the optimal mix of renewable and negative emission technologies for each region under different scenario pathways with the goal to maximize the profit while achieving the Paris agreement target.

The multi-scale studies address the techno-economic potential of the combined use of renewable energy and negative emission technologies from the local and practical operation aspect on the one hand and provide broader insight into the global applicability considering geographical difference on the other. The result regional study showed that negative emission and 100% renewable penetration were achievable with positive and competitive profitability for a rural case but not achievable for the urban case. Through the global analysis, it was found that most countries were decided to be profitable locations for the proposed system when net present value is maximized. Negative emission was possible to be achieved in the majority of the countries if greenhouse gas emission was minimized, but it may lead to a dramatic increase in cost compared to the optimal NPV scenario. The trade-off between the economic and environmental benefits was also discussed in the study.

Copyright © American Institute of Chemical Engineers. All rights reserved.

630027 Toward Future Energy Generation Systems: Multi-Scale Optimization with Market Interactions

Abstract:

There has been growing interest within power systems communities towards the development of Integrated Energy Systems (IES) [1]. Such generation systems have the flexibility to integrate diverse energy sources (e.g., electricity, heat, steam, or chemicals) in hybrid configurations to facilitate new renewable integration, curtailment of emissions, and reduction in grid operating costs [2]. A key advantage afforded by IES flexibility is the means to participate in electricity markets by offering energy and regulation services. Because an IES can dynamically apportion its energy output (e.g. using thermal and/or electrical storage), it can provide responsive generation to grid operators and help to improve grid reliability [3]. Consequently, the design and optimization of an IES must consider both technical performance of the energy system and its economic viability within dynamic energy markets.

Fundamental challenges toward the development of an IES include accounting for multiple energy market time-scales (e.g. real-time, day-ahead, and long-term capacity markets), and exogenous uncertainty that arises due to decisions as a market participant that affect the entire system [4]. The effects of exogenous uncertainty can be particularly important to capture. For instance, an IES design decision like the size of storage will impact how it bids into the market (e.g., ramping rate), which impacts both (1) the market revenues of the entire system and (2) how the IES is asked to respond from the system operator. Market prices are also often set by the marginal resource, i.e., the resource with the highest cost out of those that are dispatched. As a result, slight perturbations in the parameters of an IES that are marginal (such as generation) can induce large fluctuations in the market, both for the IES and its neighbors. Accepted modeling approaches typically use "price-taker" assumptions [5] that do not capture the exogenous uncertainty inherent to designing new generation systems, which necessitates the development of new optimization formulations.

This talk discusses novel optimization-based approaches that capture market interactions (i.e. exogenous uncertainty) to enable tractable solution approaches for IES design problems. We first discuss key IES design parameters from an energy market standpoint based on current power dispatch formulations [6], and we then present a detailed analysis of market simulations [7] over a candidate power network [8] that predict long-term revenues and operational profiles. We then discuss a surrogate modeling framework [9] that captures market revenue and dispatch as tractable algebraic functions within our optimization framework. We show that our developed surrogates well-approximate the high-fidelity simulation data [10,11], and that they confirm our intuition with respect to key design parameters. We lastly introduce a complete IES conceptual design problem that combines our market-based surrogate models with detailed plant physics in a stochastic programming framework. Our results demonstrate the advantages gained by capturing exogenous market uncertainty versus using standard modeling assumptions when designing new generation systems. We finally close with discussion towards developing fully dynamic conceptual design problems that include market participation.

[1] M. O. Malley et al., “Energy Systems Integration : Defining and Describing the Value Proposition”, 2016

[2] D. Arent et al., "Multi-input, Multi-output Hybrid Energy Systems," Joule, 2021

[3] A. W. Dowling and V. M. Zavala, “Economic opportunities for industrial systems from frequency regulation markets,” Computers and Chemical Engineering, 2018

[4] F. Sorourifar, V. M. Zavala, and A. W. Dowling, “Integrated Multiscale Design , Market Participation , and Replacement Strategies for Battery Energy Storage Systems,” IEEE Transactions on Sustainable Energy, 2020

[5] A. W. Dowling, R. Kumar, and V. M. Zavala, “A multi-scale optimization framework for electricity market participation,” Applied Energy, 2016

[6] B. Knueven, J. Ostrowski, J. P. Watson, "On Mixed-Integer Programming Formulations for the Unit Commitment Problem", INFORMS Journal on Computing, 2020

[7] J. P. Watson, B. Knueven, R. Concepcion, D. Melander, A. Short, P. Zhang, D. Woodruff, Prescient, Computer software, Available: https://github.com/grid-parity-exchange/Prescient

[8] C. Barrows et al., “The IEEE Reliability Test System: A Proposed 2019 Update,” IEEE Transactions on Power Systems, 2020

[9] A. Cozad, N. V Sahinidis, and D. C. Miller, “Learning Surrogate Models for Simulation-Based Optimization,” AIChE Journal, 2014

[10] E. Plischke, E. Borgonovo, and C. L. Smith, “Global sensitivity measures from given data,” European Journal of Operational Research, 2013

[11] A. Saltelli, P. Annoni, I. Azzini, F. Campolongo, M. Ratto, and S. Tarantola, “Variance based sensitivity analysis of model output . Design and estimator for the total sensitivity index,” Computer Physics Communications, 2010

Copyright © American Institute of Chemical Engineers. All rights reserved.

633005 Magnesium Oxychloride Formation Kinetics and Enhanced Water Stability for Sustainable Building Materials Applications

Abstract:

Magnesium oxychloride (MOC) is a ceramic material with significant fire resistant properties and growing potential as a light-weight, structurally strong alternative building material. Despite many advantages, MOC has limitations as a construction material, particularly water stability. While researchers have explored solutions to these shortcomings, fundamental understanding of this material is still urgently needed. Thus we have investigated the fundamentals of the formation cure reaction and mechanisms of water stability enhancement reactions. The formation cure reaction kinetics for magnesium oxychloride 5-phase was monitored from 35 to 55°C using time-resolved quantitative x-ray diffraction and differential scanning calorimetry (DSC). The reaction was characterized as a two-step process: dissolution of magnesium oxide into a gel state followed by crystallization of magnesium oxychloride. Assuming first-order kinetics for both MgO dissolution and MOC crystallization, a kinetic model predicts 42.4 kJ/mol and 26.1 kJ/mol for dissolution and crystallization activation energies respectively. Alternatively, the Avrami nucleation and growth model was fit to DSC measurements predicting diffusion controlled, one-dimensional growth with an activation energy of 72.4 kJ/mol, accounting for both dissolution and crystallization.

For the water stability enhancement, two methods of water stability enhancement have been explored: chlorartinite formation from CO_2 exposure and phosphoric acid addition. The conversion of magnesium oxychloride to chlorartinite by CO_2 forms a protective, semi-insoluble chlorartinite layer on the surface of the magnesium oxychloride crystals, which improves water stability. Phosphoric acid (2.5 to 10 wt. %) was added to the MOC slurry before the cure reaction. Additions of 2.5 wt. % and above had positive impacts on the water stability, preserving ~50 wt. % crystalline MOC after the water stability test, but has a significant impact on the MOC formation kinetics. This is achied via. the formation of an amorphous phase on the MOC crystal surface that contains structural motifs related to insoluble MgHPO4·3H2O (newberyte) and Mg2P2O7·3.5H2O (magnesium pyrophosphate) phases. Results from this work are significantly impactful for the use of MOC in more sustainable alternatives for conventional materials in residential and commercial building applications.

Copyright © American Institute of Chemical Engineers. All rights reserved.

624619 Optimization of TEG Dehydration Process in Natural Gas Processing Under Metamodel Uncertainty

Abstract:

Optimization of TEG Dehydration Process in Natural Gas Processing under Metamodel Uncertainty

Rajib Mukherjee[1,2], Urmila M Diwekar[2*]

[1]*Department of Chemical Engineering, The University of Texas Permian Basin, Odessa, TX 79762, USA*

[2]*Vishwamitra Research Institute, Crystal Lake, Illinois, 60012, USA*

Abstract

Natural gas processing involves removal of acidic gases followed by dehydration that is mainly carried out through absorption in tri-ethylene glycol (TEG). The dehydration process is accompanied by the emission of volatile organic compounds, including BTEX. There are several process parameters associated with major equipment that may impact BTEX emission. In our previous work, a multi-objective optimization is performed for optimal operating conditions identification related to the process parameters to mitigate BTEX emission using data driven metamodeling using machine learning followed by optimization with metaheuristic technique [1]. The metamodel is generated with data obtained from process simulation carried out in ProMax® process simulator. The metamodels are created using limited samples which introduces metamodeling uncertainty [2]. Previous metamodel-based robust designs treated a metamodel as the real model and ignore the influence of metamodeling uncertainty. Thus, the performance of the resulting optimized process parameters may have shortcomings due to the lack of adequately accounting for the uncertainty introduced by the metamodel. In the present work, the bias of metamodel uncertainty has been addressed for parameter optimization. An algorithmic framework is developed to address the challenge of parameter optimization under metamodel uncertainties. We used the better optimization of nonlinear uncertain systems (BONUS) algorithm to solve the problem. BTEX mitigation is used as the objective of the optimization [3]. Our algorithm allows the determination of the optimal process parameters for BTEX emission mitigation from TEG dehydration process under metamodel uncertainty.

Keywords: Natural gas processing, TEG dehydration, BTEX mitigation, metamodeling uncertainty, support vector regression (SVR), BONUS algorithm

[1] Mukherjee, R., & Diwekar, U. M. (2021). Multi-objective Optimization of the TEG Dehydration Process for BTEX Emission Mitigation Using Machine-Learning and Metaheuristic Algorithms. ACS Sustainable Chemistry & Engineering, 9(3), 1213-1228.

[2] Zhang, S., Zhu, P., Chen, W., & Arendt, P. (2013). Concurrent treatment of parametric uncertainty and metamodeling uncertainty in robust design. *Structural and multidisciplinary optimization*, *47*(1), 63-76.

[3] Sahin KH, Diwekar U, M. (2004) Better Optimization of Nonlinear Uncertain Systems (BONUS): a new algorithm for stochastic programming using reweighting through Kernel density estimation. Ann Oper Res 132:47–68

Copyright © American Institute of Chemical Engineers. All rights reserved.

633725 Mapping Environmental and Economic Analysis of Decentralized Cogeneration Energy Management Centers

Abstract:

In the midst of a global conversation on how to reduce GHG emissions, Ontario's electrical grid stands out due to its extensive use of low carbon-intense processes such as nuclear and hydropower that acts as baseload energy sources. However, for peak electricity demands that require a quick ramp up of production, natural gas (NG) turbines are deployed [1], which causes the carbon intensity of Ontario's electrical grid to vary hourly [2]. Another intensive and permanent Canadian energy requirement is heating. Currently, it is provided mostly by NG -fired boilers [3]. The pursuit for better NG utilization is the motivation for this study, which proposes a design of combined heat and power energy management center (CHP-EMC).

CHP-EMC is composed of an NG-powered internal combustion engine connected to an electrical generator (NG-CHP), a hot water tank storage, an NG boiler, and a cooling tower to keep the water temperature within acceptable bounds. EMC is connected to a water loop similar to the ones used in district heating (DH) by a heat exchanger and responds to a community's heating demand using an hourly lumped-demand approach. The simulations are developed in TRNSYS, which is a capable heat transfer and fluid dynamic simulator [4]. The storage tank uses a built-in dynamic model whilst the rest of the components operate in pseudo steady-state. These components are connected by pipe models that account for time delay and flowrate.

The equipment presence and its sizes are decided by minimizing the natural gas consumption while simultaneously maximizing thermal and electrical generation. The optimization strategy used is a black-box approach using modeFrontier software [5]. This optimization and data visualization software can be coupled with any other software that uses input/output files -such as TRNSYS- and it executes several simulations guided by the optimization strategy and its constraints. For this work, classic Multi-Objective Genetic Algorithm (MOGA) is chosen for design variables (including the presence of specific components) and the pareto points are analyzed in more detail. TRNSYS simulation has internal simple controllers that take care of schedule and operational constraints. The result of the optimization aims to create an EMC flexible enough to endure deviations from expected demand profiles.

The emissions of proposed EMC designs are compared to the status quo (electricity supplied by electrical grid and thermal demand supplied by NG boilers) for a year of operation. Total emissions are calculated for each possible value of gCO_{2eq}/kWh to better represent possible scenarios of carbon intensity. The results are summarized in an easy-to-read emissions vs technology map that defines the conditions under which the EMC is less pollutant than status quo (SQ). The optimal design and selection of the EMC (vs. SQ) system is strongly dependent on the carbon intensity of the electric grid during peak-load times, the electric demand profiles, and the cooling demand profiles over the course of its operation. These are best characterized by the combination of greenhouse gas emissions per kWh during peaking operation, the ranges of heating and electricity demand over the course of the year (the maximum minus the minimum), and the kurtosis of the demands.

Following this analysis, an economic evaluation of both systems is carried out, demonstrating the cost of carbon avoided (CCA). CCA is a metric that compares CO_2 and cost of two technological strategies and calculates the cost per ton of GHG emissions potentially reduced [6]. These results are compared to possible carbon tax values, in a similar way to carbon intensity. A second map, cost vs technology, is then developed for easy visualization of the impact of carbon taxes on EMC and SQ operation costs.

The goal of this work is to apply the benefits of decentralized EMCs to reduce GHG emissions from both heating and electricity generation. The results are generalizable and can be applied to any location due to its methodology in representing carbon intensity and carbon tax.

References

[1] IESO, "Power Data," 30 04 2021. [Online]. Available: https://www.ieso.ca/power-data. [Accessed 30 04 2021].

[2] L. Pereira and I. D. Posen, "Lifecycle greenhouse gas emissions from electricity in the province of Ontario at different temporal resolutions," *Journal of Cleaner Production,* vol. 270, no. 122514, 2020.

[3] S. Canada, "Primary heating systems and type of energy," 03 04 2021. [Online]. Available: https://www150.statcan.gc.ca/t1/tbl1/en/tv.action?pid=3810028601. [Accessed 28 04 2021].

[4] TRNSYS, 2019. [Online]. Available: http://www.trnsys.com/index.html. [Accessed 30 04 2021].

[5] ESTECO, "modeFrontier," 2021. [Online]. Available: https://engineering.esteco.com/modefrontier/. [Accessed 30 04 2021].

[6] K. Gillingham and J. H. Stock, "The Cost of Reducing Greenhouse Gas Emissions," *Journal of Economic Perspectives,* 2018.

Copyright © American Institute of Chemical Engineers. All rights reserved.

627430 Development of an Interactive Software Tool for Designing Industrial Solvent Recovery Processes

Abstract:

Abstract

In 2017 the chemical industry was reported to be the world's second-largest manufacturing sector (United Nations Environment Programme 2019). Solvents are used in most operations in chemical and pharmaceutical industries as a reaction medium, selective dissolution and extraction medium, and dilution agents. Thus, there is a sizable amount of solvent waste generated due to process inefficiencies such as inefficient mixing, and inappropriate process equipment being utilized. There are several options for solvent waste disposal, however, the three most common are on-site disposal, off-site disposal, and incineration (J. Raymond, Stewart Slater, and J. Savelski 2010). Even though these methods of solvent waste disposal carry considerable negative environmental impacts they are still widely used since they are readily available and provide an easy method to dispose of large quantities of waste. Solvent recovery is typically not used because of the potential difficulties in achieving a required purity to meet the stringent guidelines imposed on the products in pharmaceuticals and specialty chemical industries as well as the additional infrastructure and investments needed. To this end, this problem needs to be studied carefully by involving all aspects in regard to capital needs, environmental benefits, comparison to traditional disposal methods, while achieving the required chemical purity.

Thus, designing a solvent recovery method is a multi-objective problem focused on minimizing capital investments of the new separation technologies while maximizing the amount of solvent recovery matching the desired purity requirements. This task is further complicated because there is a lack of a guidance system to assist professional engineers in the modeling and design of solvent recovery processes. Such a guidance system or tool that could predict the value of solvent recovery and reuse for industry users, can generate an environmentally friendly trend to shift away from using incineration, on-site disposal, and off-site disposal as primary solvent waste handling methods. As seen in Figure 1A, not only will this tool provide a greener alternative, but also aid in solvent recycling into the chemical plant, which is economically favorable. To this end, if we can develop a user-friendly tool that allows engineers to easily access solvent recovery options, we can provide an economical and environmentally favorable strategy.

To account for all possible waste streams a robust framework is needed. This consists of a maximal process flow diagram that encompassed multiple stages of separations and multiple technologies within those stages. Together this process flow diagram develops the superstructure that provides multiple technology pathway options for any solvent waste stream. We established the different stages in the superstructure as Solid Removal, Recovery, Purification, and Refinement by reviewing the literature on existing materials recovery and separation processes (Yenkie, Wu, and Maravelias 2017; Biegler 1997; Geankoplis 2003). Common technologies are placed in different stages depending on the component they can separate in terms of their physical and chemical properties. A comprehensive chemical database is created using sources, such as the Design Institute for Physical Properties (DIPPR), to store all relevant chemical and physical properties ("DIPPR" 2012). Using General Algebraic Modeling Systems (GAMS) to model the various separation units and stream flows as a multi-objective optimization problem, the amount of solvent recovered can be optimized while minimizing the environmental impacts and total cost of operation (Chea et al. 2020; Yenkie et al. 2016). With the GAMS code as the backbone, a Graphical User Interface (GUI) is created to provide a user-friendly tool to the chemical industry. Figure 1B shows how the GUI connects the superstructure modeled in GAMS model, the chemical database, and user inputs to allow for anyone in the industry to find an optimal technology pathway for their solvent waste.

The GUI will allow for a user to input chemical species and parameters that are specific to the process that is being observed. By connecting MATLAB to the GAMS code, it is possible to model a solvent recovery process solely in the GUI with these input parameters. This tool reports economic, chemical, and environmental factors for the user to consider. The economic factors include an annualized cost breakdown, utility costs, consumable costs, labor costs, and overhead costs. These costs can be seen for the whole selected recovery pathway with details of the individual technologies. The chemical factors report the amount of solvent the predicted pathway can recover and its purity level. The environmental factors show the estimated emissions from the technology pathway and compare these emissions against those for incinerating the waste stream. Thus, this tool is a powerful option for those in the industry who want to look for a cost-effective way of recycling their solvents from the waste streams.

Keywords: Solvent Recovery, Recycling, Graphical User Interface

Acknowledgments

This research was supported by the United States Environmental Protection Agency's Pollution Prevention (P2) Program (NP96259218). The authors thank the Department of Chemical Engineering at Rowan University for their continued assistance with acquiring the tools required for this research.

References:

Biegler, Lorenz. 1997. *Systematic Methods of Chemical Process Design*. 1st ed. Prentice Hall PTR.

Chea, John D., Austin L. Lehr, Jake P. Stengel, Mariano J. Savelski, C. Stewart Slater, and Kirti M. Yenkie. 2020. "Evaluation of Solvent Recovery Options for Economic Feasibility through a Superstructure-Based Optimization Framework." *Industrial & Engineering Chemistry Research* 59 (13): 5931–44. https://doi.org/10.1021/acs.iecr.9b06725.

"DIPPR." 2012. The American Institute of Chemical Engineers. March 29, 2012. aiche.org/dippr.

Geankoplis, Christie J. 2003. *Transport Processes and Unit Operations*. 4th ed. Englewood Cliffs, NJ: P T R Prentice-Hall, Inc.

2010. Raymond, Michael, C. Stewart Slater, and Mariano J. Savelski. 2010. "LCA Approach to the Analysis of Solvent Waste Issues in the Pharmaceutical Industry." *Green Chemistry* 12 (10): 1826–34. https://doi.org/10.1039/C003666H.

United Nations Environment Programme. 2019. "Global Chemicals Outlook II - From Legacies to Innovative Solutions - Synthesis Report (2019)." Global Chemicals Outlook II. United Nations Environment Programme. https://wedocs.unep.org/bitstream/handle/20.500.11822/27651/GCOII_synth.pdf?sequence=1&isAllowed=y.

US EPA, OEI. 2013. "TRI Data and Tools." Data and Tools. US EPA. March 3, 2013. https://www.epa.gov/toxics-release-inventory-tri-program/tri-data-and-tools.

Yenkie, Kirti M., WenZhao Wu, Ryan L. Clark, Brian F. Pfleger, Thatcher W. Root, and Christos T. Maravelias. 2016. "A Roadmap for the Synthesis of Separation Networks for the Recovery of Bio-Based Chemicals: Matching Biological and Process Feasibility." *Biotechnology Advances* 34 (8): 1362–83. https://doi.org/10.1016/j.biotechadv.2016.10.003.

Yenkie, Kirti M., Wenzhao Wu, and Christos T. Maravelias. 2017. "Synthesis and Analysis of Separation Networks for the Recovery of Intracellular Chemicals Generated from Microbial-Based Conversions." *Biotechnology for Biofuels* 10 (1). https://doi.org/10.1186/s13068-017-0804-2.

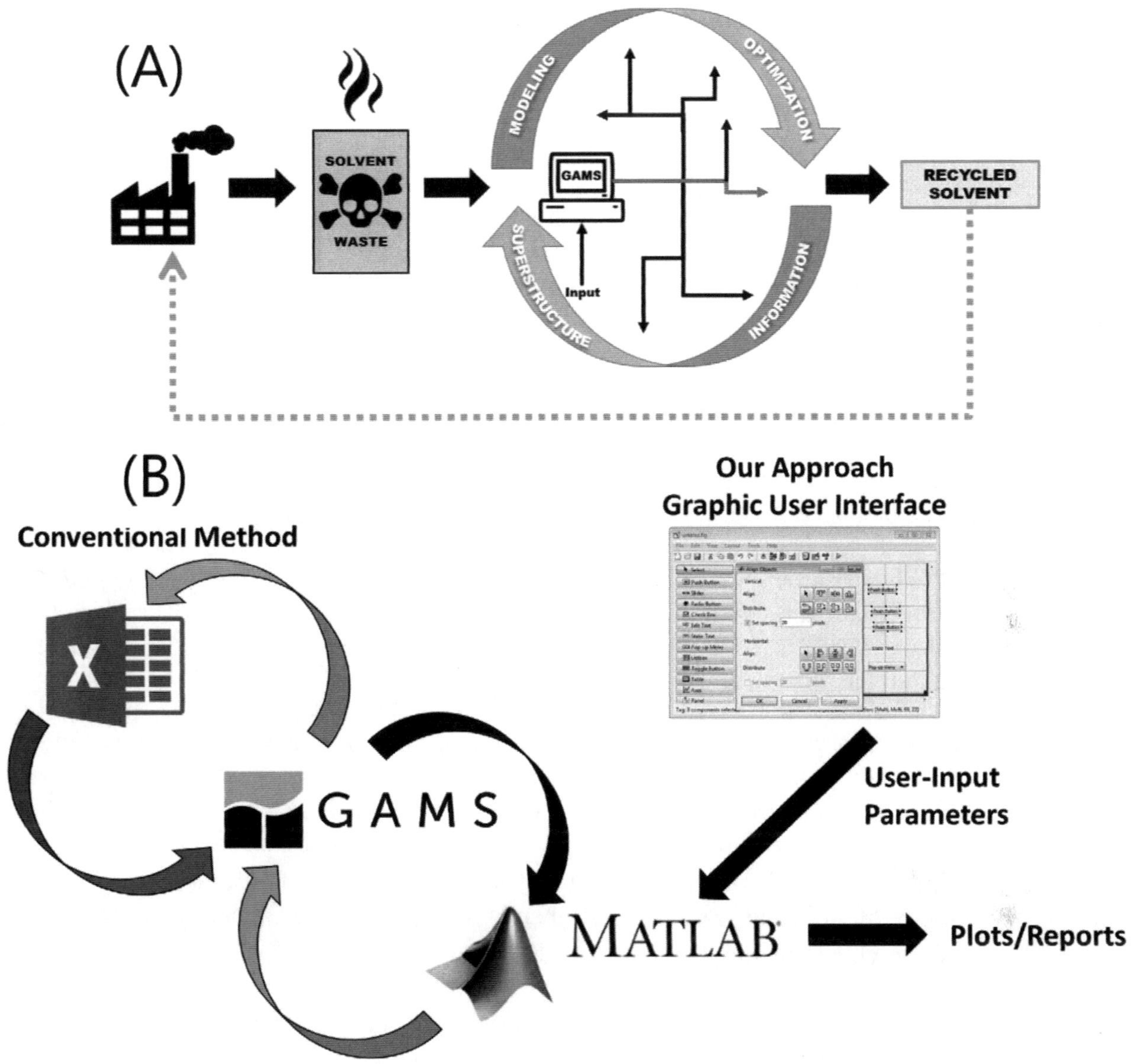

Figure 1(A) Proposed Process for Modeling Solvent Recovery for use in an Industrial setting.
(B) Flow chart of Information Processing of the Proposed Computational Software Tool

Copyright © American Institute of Chemical Engineers. All rights reserved.

625002 High Flux CO2 Selective Membranes for Renewable Natural Gas and CO2 Capture

Abstract:

Reducing carbon emissions is key to preventing further climate change and in correcting the damage that has already been done. Compact Membrane Systems is developing a carbon dioxide selective membrane designed to operate at low pressure with minimal pretreatment. Results from two applications will be discussed: removal of carbon dioxide from flue gas and upgrading of biogas to renewable natural gas.

According to the World Biogas Association, anerobic digestors for biogas production can reduce greenhouse gas emissions by 10% in 2030. Separating carbon dioxide from methane is a key challenge for the renewable gas industry. Biogas is produced largely on farms or small waste centers, not at large centralized plants. While there are existing technologies to separate methane from carbon dioxide, they require multiple stages and high pressures, and are sensitive to process poisons. An efficient and cost-effective solution is needed to bring down barriers that previously prevented small scale operations from transforming their biogas into usable fuel. CMS has developed a membrane aimed at allowing users to capture, separate and upgrade existing biogas streams that would otherwise be flared or released into the atmosphere.

Results from the first field demonstration of Optiperm™ *biogas* will be shared. Working with Energy Projects Manager Matt Steiman at the Dickinson College Farm, a small spiral wound module system has been installed. The system takes raw biogas feed directly from the digester with no pretreatment. The pilot is intended to demonstrate successful operation under field conditions for an extended period of time. Optiperm™ *biogas* technology will be tested for a period of at least 2 months to demonstrate stability in the field and validate performance measured in the CMS laboratory. Optiperm™ *biogas* upgrading streams from 60% to 90% methane in a single stage.

Finally, initial results for utilizing the CMS membranes to capture carbon dioxide from flue gas, including the effect of operating conditions and initial system modeling, will be shared. This application is earlier in development, but results thus far have shown a very high flux membrane that could result in substantially lower capture costs than the current state of the art.

Copyright © American Institute of Chemical Engineers. All rights reserved.

AICHE Fall 2021

Optimizing Energy Efficiency of Ammonia Production via Electrochemical Reaction and Haber-Bosch process.

Gbemisola Ojo and Kyle Camarda
[1]The University of Kansas, Department of Chemical & Petroleum Engineering
4132 Learned Hall, 1530 W 15th Street, Lawrence, KS 66045

Ammonia (NH_3) synthesis is crucial to the production of fertilizers, and accounts for nearly 2% of total global energy consumption. Therefore, finding methods to produce ammonia employing only renewable energy and sustainable feedstocks is paramount. This project uses dynamic simulation to design a 100% renewable ammonia synthesis plant, using only air and water as raw materials. Energy is supplied from a solar source, which is supplemented with industrial batteries. The intermittent nature of solar energy production requires the development of a dynamic production profile for the plant.

In this work, an electrolytic cell powered by a time-varying solar source is employed for hydrogen production. Air is separated via a membrane to provide a nitrogen source, and then the Haber-Bosch reaction is used to produce the product. The degrees of freedom in this problem are the number of batteries purchased, as well as the dynamic production and charging schedules. To determine the optimal production profile, a mixed integer non-linear programing model is formulated and solved. Results demonstrate how varying levels of production and energy storage usage correspond to different levels of energy availability, and show the feasibility of completely sustainable ammonia production, using no raw materials other than water and nitrogen, and creating zero emissions.

629230 Computational Prediction of Polymer Solubilities and Its Application in Multilayer Plastic Film Recycling

Abstract:

Billions of pounds of multilayer plastic films are produced annually as packaging materials. These films are made of distinct polymer layers, which makes it a challenging task to recycle them. Recently, we developed a strategy called solvent-targeted recovery and precipitation (STRAP) that can deconstruct and recycle such films through a series of solvent washes. It utilizes the solubility difference of the polymers to selectively dissolve them in carefully chosen solvent systems. One key step in designing a STRAP process for a target film is to identify selective solvent systems and operating temperatures, which could be prohibitively time-consuming using experimental screening alone. Here, we report a computational scheme for polymer solubility prediction and solvent selection that employs Hansen solubility parameters, molecular dynamics simulations, and the COnductor-like Screening Model for Realistic Solvents. These tools take molecular interactions and conformational information into consideration to yield quantitative, temperature-dependent solubility predictions of polymers in pure solvents and solvent mixtures. We demonstrate that this computational scheme can efficiently screen appropriate solvents to guide the design of solvent-mediated processes for multilayer plastic film recycling.

Copyright © American Institute of Chemical Engineers. All rights reserved.

631196 Calibration of a CFD-DEM Fluidised Bed Simulation for Plastic Recycling Using Positron Emission Particle Tracking

Abstract:

Industrial-scale plastic recycling using pyrolysis in fluidised beds is an emerging technology aiming to address the ever growing plastic waste crisis. Whilst we developed a basic understanding of fluidised beds with a single solid phase, such as the fluid catalytic cracking in modern oil-refineries which have been used for decades, binary systems which introduce a non-identical second solid species often lead to difficulties in reactor design. Uneven particle distributions, segregation and hot-spots lead to not only product issues but also safety issues which need to be understood and removed.

The Discrete Element Method (DEM) does, when calibrated properly, reproduce the behaviour of moving granular material very well. Coupled with the well developed computational fluid dynamics (CFD) it provides a powerful tool in predicting the behaviour of a fluidised granular system such as fluidised beds.
These systems however need to be calibrated and even though the conventional techniques of calibration for DEM models such as shear cells or angle of repose measurements can lead to quantitative granular properties, the characterisation and calibration of the fluid phase is currently much less precise.

We use positron emission particle tracking, a Lagrangian particle tracking technique, to directly compare coupled DEM-CFD simulations of fluidised beds with real world fluidised beds. The rich, high-resolution data yielded by PEPT allows us to fine tune drag models and particle parameters in a significantly more precise manner than is possible using conventional measurements such as pressure drop alone.

Copyright © American Institute of Chemical Engineers. All rights reserved.

631881 Use of Waste Plastic in Asphalt Binder: Theoretically and Experimentally Justified Identification of Compatible Blends

Abstract:

Previous attempts to add waste plastic to asphalts were not met with great success primarily due to the incompatibility of common plastic and asphalt cement, which resulted in poor performance. If the compatibility between waste plastic and asphalt binder can be systematically understood and improved, then large amounts (millions of tons annually) of previously land-filled waste plastic can be repurposed in asphalt binder. Here, we fundamentally investigate the compatibility of waste plastic and asphalt using a combination of atomistic simulation, coarse-grained simulation, and experimental validation to provide guidelines for predicting the compatibility between asphalt and waste plastic. Waste plastics such as HDPE, LDPE, PVC, PET, and PS are considered, along with some approaches to improve compatibilitzation such as crosslinking with ground tire rubber, aminolysis of PET, and wax residue addition. These materials and approaches are considered in terms of fundamental polymer science concepts.

Copyright © American Institute of Chemical Engineers. All rights reserved.

625807 Application of Waste Hay-Derived Biochar for Adsorption of Cyanobacterial Toxin in Water

Abstract:

The biochar-driven adsorption of microcystin was investigated as a promising remediation technology. The objective of present study is to investigate alfalfa-derived biochars (AF biochars) for adsorption of microcystin-LR (MC-LR) in water for the first time. The relationships between properties of AF biochars produced at various temperatures (350-650°C) and adsorption capacities of MC-LR were evaluated using a statistical analysis. The pyrolysis temperature led to a highly increase in surface area, aromaticity, and adsorption capacity of MC-LR. Among the AF biochars, AF650 had much higher adsorption capacity for MC-LR because of mechanisms such as electrostatic, π-π, and hydrophobic interactions supported by its positively charged surface, high BET surface area, and high hydrophobicity. The adsorption kinetic and isotherm models fitted pseudo-second order and Langmuir for MC-LR onto AF650, respectively. The AF650 possessed highly adsorption capacity for MC-LR but lower adsorption capacity than those of activated carbon.

Copyright © American Institute of Chemical Engineers. All rights reserved.

631666 Electro-Enhanced Removal of Heavy Metal Ions from Aqueous Solutions By Capacitive Deionization Process Using Biochar-Based Electrodes

Abstract:

The increased level of heavy metal contamination in wastewater is a major concern due to its potential impacts on human health and the environment. Most heavy metal ions are highly persistent environmental pollutants and exhibit carcinogenic and mutagenic properties. Therefore, it is imperative to develop efficient methods to eliminate heavy metal ions from wastewater before being discharged into the environment. Among various available wastewater treatment methods, capacitive deionization (CDI) has received great attention lately due to its energy-efficient, low cost, and environment-friendly technology. The CDI process is based on the electrosorption or polarization-induced adsorption of ions on the surface of porous electrodes. The surface area and chemistry of the electrode material used in the CDI process are the key parameters that control the removal of ions via electrosorption. In this study, activated biochar-based electrodes were used to study their efficiency towards the removal of heavy metal ions (Pb^{2+} and Co^{2+}) from aqueous solutions. The morphological, structural, and electrochemical properties of prepared biochars were characterized by using BET surface area and pore size distribution analyzer, FTIR, SEM, Raman spectroscopy, cyclic voltammetry, and others. The kinetics of heavy metal ion removal was studied using different models to elucidate the mechanisms involved in the electrosorption of metal ions on the electrode surface. The effect of applied electric voltage, pH, and concentration of the metal ion solution on the electrosorption performance of the electrodes was also determined.

Copyright © American Institute of Chemical Engineers. All rights reserved.

632655 The Impact of Inorganics on the Production and Use of Biochars Made from Household Biowaste

Abstract:

Biochar can be used as an alternative fuel source or as a soil amendment and potential negative emissions technology for Biomass Carbon Removal and Storage (BiCRS). We have investigated the production and use of biochar from various biowaste sources such as nut shells, coffee grinds and corn cobs. For eight different feedstocks we have studied the impact of ash content on the pyrolysis process and on the reactivity of the subsequent char. Slow pyrolysis was conducted in a tube furnace with flash heating to temperatures between 400-500°C and the subsequent chars were measured for CO_2 gasification reactivity using Thermal Gravimetric Analysis at temperature from 700-900°C. We have found that ash content in the feedstocks was negatively correlated with char yield and positively correlated with CO_2 gasification reactivity for the feedstocks investigated. The impact of the inorganics on reactivity was further demonstrated by leaching studies on four feedstocks: corncobs, pistachio shells, walnut shells, and pumpkin shells. We found that biochars leached of their inorganics had gasification rates significantly slower than their non-leached counterparts for three of the four feedstocks studied (up to three times slower for walnut shell biochars). As an additional study, we have developed a 2.5 L semi-batch pyrolysis reactor fueled by propane and synthesis gas to produce biochar from household biowaste. We will also present design iterations of this reactor and biochar yields, emissions, residence times, and temperature profiles for various feedstocks.

Copyright © American Institute of Chemical Engineers. All rights reserved.

621594 Over Half a Century of Protecting the Environment through Various Policies, Administrations, Successes, & Failures; An Overview of 50+ Years of EPA Regulations and Standards

Abstract:

Since the establishment of the U.S. Environmental Protection Agency (EPA) on December 2nd, 1970, the Agency has implemented a vast number of regulations to benefit the American people and the environment. Over the last 51 years, the EPA and its dedicated workers have endeavored to fulfil its mission to protect human health and the environment; the Agency has sustained its critical function through 51 years and 10 different administrations (6 Republican and 4 Democratic presidents). The change of administration leads to change and priorities, and at times new or rolled back regulations. One such case is the Bush-era fuel standard increasing the fuel standards (established in 1975) to a corporate average of 35 miles per gallon by 2020, further increased by the Obama-era fuel efficiency standard of 54.5 mpg by 2025, then replaced by the Trump-era standard requiring a small 1.5% increase in average fuel economy compared to the pre-Obama standards, and then upon the inauguration of the Biden-era, auto makers themselves began to support higher fuel efficiency standards as the Administration foreshadowed increasing the standards akin to the Obama-era. This presentation aims to provide an overview of current U.S. environmental regulations, where we currently are, where we have come from, and where we are likely to go in the future.

Copyright © American Institute of Chemical Engineers. All rights reserved.

628185 Decarbonising Residential Heating: Cost-Benefit Analysis to Support Policy Design

Abstract:

The UK's legally binding targets to achieve "net zero" greenhouse gas emissions by the year 2050 may require full decarbonisation of residential heating. This will involve deploying low-carbon heating technologies (such as heat pumps, electric storage heaters, and/or various hydrogen heating technologies) across the country, including in the 24 million homes that currently use natural-gas-based central heating systems. A key challenge of this transition is that the capital and operating costs of low-carbon heating systems are often greater than current natural-gas-based systems. In addition, the UK has a wide variety of building stock, which defies a one-size-fits-all solution.

The aim of this study is to assess the suitability of various policy instruments to promote the uptake of low-carbon heating technologies. A cost-benefit model is being developed based on housing stock data from the English House Survey, performance data from field trials, and cost estimates from the literature. The model will be used to calculate the lifetime cost and levelised cost of heat for a wide range of households and the results analysed to consider how the costs of different low-carbon heating options vary with characteristics such as building type, number of occupants, thermal efficiency, and geographical location. In addition, an uncertainty analysis will used to evaluate the impact of different fuel prices, outdoor temperatures, and personal preferences and to identify the household types likely to be impacted most by these uncertainties.

The results of this study will be presented and discussed in relation to current policies designed to promote low-carbon heating. This study will consider what level of government support will be sufficient to promote low carbon heating over the current natural-gas-based heating systems and will identify if a targeted approach based on geographic location, building type, tenure type, or other characteristics would be more appropriate.

Copyright © American Institute of Chemical Engineers. All rights reserved.

625673 Improving the Economics of Industrial Battery Storage: A Proactive Policy and Management Approach

Abstract:

Historically, implementation of battery energy storage has been proportionally higher in commercial and residential systems than industrial systems [1]. Economic and operational constraints of industrial facilities make battery installation close to infeasible. Industrial energy usage accounts for about 25% of total energy usage in the US. As the deployment of battery storage increases, lack of implementation in the industrial sector creates a gap, bridged by adapting control and policy strategies to constraints of industrial users. Projections indicate that battery prices will continue to fall, making it even more important to find ways to implement batteries in the industrial sector [2].

This research aims to show that the utilization of electric battery storage to decrease the energy costs and peak power usage of industrial energy users can be made attainable by identifying and closing the policy gap between affordability and utility of industrial battery installations. Similar studies have been done to improve economics and management of storage in residential installations [3]. Other studies have looked at regulatory design of energy storage coupled with renewable sources such as solar and wind [4]. Though these efforts have furthered the penetration of battery storage in the residential and commercial sectors, this same penetration has not been seen in the industrial sector from lack of analysis and specificity to the needs of industry. Policies specific to large electrical users in the industrial sector have also not been explored.

Load data from actual industrial facilities is used spanning a full year of typical manufacturing operations in which the daily electrical peak is about 2200 to 2500 kW. Two models are used, a battery model utilizing the facility load data to look at battery dynamics and effects of system load, and a financial model to analyze overall cost and effect of policy on project economics. The study identifies policies that are currently in place in parts of the US or similar policies that would allow batteries to be financially viable for industrial facilities.

[1] EIA, "Battery Storage in the United States : An Update on Market Trends," Washington, DC, 2020.

[2] W. Cole and A. W. Frazier, "Cost Projections for Utility-Scale Battery Storage," 2019.

[3] M. N. Sheha and K. M. Powell, "An economic and policy case for proactive home energy management systems with photovoltaics and batteries," *Electr. J.*, vol. 32, no. 1, pp. 6–12, 2019, doi: 10.1016/j.tej.2019.01.009.

[4] O. Zinaman, T. Bowen, and A. Aznar, "An overview of behind-the-meter solar-plus-storage regulatory design: Approaches and case studies to inform international applications (Report no. NREL/TP-7A40-75283)," *NREL Rep.*, no. March, pp. 1–76, 2020.

Copyright © American Institute of Chemical Engineers. All rights reserved.

631890 Continuous Flow Catalytic Hydrogenation of Groundwater Nitrate Using Hydrogen-Permeable Hollow Fiber Membranes Decorated with Palladium-Indium Nanoparticles

Abstract:

Nitrate is the most commonly occurring anthropogenic groundwater pollutant in the US, due largely to over application of nitrogen-based fertilizers. In drinking water, nitrate consumption can cause oxygen deficiency in the bloodstream; an especially dangerous occurrence for infants. Accordingly, the US EPA has imposed an MCL of 10 mg NO_3-N/L. Existing technologies to treat nitrate focus on separating nitrate from water using ion exchange or reverse osmosis that produce brine waste streams that require disposal. Also, a few full-scale treatment systems currently operate using biological denitrification for drinking water treatment wherein nitrate is converted to innocuous by-products.

Non-biological catalytic hydrogenation rapidly converts nitrate to nitrogen gas using hydrogen gas (H_2) and a solid-phase catalyst. Unlike biological systems, these abiotic catalytic hydrogenation can be turned on or off with near zero start-up time. However, 80% of publications regarding catalytic hydrogenation treatment option focus on material development and employ batch slurry reactor with a catalyst in suspension and uncontrolled hydrogen flow, which is not scalable for drinking water treatment.

The presented technology uses hollow fibers as a method of controlling hydrogen delivery in a continuous flow reactor. This method improves the efficiency of hydrogen delivered, which results in an order of magnitude decrease in required hydrogen flow rate and cost. Previously studied nitrate reduction systems have incorporated catalysts in suspension, but those configurations risk metal nanoparticles carrying over into drinking water. To resolve this technology barrier, the hollow fibers also act as a surface attachment point for the catalyst. After analyzing various catalyst coating methods, we have found a method to irreversibly attach nano-catalysts to the H_2-permeable hollow-fibers. The research focuses on a bimetallic Palladium-Indium catalyst because it has the highest selective reduction of nitrate to nitrogen gas without leaching. The parameters of the synthesized Pd-In catalyst and the reaction result in up to four times the nitrate reduction activity as similar catalysts while still achieving nitrogen selectivity above 95%. Reactor parameters and the corresponding effects on activity and selectivity of the catalyst will be discussed; particularly relating to scale up from batch to continuous flow. The reactor system is operated for extended periods to determine any sources of long term catalyst deactivation. This presentation focuses on the development of a scalable technology for groundwater nitrate treatment with high selectivity to desired nitrogen gas and minimized catalyst deactivation over time.

Copyright © American Institute of Chemical Engineers. All rights reserved.

624337 Naa Zeolite-Coated Meshes with Tunable Hydrophilicity for Oil-Water Separation

Abstract:

NaA Zeolite-Coated Meshes with Tunable Hydrophilicity for Oil-Water Separation

Ghader Mahmodi [a], Shailesh Dangwal [a], Payam Zarrintaj[a], Mengfan Zhu [b], Yu Mao [b], David N. McIlroy [c], Mohammad Reza Saeb [d], Vahid Vatanpour [e], Joshua D. Ramsey [a], and Seok-Jhin Kim [a]*

[a] School of Chemical Engineering, Oklahoma State University, 420 Engineering North, Stillwater, OK, 74078, United States

[b] Department of Biosystems and Engineering, Oklahoma State University, 213 Agricultural Hall, Stillwater, OK, 74078, United States

[c] Department of Physics, Oklahoma State University, 145 Physical Sciences Bldg., Stillwater, OK, 74078, United States

[d] Department of Resin and Additive, Institute for Color Science and Technology, P.O. Box 16656118481, Tehran, Iran

[e] Department of Applied Chemistry, Faculty of Chemistry, Kharazmi University, Tehran, Iran

NaA zeolite was fabricated on a stainless steel mesh by secondary growth method through optimizing aluminum/silicon ratio (ASR) to effectively separate oily water. The NaA zeolite-coated mesh with the highest superoleophobic property had an ASR of 1.21 and an oil contact angle of 163.7^{o}. The highest membrane flux obtained was 13,513 L m^{-2} h^{-1} for an ASR = 0.91 and the highest separation efficiency was 99.5% for an ASR = 1.21. NaA zeolite meshes were recycled and reused for 15 cycles by rinsing the membrane with DI water between each test. The oil rejection rate of the mesh for ASR = 0 decreased approximately 2.7% after 4 cycles of separation, but the oil rejection rate of the meshes for ASRs = 0.3-182 did not change with 4 cycles of oil water separation, which is attributed to the uniformity of the NaA zeolite coating and its thermal stability. NaA zeolite meshes were also used for separation of various oils, such as n-hexane, mineral oil, olive oil, and diesel. The oil rejection rates were higher than 98.3% in all cases, which demonstrates that NaA zeolite mesh is very promising technology for oil/water separation applications.

Copyright © American Institute of Chemical Engineers. All rights reserved.

621345 Cobalt-Based ZIF Coordinated Hybrids with Defective TiO2-x Forboosting Visible Light-Driven Photo-Fenton-like Degradation of Bisphenol a

Abstract:

Designing heterostructure of photocatalyst as an efficient approach to boost visible light-driven photocatalytic degradation, we prepared a series of cobalt-based ZIF coordinated with defective TiO_{2-x}, denoted as B-TiO_{2-x}@ZIF-67 composites, through wrapping defective B-TiO_{2-x} on ZIF-67 for promoting photocatalytic degradation efficiency of biphenyl A. The B-TiO_{2-x}@ZIF-67 composites displayed superior photocatalytic performance to pure TiO_{2-x} or ZIF-67 because of faster separation of photogenerated charge carriers and more suitable redox potentials. Such a novel photo-Fenton-like system composed of B-TiO_{2-x}@ZIF-67/H_2O_2/visible light accelerated the peroxidative degradation of biphenyl A up to a removal efficiency of 95.30%, which is also higher than that of photocatalysis or Fenton-like reaction alone. In addition, the degradation efficiency of biphenyl A is unchanged after catalyst reuse of four cycles. Integrating the trapping experiments and electrochemical analysis, we found the oxygen vacancy on BeTiO_{2-x} capturing the electrons to promote the separation of photogenerated charges, meanwhile the Co(II) in the composite decomposed hydrogen peroxide (H_2O_2) to produce more OH radical. Both of them mutually boosted the removal efficiency. Finally, feasible degradation pathways of biphenyl A were proposed based on the assay of LC-MS spectrometry. This strategy offers a novel insight into fabrication of Co-ZIF-based TiO_{2-x} materials and application to visible light-driven photocatalytic and Fenton-like degradation reaction.

Copyright © American Institute of Chemical Engineers. All rights reserved.

625247 Photocatalytic Removal of Waterborne Pollutants By Electrospun Nanofibers of Graphitic Carbon Nitride and Titanium Nitride

Abstract:

In this work, we have fabricated the of composite plasmonic nanofiber mat consisting of graphitic carbon nitride (g-C_3N_4) and titanium nitride (TiN) nanoparticles for photocatalysis. The nanofiber mat was fabricated by electrospinning g-C_3N_4, TiN nanoparticles and polyvinylidene fluoride (PVDF) and it demonstrated well flexibility without losing the structure. The mat was successfully characterized with SEM, TEM, AFM, FTIR, XPS, XRD and PL spectroscopy. Hexavalent chromium, which is the toxic state of chromium has been successfully reduced by this nanofiber mat under simulated solar light and in presence of trace amount of formic acid. The nanofiber mat also demonstrated well cyclability for hexavalent chromium reduction.

Copyright © American Institute of Chemical Engineers. All rights reserved.

629288 Synthesis of Imidazole and Amine Based Deep Eutectic Solvents for Low Concentration so_2 Capture

Abstract:

While the use of fossil fuels for energy production has improved the quality of human life, the environmental problems caused by pollutant emissions are a major drawback to their usage. In particular, the sulfur content in fossil fuels is oxidized in the combustion process and released in the form of sulfur oxides (SO_2 and SO_3), environmental pollutants that adversely affect air quality and the human body. Therefore, industrial facilities where sulfur oxides are mainly discharged should make efforts to reduce them.

Flue gas desulfurization (FGD) is generally used to remove SO_2, which uses lime ($Ca(OH)_2$) or limestone ($CaCO_3$) to produce calcium sulfite, which further reacts with oxygen to produce gypsum ($CaSO_4 \cdot 2H_2O$). However, the disadvantage is that the reserve of high purity natural limestone required as raw material is limited. As a solution to this issue, wet absorption methods that absorb acidic gases using liquid absorbents are usually considered as an alternative technology. Amine type absorbents are generally considered good absorbents for acidic gases, but they are not suitable as SO_2 absorbents due to high energy required for their regeneration and that they are highly volatile. New absorbents are constantly studied with the aim of overcoming the current difficulties, and recently a new absorbent called Ionic liquids (ILs) emerged. They present stable physicochemical properties such as low vapor pressure and low corrosivity, but due to the complexity of the production process and the high cost of the materials their practical application on the industry does not seem very likely. As a new alternative, Deep eutectic solvent (DES) for acidic gas absorption have been explored since the year 2003. DES are made from the synthesis of a Hydrogen bond acceptor (HBA) and a Hydrogen bond donor (HBD) and is characterized by the drastic decrease of the melting point of the mixture compared to their individual components and are often liquid at room temperature. Depending on their characteristics, DES may be considered as an absorbent for the wet absorption method to absorb acidic gases. DES, like ILs, has low vapor pressure and stable chemical properties, but have the advantage of an easy manufacturing process and low costs. Therefore, DES was selected as the absorbent for wet absorption in this study due to its advantages over the available alternatives.

An important focus of the study is the low concentration SO_2 absorption capacity of DES. Since most industrial processes emit low-concentration SO_2 research on low-concentration absorption must be conducted to achieve practical results. In general, DES physically absorbs SO_2, process suited for high concentration gases, and therefore cannot effectively absorb low concentrations of SO_2. It is important to synthesize DES by selecting HBDs and HBDs that can chemically absorb SO_2. Thus, Imidazole, which has been reported to be able to chemically absorb SO_2 according to its structural properties and types of functional group, was used as HBA. Another focus of research is the use of amine-based materials such as MEA, DEA, and MDEA as HBD.

Gas-analyzer was used to analyze the absorption capacity of DES, and the experiments and analyses were conducted by synthesizing HBA and HBD at a ratio of 1:2, 1:3, 1:4 to find the optimal ratio of the target absorbent. The gas was injected into the absorbent with the composition of SO_2 5,000 ppm and 99% N_2. The absorption mechanisms were analyzed with Fourier-transform infrared spectroscopy

(FT-IR) and Nuclear Magnetic Resonance (NMR). In addition, the density and viscosity of absorbents, which are factors that affect absorption efficiency, were analyzed by mole ratio using a densitometer and viscometer, and Thermogravimetric Analysis (TGA) was performed to analyze thermal stability.

Copyright © American Institute of Chemical Engineers. All rights reserved.

632255 Finding Properties Effecting Selectivity of SeO_4^{2-} Removal from Water Containing Se/S/P Oxo-Anions Using Adsorption on Metal Oxide: DFT Study with Hybrid Solvent Method.

Abstract:

Selenium is highly soluble in water as selenate and selenite and is toxic to human and aquatic life even at very low concentration (TWA 0.2 mg/m^3). Adsorption is a promising remediation strategy; however existing metal oxides have low selectivity towards Se and As oxoanions. Here we use first principal calculations to study the facet and material dependent behavior of Se(IV), Se(VI), S(VI), P(III) and P(V) adsorption. We delineate material properties affecting the adsorption characteristics such as valance electrons, and bond lengths, and surface properties such as interfacial water network, oxo-anion surface configuration, pH, etc. We find that the water network has significant control over adsorption behavior. Since number of H on the surface and oxo-anion directly control the water network, Al_2O_3 is not selective adsorbent towards selenate as compared to sulphate. The adsorption is highest at low pH, as the surface is concentrated with excess H. The facet dependent adsorption is also related to changes in water network over the surface. The adsorbents electronegativity also affects the adsorption energy of oxo-anion. The hematite binds Se(VI) and Se(IV) more strongly than Al_2O_3 surfaces because the Fe atoms are able to donate electrons to oxoanion-surface bonds . The adsorbent material has stronger effect on oxo-anions in inner sphere configuration. The potential adsorbent material, therefore, needs to bind the oxo-anion in inner sphere configuration, for highly selective adsorption.

Copyright © American Institute of Chemical Engineers. All rights reserved.

622858 A Comparative Study of the Thermal and Hydrothermal Aging Effect on Cu-SSZ-13 for the Selective Catalytic Reduction of NOx with NH_3

Abstract:

A comparative study of the thermal and hydrothermal aging effect on Cu-SSZ-13 for the selective catalytic reduction of NOx with NH_3

Huawang Zhao [a], Xiaomin Wu [a], Zhiwei Huang [a], Ziyi Chen [a], Guohua Jing [a]*

a Department of Environmental Science & Engineering, College of Chemical Engineering, Huaqiao University, Xiamen, Fujian 361021, China

Corresponding author: zhoujing@hqu.edu.cn

Abstract: In this work, the characterizations of Cu-SSZ-13 after hydrothermal aging (HTA) and thermal aging (TA) at different temperatures (750, 800, and 850 °C) are compared, and the differences between those two serious aged samples are analyzed. With this data, the effect of steam on catalysts deactivation during hydrothermal aging is analyzed. The TA at 750 and 800 °C causes the dealumination and the agglomeration of Cu^{2+} ions to CuO, resulting in the activity loss of Cu-SSZ-13. The presence of steam upon HTA at 750 and 800 °C aggravates the catalyst deactivation by increasing the Al detachment and the Cu^{2+} agglomeration. The structure and cupric state are almost the same in the Cu-SSZ-13 after TA and HTA at 850 °C, respectively, indicating that the steam has little influence on the deactivation. The formation of $CuAl_2O_4$ spinel is the primary reason for the deactivation after both HTA and TA at 850 °C, probably attributed to the strong interaction between Cu^{2+} ions and framework Al sites at high temperatures.

Copyright © American Institute of Chemical Engineers. All rights reserved.

624707 Characterization of Lethal Toxicity of Engineered Nanoparticles on Isochrysis Galbana Via Guts-Derived Fate-Transport-Effect Model

Abstract:

Ecotoxicological modeling is a crucial technique used in environmental risk assessments. Ecotoxicological models are used to determine the potential harms a proposed ecotoxicant can have on an ecosystem of interest. Fate ecotoxicological models make predictions on the distribution of ecotoxicants in an ecosystem. Toxicokinetic/toxicodynamic models make predictions on the effects of ecotoxicants on biota in an ecosystem. Models which link these two fields of ecotoxicological modeling are fate-transport-effect models. Dynamic biogeochemical mass-balances solved via numerical methods offer a robust approach to fate-transport-effect ecotoxicological modeling. We propose a biogeochemical modeling framework to bridge the gap between commonly used fate-transport models and toxicokinetic-toxicodynamic models using an extension of the GUTS toxicodynamic modelling framework. This model is then used to characterize the lethal toxicological effects of several metal engineered nanoparticles on a common marine alga, *Isochrysis galbana*. The model is also retroactively applied to two studies, one on the sub-lethal effects of engineered nanomaterials on *Isochrysis galbana*, another on the accumulation of 17α-ethinylestradiol in *Desmodesmus subspicatus*. In these analyses, we see the capacity of the model to descriptively characterize the population dynamics of microbial species, the accumulation of two classes of ecotoxicants in those microbes, and the toxicological effects of these ecotoxicants on the microbes.

Copyright © American Institute of Chemical Engineers. All rights reserved.

624921 Techno-Economic Analysis and Life Cycle Assessment of Contaminant Removal from Landfill Gas for Electricity Generation.

Abstract:

Landfill gas (LFG) is increasingly used and researched as a feedstock for a variety of traditional and proposed Waste-to-Energy (WTE) technologies, which includes electricity generation, compressed natural gas, or liquid hydrocarbon fuels. In these various scenarios, contaminants in the LFG can have substantial economic and environmental consequences in the WTE processes. There are several classes of contaminants contained in varied quantities in LFG such as siloxanes, sulfur compounds, chlorides, halogenated volatile organic compounds (VOCs), nitrogen, oxygen, carbon monoxide, alkanes/alkenes and mercury compounds (Andriani *et al* 2013). The two most important classes of contaminants requiring further treatment are the sulfur species and siloxanes, as these contaminants lead to substantial processing challenges and irreversible damage even at low concentrations and require removal technologies. The EPA has regulated the emission of harmful gases like hydrogen sulfide from landfills since it oxidizes to sulfur dioxide in the atmosphere (Public Lab n.d.).

Damages caused by hydrogen sulfide includes corrosion of equipment such as pipelines, compressors, engines and storage tanks; poisoning of fuel cells and catalysts; conversion to harmful sulfur dioxide during combustion of LFG composed of H_2S, thereby raising environmental concerns (Andriani *et al* 2013). Siloxanes are an emerging component in many consumer products that are landfilled and have a high enough vapor pressure such that a substantial amount is contained in the LFG. Siloxanes decompose to silica causing equipment damage that results in process downtime and reoccurring maintenance costs. The current purification techniques available are too expensive, in that it costs less to repair damaged engine parts than to adopt the current gaseous siloxane scrubbing technology. In order to accelerate adoption of waste to energy processes, a desire for more economical ways for removing siloxanes from LFG exists.

The goal of this research project is to conduct the techno-economic analysis (TEA) and life cycle assessment (LCA) of LFG cleaning for the generation of electricity using internal combustion engines, solar turbines and microturbines. The result from this research will identify the process unit with the highest economic and environmental impact in the LFG to energy process, thereby informing decision making when improving the entire process. Literature available have either focused on just the TEA of the removal of key contaminants from LFG (Hill 2014, Kuhn *et al* 2017, Tansel and Surita 2019) or the environmental assessment of biogas upgrading technologies that involves the removal of CO_2 for applications such as biomethane for gas grid injection, and compressed natural gas for transportation fuels (Cozma *et al* 2013, Lombardi and Francini 2020, Lorenzi *et al* 2019, Starr *et al* 2012). Although LFG is being converted into several end products (electricity, biomethane, bio-CNG, liquid fuels), about 70 percent of currently operational LFG energy projects in the United States generate electricity (EPA 2021). For this reason, only the assessment of contaminant removal from LFG for electricity generation was conducted in this study.

In this work, a natural zeolite (clinoptilolite), which costs as low as ~\$100/ton, as compared to activated carbon that is factors of 3+ more expensive, and four other low-cost materials (diatomaceous earth, crushed glass, biochar and hydrochar), were characterized using various techniques such as N_2 physisorption, x-ray diffraction, CO_2 adsorption and water vapor sorption analysis. Clinoptilolite showed the most favorable characteristics for siloxane adsorption among all samples characterized and hence was used in further analysis. TEA and LCA were conducted for siloxane removal units with clinoptilolite as an adsorbent and then compared with the current state of technology, activated carbon and another low-cost material that had been tested for siloxane adsorption in literature - 13X zeolite. The TEA results revealed that clinoptilolite as an adsorbent for siloxane removal is too expensive as compared to activated carbon and 13X zeolite. Sensitivity analysis was conducted to study the effect of the adsorption capacity of activated carbon on the total annual cost. The result from the sensitivity analysis revealed that 13X zeolite becomes a more cost-effective adsorbent for siloxane removal from LFG when compared with activated carbon that possesses an adsorption capacity less than 500 mg D4/g media. Assessment was also conducted to find out the effect on the total annual cost (TAC) for facilities with and without siloxane removal units installed. It was found that it becomes economically feasible for facilities with LFG siloxane concentration of 15 mg/m^3, to install a siloxane removal unit for biogas flowrates above 1300 scfm. The result from the life cycle assessment revealed that the CO_2 emissions from all siloxane removal units analyzed are approximately equal and therefore, only the cost of siloxane adsorption systems can be used in making decisions on adoption.

Fe oxide-based media is the most common adsorbent for H_2S. LFG is usually cooled down to about 5^0C to remove moisture. Activated carbon is usually used as the adsorbent in the halide removal unit. The TEA and LCA of the process to remove all contaminants from LFG prior to electricity generation will be conducted. Comparisons will be made between impact generated from engines with varied contaminant tolerance limits. Sensitivity analysis will be conducted to understand the impact of varying certain factors.

References

Andriani, Y., Morrow, I.C., Taran, E., Edwards, G.A., Schiller, T.L., Osman, A.F., and Martin, D.J. "In vitro biostability of poly(dimethyl siloxane/hexamethylene oxide)-based polyurethane/layered silicate nanocomposites," Acta Biomater. 9: 8308-17 (2013).

Cozma, P., Ghinea, C., Mămăligă, I., Wukovits, W., Friedl, A., and Gavrilescu, M. "Environmental Impact Assessment of High Pressure Water Scrubbing Biogas Upgrading Technology," CLEAN - Soil, Air, Water. 41: 917-27 (2013).

EPA "Landfill Methane Outreach Program (LMOP)," (2021).

Hill, A. Conduct a Nationwide Survey of Biogas Cleanup Technologies and Costs; 2014.

Kuhn, J.N., Elwell, A.C., Elsayed, N.H., and Joseph, B. "Requirements, techniques, and costs for contaminant removal from landfill gas," Waste Manag. 63: 246-56 (2017).

Lombardi, L., and Francini, G. "Techno-economic and environmental assessment of the main biogas upgrading technologies," Renewable Energy. 156: 440-58 (2020).

Lorenzi, G., Gorgoroni, M., Silva, C., and Santarelli, M. "Life Cycle Assessment of biogas upgrading routes," Energy Procedia. 158: 2012-8 (2019).

Public Lab "Hydrogen Sulfide Regulations " (n.d.).

Starr, K., Gabarrell, X., Villalba, G., Talens, L., and Lombardi, L. "Life cycle assessment of biogas upgrading technologies," Waste Management. 32: 991-9 (2012).

Tansel, B., and Surita, S.C. "Managing siloxanes in biogas-to-energy facilities: Economic comparison of pre- vs post-combustion practices," Waste Management. 96: 121-7 (2019).

Copyright © American Institute of Chemical Engineers. All rights reserved.

632795 Leaching and Fixation of Heavy Metals in Municipal Solid Waste Incineration Fly Ash (MSWI FA) Based on CO_2 Uptake Process

Abstract:

Due to the massive production by development in industries and economy, lots of municipal solid waste (MSW) have been produced daily and most of them are treated by utilization, incineration and landfill. For some developed nations in Europe, researches to utilize MSW incineration (MSWI) ash as a construction materials have been conducted since their physical properties are similar to that of aggregate. Except for the utilization, most of the produced MSW is treated by incineration and landfill. However, for the nations whose land area is confined, landfill is not preferred since it requires large area due to the volume of MSW itself. To decrease the volume of MSW, incineration can be considered, however, the problem occurs in the treatment of MSWI ashes. When MSWI ashes are buried, hazardous components such as heavy metals and other metal components can be leached out resulting in the pollution of underground water. Since underground water can flow to the far distance, environmental and biological problem is not limited to the land close to the landfill area.

To prevent negative impact on the environment, stabilizing method of MSW ash have been studied and applied. Representative ways are thermal treatment, solidification and washing treatment. In thermal treatment method, MSWI ash is heated to the high temperature in the range of 1500 – 1700 K resulting in the stabilization of the heavy metal and salts whose vapor pressure is relatively high. One of the merit of this method is that the hazardous organic component can be treated in the same time by the thermal oxidation in that temperature range. However, it costs a lot to maintain high temperature condition. For solidification process, some components in MSWI ash are fixed by the reaction with waste cement or the slag. The merit of this method is the prevention of hazardous component leaching by forming the compounds whose binding force is higher than that in the ash. Also, due to the reaction, the mass after the solidification is increased make it harder to transport to the landfill site. In the washing treatment, hazardous materials in soluble phase are dissolved into the solvent, mainly water. However, it requires the further treatment of wastewater containing dissolved components. In this study, we discussed a new ways to treat MSWI fly ashes which contains shows higher heavy metal contents compared to the MSWI bottom ashes.

Among various kinds of methods for the stabilization of heavy metals in municipal solid waste incineration fly ash (MSWI FA), acid leaching is the most prevalent. However, the process of heavy metals leaching in MSWI FA by various acids must be compared. In this study, we extracted the heavy metals in MSWI FA using 0.5 M solutions of strong acids (HCl, H_2SO_4 and HNO_3), as well as weak acids (NH_4Cl, $(NH_4)_2SO_4$ and NH_4NO_3). Solidification by CO_2 injection occurred in all the leaching solutions. The leaching test was performed for solids such as treated MSWI FA and the precipitates produced in the solidification process, compared with untreated MSWI FA. The solids were analyzed by using X-Ray fluorescence spectroscopy, X-Ray diffraction, and scanning electron microscopy. The leaching solutions were analyzed by inductively coupled plasma-optical emission spectroscopy. Using weak acid solutions, Zn, Pb, and Cu were leached 15 - 70 times, two times, and 20 - 150 times, respectively, more than those leached strong acids. Except H_2SO_4 and $(NH_4)_2SO_4$ solutions, all the acids reduced the mass of MSWI FA and extracted heavy metals. The crystalline precipitates were the carbonates. HNO_3 solution was found to be the most appropriate solvent for the proposed method.

Copyright © American Institute of Chemical Engineers. All rights reserved.

627119 Life Cycle Comparison of Battery Recycling and Raw Material Extraction

Abstract:

Background

Demand for lithium-ion batteries (LIBs) has skyrocketed in the last decade, causing concern for the rapid depletion of precious metal sources that constitute these products and the environmental impacts of resource extraction. In addition, battery disposal can lead to environmental contamination and economic losses from discarding valuable materials. Battery recycling is being investigated as a solution to sustain the growing demand for LIBs and to address the problems resulting from spent LIB disposal. However, selective component separation and recovery can still be energy-intensive, require harsh chemicals, and cause environmental pollution. Therefore, a life cycle comparison between battery recycling and raw material extraction is essential to ensure adoption of the most sustainable processes going forward. In this work, we conducted a comprehensive evaluation of battery recycling through three objectives: (1) combine data from literature and government databases to characterize and assess life cycle criteria for raw material extraction; (2) analyze the unit processes in a case study to assess life cycle criteria for battery recycling; and (3) compare the results of the first two objectives and produce a final conclusion with recommendations for future research.

Methods

We performed a gate-to-gate life cycle assessment (LCA) of battery recycling and raw material extraction using primary data and the Argonne National Laboratory Greenhouse gases, Regulated Emissions, and Energy use in Technologies (GREET) model. The functional unit employed in the analysis was a 1 kWh lithium nickel manganese cobalt oxide (NMC) battery. The LCA criteria evaluated were global warming potential (CO_2 eq), ozone depletion (CFC-11 eq), primary energy demand, water consumption, and the National Ambient Air Quality Standards (NAAQS) six common air pollutants set by the Environmental Protection Agency (EPA): ozone, carbon monoxide, lead, sulfur dioxide, nitrogen dioxide, and particulate matter. We extracted information regarding raw material extraction for each battery component from peer-reviewed journals and government databases.

Results and Implications

Our review of raw material extraction revealed variations in life cycle results based on the location of extraction, the method of extraction (e.g., mining vs. brines), the type of ore being processed, and the metals being targeted. To facilitate a detailed analysis, the raw extraction assessment was broken down into each NMC battery component, and the results were then combined for an overall evaluation. Variables were built into the model to allow for a diversity of inputs so that multiple scenarios of raw extraction could be compared against recycled materials. The comparison of life cycle metrics for each process indicated battery recycling as a feasible option to supplement virgin materials from mining and brine extraction and as a solution to the growing volume of batteries reaching end-of-life. However, several opportunities for development can improve the case of recycling as a substitute for raw extraction. To expand the reach of this work, we proposed quantitative performance metrics specifically relating to environmental impacts for evaluating other battery recycling methods. This study will guide developments in battery recycling technology, provide sustainability metrics for performance evaluations, and contribute to the movement towards a circular battery economy.

Copyright © American Institute of Chemical Engineers. All rights reserved.

629412 Modeling Bioplastic Production from Astronaut Organic Wastes for Resource Recovery Using Coupled Anaerobic-Aerobic Bioreactors for Long-Duration Space Operations

Abstract:

The National Aeronautics and Space Administration (NASA)'s Artemis program centers on sustained lunar surface exploration and technology validation. Incorporated into the program are extended lunar missions. NASA and its partners intend to develop an initial base camp at the lunar South Pole, which includes a lunar habitation module and in-situ resource utilization systems. Over time, NASA will conceivably establish a permanent presence, which will require enhanced environmental control and life support systems. Additionally, as more astronauts populate a permanent lunar base, questions concerning increased astronaut wastes need to be answered with sustainable, cradle-to-cradle solutions that recover critical resources such as water, oxygen, and nutrients in-situ. Several biological approaches for recovering and reusing resources have been explored; however, there is no clear solution and significant research in this area is required.

This project proposes and models a cradle-to-cradle approach for converting astronaut waste streams (urine, fecal matter, and grey water) into usable products, which can be subsequently recycled in situ. Specifically, our team first modeled an anaerobic-aerobic bioreactor system to convert the chemical energy in organic wastes to methane and biodegradable bioplastics. The anaerobic portion of the modeled bioreactor system consisted of an upflow anaerobic sludge blanket (UASB) and an anaerobic membrane bioreactor (AnMBR) for conversion of organic material to methane. Second, the liquid effluent from the anaerobic bioreactor system was routed to a distillation unit for separation of potable water from a concentrated nutrient stream (consisting of residual nitrogen, phosphorous, and trace metals). Third, the methane-rich biogas produced from the anaerobic bioreactors was routed to an aerobic bioreactor for suspended growth of methanotrophic proteobacteria. Certain strains of methanotrophs (we modeled strain *Methylocystis parvus* OBBP) can accumulate polyhydroxybutyrate (PHB), which can be harvested and converted to a biodegradable bioplastic. Fourth, residual methanotrophic biomass was considered for use as a food stuff, while wasted anaerobic sludge was considered as a fertilizer. Once each product (bioplastic, water, fertilizer, nutrient stream) is used, each can conceivably be digested anaerobically with urine, fecal matter, and grey water, thereby creating a cradle-to-cradle system for resource recovery.

To estimate bioplastic production, our team first conducted laboratory work to validate and strengthen existing knowledge concerning the anticipated chemical constituents of astronaut wastes using ersatz formulas. Three different waste permutations were explored: (1) fecal matter only; (2) fecal matter and urine; and (3) fecal matter, urine, and grey water. Results of chemical oxygen demand (COD) and ammonia provided an estimation of the total chemical energy in each waste permutation. Published anaerobic bioreactor conversion efficiencies were used to model biogas and methane production. Further, studied values for methanotrophic yield were used to estimate biomass production (grams volatile suspended solids, VSS) and PHB yield. Monte Carlo Simulation was employed to increase confidence in biomass and PHB yield results. Initial modeling results suggest that a lunar base consisting of five astronauts may produce between 30 and 430 grams of intracellular PHB per day. Production rates for potable water, methanotrophic biomass, concentrated nutrients, waste anaerobic sludge, as well as oxygen and energy requirements, were also determined. While further refinement of our team's modeling approach is required, initial results suggest that an anaerobic-aerobic bioreactor system creates a cradle-to-cradle approach for resource recovery and may be viable for a larger-scale lunar base.

Copyright © American Institute of Chemical Engineers. All rights reserved.

631461 Optimizing Chemical Recycling of Waste PET/Polyolefin Streams through Agile Life Cycle Assessment (LCA)

Abstract:

Restorative and sustainable future of the petrochemical resources calls for a circular economy of plastics. Conventional end-of-life treatments of plastics require either huge land area for landfilling or produces noxious and harmful gases via incineration. To exploit the full potential of the waste plastics, plastic to fuel upcycling path is a solution to this challenge. PET and PP have myriad of applications, mostly in making containers, packaging, and clothing. LCA can be used to identify the best environmentally sustainable plastic waste management option(s). In this study, we present a multidisciplinary modeling framework that integrates kinetic model, process design, and LCA to evaluate the performance, energy demand, and environmental implications of plastic upcycling pathways. The experiments and microkinetic modeling that investigates the interaction effects in mixed plastic pyrolysis is leveraged and the results on yields, molecular weight ranges, and kinetics is used to inform the LCA parameters. Further, by linking technical performance metrics through mechanistic modeling, predictive design decisions can be implemented to optimize the process on both energy use and carbon emissions of recycling a mixture of PET and PP waste feedstocks. The impacts of individual process and decision variables on overall systems level sustainability can also be characterized.

Copyright © American Institute of Chemical Engineers. All rights reserved.

625954 Evaluation of *Galdieria Sulphuraria* for the Bioremediation of Produced Water

Abstract:

Produced water (PW) is the largest volume of wastewater in oil and gas industries and accommodates 80% of the waste and residuals generated during oil and gas exploration. In the United States alone, onshore oil and gas extraction operations generate an estimated 24 billion barrels of PW annually, making it the largest waste stream associated with the upstream development of petroleum hydrocarbons. Produced water has a complex composition, containing inorganic and organic components as well as heavy metals, dissolved gases, treating chemicals, radionuclides, scaling products, and microorganisms. Currently, most of the PW is reinjected into disposal wells, which incurs transportation problems and induces seismicity issues. In its native state, PW is not considered usable for land applications due to environmental concerns. Moreover, the cost of treating one barrel of PW is \$ 0.775 USD, whereas the reinjection cost is \$ 0.75 – 8.00 USD per barrel. These facts act as the driving forces for the treatment of PW.

The utilization of PW as an algal growth medium has a great potential to eliminate chemicals from the PW and minimize the large volumes of freshwater needed for cultivation. Additionally, algae's ability to utilize certain pollutants as nutrient sources has proven beneficial to remediate PW contaminated by petroleum hydrocarbons, complex organic chemicals, and solvents. However, significant dilution of PW is often required in algal-based systems due to complex chemical contaminants present in PW. In light of these facts, the current work has investigated the potential of cultivating *Galdieria sulphuraria* in PW using multiple dilutions; 0% PW, 5% PW, 10% PW, 20% PW, 50% PW, and 100% PW. The current results indicated that *G. sulphuraria* has a high potential of growth in up to 50% PW (total dissolved solids of up to 55 g L^{-1}) with a growth rate of 0.72 ± 0.05 g L^{-1} d^{-1} and can achieve a final biomass density of 4.28 ± 0.16 g L^{-1} in seven days without the need for additional micronutrients. Additionally, the algal strain showed the potential of removing 99.6 ± 0.2% ammoniacal nitrogen and 74.2 ± 8.5% phosphate phosphorus from the PW. Overall, the current work highlights the PW bioremediation in an eco-friendly and low-cost approach using algae and brings attention to utilizing PW to grow biomass that can be processed to generate useful bioproducts.

Copyright © American Institute of Chemical Engineers. All rights reserved.

622103 Rapid and Real-Time Mixed-Plastic Waste Analysis Using Infrared Spectroscopy and Machine Learning

Abstract:

The production of plastics continues to climb but their recycling rate remains quite low. In 2015, nearly 381 million tons of plastic were produced, but only 20% of the world's plastic waste was recycled [1]. Mixed plastic waste (MPW) is putting unprecedented pressure on landfills [2] and it is critical to convert MPW into clean and usable materials to enable a more circular system. A fundamental challenge that hinders this is the difficulty in analyzing the composition of MPW; such information can enable the development of more appropriate treatment strategies.

Traditionally, offline infrared (IR) spectroscopy has been used to identify the chemical components in MPW [3]. Offline IR spectra are easy to identify but are labor- and time-intensive to obtain; as such, this method has limited use in real-time plastic composition assessment. In contrast, online IR spectroscopy allows for rapid detection of plastic composition with a single scan time of 11 microseconds [4]. However, the spectra obtained by this method can be highly noisy (due to interference).

Recently, machine learning methods such as convolutional neural networks have been used to analyze spectral data (in the context of soil characterization) and they have been shown to provide high predictive accuracies and to be resistant to noise [5]. Conventionally, IR spectra are treated as one-dimensional vectors that are analyzed using one-dimensional (1D) CNNs [6]. The 1D CNN works by convolving the spectrum using filters to extract hidden patterns. Although IR spectra already carry a rich amount of information, the hidden correlations between the signal intensities at different wavenumbers are not explicitly encoded.

In this work, we propose to encode the signal correlation information using Gramian angular fields (GAFs) and analyze GAF matrices using 2D CNNs [7]. GAFs have been used to transform time-series vectors into matrices to capture temporal correlations and thus improve classification accuracy. To facilitate fast training of 2D CNN-based algorithms, we use a Piecewise Aggregate Approximation (PAA), which allows us to reduce the dimension of the field matrix [8]. We also used saliency analysis [9] to understand the most important signal intervals and correlations between signals. We show that this approach provides a robust, low-cost, and rapid method for analyzing the composition of MPW and can enable future, high-throughput recycling of plastics.

[1] Ritchie, Hannah, and Max Roser. "Plastic pollution." Our World in Data (2018).
[2] Abdel-Shafy, Hussein I., and Mona SM Mansour. "Solid waste issue: Sources, composition, disposal, recycling, and valorization." Egyptian journal of petroleum 27.4 (2018): 1275-1290.
[3] Zhu, Shichao, et al. "Plastic solid waste identification system based on near infrared spectroscopy in combination with support vector machine." Advanced Industrial and Engineering Polymer Research 2.2 (2019): 77-81.
[4] "S2050 2.0-5.0μm Fiber Spectrometer - NLIR." [Online]. Available: https://www.nlir.com/products/s2050-2-0-5-0μm-fiber-spectrometer/. [Accessed: 15-Mar-2021].
[5] Ng, Wartini, et al. "Convolutional neural network for simultaneous prediction of several soil properties using visible/near-infrared, mid-infrared, and their combined spectra." Geoderma 352 (2019): 251-267.
[6] Chen, Xiaoyi, et al. "1D convolutional neural network for the discrimination of aristolochic acids

and their analogues based on near-infrared spectroscopy." Analytical Methods 11.40 (2019): 5118-5125.
[7] Wang, Zhiguang, and Tim Oates. "Encoding time series as images for visual inspection and classification using tiled convolutional neural networks." Workshops at the twenty-ninth AAAI conference on artificial intelligence. Vol. 1. 2015.
[8] Keogh, Eamonn J., and Michael J. Pazzani. "Scaling up dynamic time warping for datamining applications." Proceedings of the sixth ACM SIGKDD international conference on Knowledge discovery and data mining. 2000.
[9] Sundararajan, Mukund, Ankur Taly, and Qiqi Yan. "Axiomatic attribution for deep networks." International Conference on Machine Learning. PMLR, 2017.

Copyright © American Institute of Chemical Engineers. All rights reserved.

625006 Efficient Biodegradation of Poly(ethylene Terephthalate) with Leaf-Branch Compost Cutinase

Abstract:

Poly(ethylene terephthalate) (PET) is one of the most widely used polyester plastics in the world, but is extremely difficult to be hydrolyzed. Accumulation of the PET in the environment at a staggering rate as discarded packaging and textiles pose a global environmental problem. Cutinases are polyester hydrolases that show a remarkable capability to PET to its monomeric units. This revelation has stimulated research aimed at developing sustainable and green cutinase-catalyzed PET biorecycling methods. The gene encoding a cutinase homolog, LCC, was cloned from a fosmid library of a leaf-branch compost metagenome by functional screening using tributyrin agar plates. LCC is particularly suited toward these ends given its relatively high PET hydrolysis activity and thermostability. With the assistance of the pelB leader sequence, more than 36% of the recombinant LCC produced by *Escherichia coli* BL21(DE3) cells was excreted into the culture medium. To improve the biodegradation efficiency, recycled PET (RPET) was cut into small pieces via grinding to decrease PET particle size and increase surface area. Different particle size of RPET were examined. As a result, the degradation rate of the LCC in the presence of smaller RPET particle size (with 60 mesh screen) was significantly improved, with a productivity of 1.2 milligrams of terephthalic acid (TPA) per milliliter over 48 hours (with enzyme concentration of 0.8 micrograms per milligram of PET). Moreover, the substrate-binding modules from a polyhydroxyalkanoate depolymerase from *Alcaligenes faecalis* (PBM) was fused with the linker (2.3 kDa) from a cellobiohydrolase from *Trichoderma reesei* to the C-terminus of LCC to improve the PET adsorption. Although PBM (6.3 kDa) binding modules have a hydrophobic nature, it was possible to express the proteins in *E. coli*. Our result exhibited that adsorption to PET by the LCC-PBM fusion enzyme was strikingly enhanced as compared with LCC. Though the PET hydrolysis performance of any LCC variants was much higher than that of the bacterial enzyme Is-PETase from *Ideonella sakaiensis* 201-F6, the degradation efficiency of LCC-PBM into TPA was not as high as LCC. This study provides us a great potential to achieve enzymatically depolymerization of PET waste contributing towards the concept of a circular PET economy.

Copyright © American Institute of Chemical Engineers. All rights reserved.

631844 The Use of Life Cycle Assessment to Identify Sustainability Characteristics of Plastic Recycling Systems: A Review

Abstract:

The extensive use of plastics in a “take, make and dispose” order is putting a strain on our environment. Conventional end-of-life treatments of plastics diminishes the potential of waste plastics to extract the maximum value from it through recycle or upcycle. Life cycle assessment (LCA) is an established tool to evaluate technologies of their environmental footprints. A standardized LCA framework enables comparison between technologies and guide the path forward. We have critically reviewed 33 research articles and scientific reports of published LCAs with a view to present an extensive state of the art of the current chemical recycling processes involved in the plastics circular economy system along with their economic and environmental impacts. The overarching goal is to compare environmental implications of technologies for waste to fuel/monomer pathways (such as pyrolysis, and/or gasification) with that of the conventional waste plastic conversion technologies such as incineration and/or landfilling. An array of different waste feedstocks (e.g., both high- and low-density polyethylene, polystyrene, and polyethylene terephthalate) with or without contaminants were analyzed. In the process, the results obtained imply large uncertainties, for example, one study showed mixed plastic incineration was a better option than gasification-pyrolysis due to lower greenhouse gas (GHG) emissions whereas a large number of other studies favored gasification and pyrolysis over incineration due to lower emissions. This work also aims to develop systematic guidance for researchers and practitioners to use LCA to inform decision-making, design, and optimization of recycling pathways and infrastructure, and to support the development of new technologies to advance the plastics circular economy.

Currently, the availability of life cycle inventory (LCI) data remains a major problem. Previous studies usually apply static data inputs (such as using Ecoinvent inventory which is mostly for European data when conducting LCA for US based scenarios), despite the large variations stemming from assumptions of capacity, process design decisions, unit operation performance, etc. To address this gap, lab-scale results coupled with systems design and scale-up must be integrated with LCA. There’s also a clear need to provide transparent characterization of uncertainty and the sensitivity of results in modeling environmental impacts of plastic recycling systems, but this is yet to become standard practice in the field, only 15% of the total papers include both sensitivity and uncertainty analysis. Furthermore, use of different impact assessment (IA) methods may present variability in the result due to differences in characterization factors across IA methods under the same impact category. For co-products handling, System expansion is most used in the reviewed papers, with the need to examine how different co-product allocation methods can impact environmental implications. Finally, while addition of TEA strengthens the feasibility of newer technologies in plastic upcycling, very few studies have incorporated both techno-economic analysis (TEA) and LCA together. To elucidate the sustainability implications of technologies via comparative systems analysis, a unified LCA framework is required along with quantifying the technical and economic implications of the system via TEA. Addressing these gaps will help practitioners to make informed and justified methodological decisions on the current upcycling technologies.

Copyright © American Institute of Chemical Engineers. All rights reserved.

630296 Comprehensive Planning of Annual Delivery Program for LNG Suppliers

Abstract:

The demand for natural gas has increased over the decade as it is the cleanest fossil fuel. However, it is economically infeasible to transport natural gas over long distances. Therefore, it is transported in liquefied form at cryogenic conditions using specialized ships known as LNG carriers. LNG supply chain planning is done on three levels namely strategic, tactical and operational planning. Strategic planning involves making long term decision that are effective over 20-30 years of duration. For e.g., selection of a contract portfolio, investment decisions associated with both liquefaction and regasification terminals, fleet composition and size etc. Tactical planning involves decision making which are effective over a horizon of 12 - 18 months. Creation of an Annual Delivery Program (ADP) by the supplier is an important tactical decision. Operational planning involves decision making which are effective over a horizon of 1 – 3 months. Rescheduling due to unforeseen events such as equipment breakdown or ship delays due to bad weather etc. are some examples of operational decisions. In this paper our focus is on tactical planning. LNG is procured by customers through Long Term Contracts (LTC) and spot market. LTC is an agreement signed between a supplier and a customer usually spanning up to 20-25 years of duration. So, with increasing demand, fleet sizes, customers around the world and other complicated decisions to make it becomes important for the supplier to plan deliveries to the customers at the accepted delivery dates throughout the supply period. To plan for deliveries for a coming year an Annual Delivery Program (ADP) is created by the supplier. ADP is the list of scheduled voyages which contains information about the long-term contract served, ships nominated to serve the contracts, regasification terminals nominated to receive delivery, date of ship loading etc. In this paper we present an LNG supply chain planning problem with an objective of creating an ADP to maximize the revenue generated by selling LNG in spot market and minimize the cost incurred by the supplier in delivering LNG to the long-term customers.

Currently, most of the literature present on creating ADPs is to satisfy the LTC demands only considering the inventory and berth management at the supplier's port. Rakke et al. (2011) developed a rolling horizon heuristic (RHH) for creating LNG annual delivery programs. A Mixed Integer Programming (MIP) formulation was proposed[1]. Stålhane et al. (2012) developed a construction and improvement heuristic (CIH) to create ADPs[2]. Their models allowed under supply of LNG and did not consider inventory and berth management at the customer's terminal. Mutlu et al. (2016) proposed a MIP model which incorporated split delivery for delivering LNG cargos. Commercial optimizers were not providing even a feasible solution to the MIP model in reasonable time. Small instances were tested but even those instances could not be solved to optimality in a 24-hour run time. So, they proposed a Vehicle Routing Heuristic (VRH) which gave cost effective solutions and outperformed commercial optimizers[3]. Sheikhtajian et. al. (2020) considered the model proposed by Mutlu et. al. (2015) and developed a model considering ship's speed as a fuzzy parameter. The aim of the study was to compare shipping expenses of split and non-split delivery strategies considering both deterministic and uncertain situations. They also proposed a meta heuristic as a solution method to solve the problem[4]. To summarize, some of the models in literature allowed under-supply neglecting inventory and berth management at the customer port, some models did not consider production and delivery of multiple grades of LNG, while others were not able to give optimal solution within reasonable time. We seek to address these gaps through our model.

In this paper, we present a Mixed Integer Linear Programming (MILP) model for delivering LNG shipments from the supplier to LTC customers and a regional spot market. The problem has one supplier and multiple customers spread around the world. The supplier produces and delivers multiple grades of LNG. Our objective is to create an ADP which maximizes profit for the supplier while considering ongoing long-term contracts and regional spot market. The supplier generates revenue by selling excess LNG in spot market. The cost incurred by the supplier in delivering LNG is the sum of transportation and penalty costs if it fails to abide by LTCs. The minimum and maximum production rates, storage capacities at the supplier and customer terminals, transportation time from supplier to customer ports and the slot wise demand at the customers terminal are known a priori. Our model allows over-supply to customers in a reasonable limit and penalizes the supplier for delivering more than the demand. In this paper, we also focus on the inventory and the berth management at the customer's port. The supplier is responsible for scheduling the heterogeneous fleet of ships. Our model explicitly accounts for loading of ships at the supplier's port, inventory of LNG at the supplier's and customer's terminal. The supplier sends excess LNG to the spot market over the planning horizon. Maintenance of ships is also incorporated in our model. The MILP model is demonstrated on various scenarios having different demands. We get an optimal ship schedule from our model which satisfies the inventory and berth constraints at the supplier and customer ports while fulfilling customers demand with reasonable over-supply. The MILP model for LNG delivery has been implemented in and solved using CPLEX. We believe that the results from this paper would help develop critical insights regarding the tactical planning aspects of the LNG supply chain from the perspective of both supplier and customer.

References

1. J. G. Rakke *et al.*, "A rolling horizon heuristic for creating a liquefied natural gas annual delivery program," *Transp. Res. Part C Emerg. Technol.*, vol. 19, no. 5, pp. 896–911, Aug. 2011, doi: 10.1016/j.trc.2010.09.006.

2. M. Stålhane, J. G. Rakke, C. R. Moe, H. Andersson, M. Christiansen, and K. Fagerholt, "A construction and improvement heuristic for a liquefied natural gas inventory routing problem," *Comput. Ind. Eng.*, vol. 62, no. 1, pp. 245–255, Feb. 2012, doi: 10.1016/j.cie.2011.09.011.

3. F. Mutlu, M. K. Msakni, H. Yildiz, E. Sönmez, and S. Pokharel, "A comprehensive annual delivery program for upstream liquefied natural gas supply chain," *Eur. J. Oper. Res.*, vol. 250, no. 1, pp. 120–130, Apr. 2016, doi: 10.1016/j.ejor.2015.10.031.

4. Sheikhtajian, Sanaz, Ali Nazemi, and Majid Feshari. "Marine Inventory-Routing Problem for Liquefied Natural Gas under Travel Time Uncertainty." *International Journal of Supply and Operations Management*7, no. 1 (2020): 93-111.

Copyright © American Institute of Chemical Engineers. All rights reserved.

631079 A New Integrated Scheduling and Optimization Framework for Holistic Refinery Supply Chain Management

Abstract:

The holistic refinery supply chain management is critically important over the whole oil industry supply chain. It covers three highly correlated sub problems: the front-end sub problem, the refinery manufacturing sub problem and the multi oil-product pipeline distribution sub problem. According to Chima (2007), there is a need to ensure that each sector along the supply-chain can respond quickly to the exact demand of its downstream customers, protecting itself from problems with suppliers and buffering its operations from the demand and supply uncertainty it faces. Since the oil companies are aiming at maximizing the profit margin by weighing benefits versus costs throughout the entire refinery supply-chain. Thus, by coordinating the management and operations of the three sub problems along the holistic refinery supply-chain, it can greatly maximize the potential benefit margin. Although there exists a lot of works explicitly and deeply exploited and studied these sectors separately, only a few works addressed these sub systems along the refinery supply-chain simultaneously. Guyonnet et al. (2009) integrated the crude oil unloading, production planning, and distribution sub models on a tactical decision level by solving each part in a sequential push or pull manner, where the planning horizon is discretized into time periods, typically a day or a week. Siwi et al. (2018) proposed a strategical planning of the petroleum and petrochemical supply chain. However, systematic studies for the integrated scheduling and optimization of the holistic refinery supply chain are still lacking.

To minimize the total operating cost along the holistic petroleum refinery supply-chain, this work developed a new integrated scheduling and optimization framework covering the crude oil unloading, transferring, manufacturing, and pipeline distribution subsystems. Note that the specifications requirement for crude oil blending and oil product blending (where nonlinearity may occur due to the blending operations) must be satisfied. Meanwhile, operating rules such as tank can either receive or discharge oil product at a time has been considered, which further closed the gap between the modeling and the realistic problem (Yu et al., 2020). Also, the inventory/capacity limits, delivery feasibility constraints and oil product demands must be satisfied, which may result in a multi-stage and large-scale MINLP model. The efficacy of the developed integrated scheduling and optimization framework is demonstrated by an industrial-scale case study.

Keywords: Integrated scheduling and optimization, MINLP, Holistic refinery supply chain, Crude unloading and transferring, Refinery manufacturing, Pipeline distribution

Copyright © American Institute of Chemical Engineers. All rights reserved.

631348 Integrated Load Shifting and Curtailment for Demand Response of Central Chilled Water Plants

Abstract:

The participation of industrial electricity users in Demand Response (DR) programs has increased in the past decade (FERC, 2019), motivated by the necessity of ensuring the electricity grid's generation-load balance and transmission reliability. Programs that promote load shifting and curtailment and the dispatching of distributed generation assets have become common products for independent system operators. While load shifting as a response to price signals has been widely studied in the past, the provision of load curtailment services from the viewpoint of the user started to gain attention in the last decade. Previous studies in this area considered processes such as cement milling (Vujanic *et al.*, 2012), aluminum smelters (Zhang and Hug, 2015) and air separation plants (Zhang *et al.*, 2015). In these works, there is a considerable variation in the load curtailment program (e.g., how far in advance the bids are placed and how far into the future they are implemented) and industrial process addressed, the combination of which results in different problem setups and solution approaches.

In this work, we study how load curtailment bids can be optimized and implemented along with load shifting and analyze its closed-loop performance under forecast uncertainty, specifically for large-scale central chilled water plants with Thermal Energy Storage (TES) used for District Cooling. First, an optimization framework is proposed that integrates the two problems of real-time load shifting (considering hourly day-ahead prices) and day-ahead load curtailment bidding and implementation. We perform a cost-benefit analysis that demonstrates the potential benefits of providing load curtailment under varying incentive profiles. Lastly, we analyze the effect of forecast uncertainty on the closed-loop performance and identify opportunities for integrating uncertainty into the optimization problem. The case-studies are performed with real data from the central chilled water plant located on the campus of the University of California, Davis.

References

FERC, 2019. 2019 Assessment of Demand Response and Advanced Metering, Technical Report.

R. Vujanic, S. Mariéthos, P. Goulart, M. Morari, 2012. Robust integer optimization and scheduling problems for large electricity consumers. Proceedings of the 2012 American Control Conference, 3108–13.

Q. Zhang, I. E. Grossmann, C. F. Heuberger, A. Sundaramoorthy, J. M. Pinto, 2015. Air separation with cryogenic energy storage: optimal scheduling considering electric energy and reserve markets. AIChE Journal 61 (5), 1547–58.

X. Zhang, G. Hug, 2015. Bidding Strategy in Energy and Spinning Reserve Markets for Aluminum Smelters Demand Response. In: Innovative Smart Grid Technologies Conference (ISGT), 2015 IEEE PowerEnergy Society, 1–5.

Copyright © American Institute of Chemical Engineers. All rights reserved.

624503 A Web-Based Application for Chemical Production Scheduling

Abstract:

While a large range of models and methods have been proposed in the literature for chemical production scheduling, very limited tools are available for researchers to try and test these methods. In this work, we develop a web-based application to facilitate the generation and solution of chemical production scheduling problems. The application offers a graphical interface through which the user can define an instance using the state-task network representation (Kondili *et al.*, 1993). The connections between the states and tasks can be created using connectors and processing data (e.g., task processing times, conversion coefficients) can be entered through the interface.

Given additional user-defined parameters (e.g., horizon, optimality gap) and choice of the objective function (e.g., profit maximization, cost minimization, makespan minimization), the application then automatically generates the corresponding mixed-integer linear programming discrete-time model. In terms of solution methods, the user can choose among three options: (1) tightening constraints, based on a constraint propagation algorithm (Velez *et al.*, 2013); (2) reformulation based on the introduction of integer variables representing the total number of executed batches of each task (Velez and Maravelias, 2013); and (3) a discrete-continue algorithm which identifies good solutions fast based on a discrete-time model and then refines the solution based on a continuous time representation (Lee and Maravelias, 2018).

A data string representing the created network is generated automatically and submitted to the NEOS server (hosted at the University of Wisconsin-Madison). The user must provide their email address to retrieve the results. When the user submits the network through the web application, the NEOS server queues the relevant network data and based on the solution method(s) chosen, appropriate constraints are activated in the optimization model solved using the CPLEX solver in the GAMS platform. The user can access the model and solution statistics as well as a visual representation of the solution in the form of a Gantt chart along with the inventory profiles, which are zipped into a folder and emailed to the user.

References

1. Kondili, E., Pantelides, C.C., Sargent, R.W.H. (1993) A general algorithm for short-term scheduling of batch operations—I. MILP formulation. *Computers & Chemical Engineering*, **17**, 211–227.
2. Velez, S., Sundaramoorthy, A., Maravelias, C.T. (2013) Valid inequalities based on demand propagation for chemical production scheduling MIP models. *AIChE Journal*, **59**, 872–887.
3. Velez, S., Maravelias, C.T. (2013) Reformulations and branching methods for mixed-integer programming chemical production scheduling models. *Industrial Engineering and Chemistry Research*, **52**, 3832–3841.
4. Lee, H., Maravelias, C.T. (2018) Combining the advantages of discrete and continuous-time scheduling models: Part 1. Framework and mathematical formulations. *Computers & Chemical Engineering*, **116**, 176-190.

Copyright © American Institute of Chemical Engineers. All rights reserved.

634119 Keynote Talk - Positive Power with Negative Emissions: Flexible Ngcc Enabled By Modular Direct Air Capture (DAC)

Abstract:

Natural gas power generation is a vital part of the U.S. energy system based on an abundant domestic energy source. In this talk a way to modify natural gas combined cycle (NGCC) plants such that they produce power with a negative carbon footprint will be described. To achieve this, we have integrated the natural gas combined cycle (NGCC) system with both conventional amine capture and direct air capture (DAC). This combination provides dispatchable operation of the combined system to deliver carbon negative electricity even when variable renewable energy (VRE) resources are scarce. During times of high VRE, the unit would reduce its electrical output and allocate more energy to the DAC system, removing more CO2 from the atmosphere driving the carbon footprint further negative. This requires manipulation of the DAC system operation to optimize steam use to enable maximum value to be derived from time varying pricing of electricity.

The NGCC plant will include conventional amine scrubbing to remove 97% of the CO_2 contained in the flue gas. The amount of DAC will depend on the balance of using steam from the heat recovery steam generation system to make power through the turbine system and its use for regenerating the DAC sorbent. The NGCC plant will also be able to ramp up and down maintaining a minimum load to allow for fast response to pricing signals. Overall this will allow for carbon negative electricity to be produced between 60-100% of the NGCC operating capacity with an additional penalty of 5-20% of the NGCC generating capacity, above that required for conventional capture, depending on how much steam is directed to the DAC plant.

Copyright © American Institute of Chemical Engineers. All rights reserved.

634120 Keynote Talk - Machine Learning of Molecular and Materials Properties at the Low-Data Limit

Abstract:

The development of next-generation chemical manufacturing technologies will immensely benefit from robust, reliable, and accurate multiscale models that capture the underlying physical phenomena at multiple scales. A challenge, to this end, is to develop fast and accurate models of molecular and material properties using a combination of experimental and ab initio techniques. In this context, recent years have witnessed a burgeoning growth in the application of data science and machine learning for predicting such properties from large, homogeneous, and often computed molecular/material property datasets. Many properties, however, are hard to measure or compute; consequently, the available data in such cases is scarce, heterogeneous (that is available from across a variety of sources), and of differing fidelity. This talk will focus on data-driven model building for such "imperfect" datasets.

Using examples from molecule design and catalysis, this talk will focus on different methods, arising from process systems engineering, that can be applied to address the aforementioned challenges. First, **active learning** techniques that balance exploration of the molecule/material space and exploitation of the current model will be discussed to learn linear (generalized group additive) and nonlinear (graph convolutional neural network) property models. The second example will focus on the concept of **transfer learning** whereby information (model structure and features) from a model of a molecular property for which data is plentiful is subsequently leveraged ("transferred") while training a model of a related property for which data is scarce. Third, we will show that, by infusing a relatively large amount of low fidelity data via **multitask learning**, the thermodynamic properties of adsorption on catalytic surfaces can be modeled to a high level of accuracy using only small amounts of high accuracy data.

Copyright © American Institute of Chemical Engineers. All rights reserved.

625515 Analysis of Plastic-Derived Fuel Oil Produced from Plastic Waste Via Slow Pyrolysis Operations

Abstract:

A viable approach for managing waste plastic accumulation in resource-constrained, rural economies is the conversion of waste plastic to plastic-derived fuel oil (PDFO) via slow pyrolysis. Utilizing localized collection, sorting, and conversion of waste plastic by the application of appropriate technology, PDFO can be readily produced in rural economies by local entrepreneurs. However, when implementing slow-pyrolysis via appropriate technology, a challenge faced by small-scale entrepreneurs is the composition of the product fuel blend. The PDFO produced can vary in composition from a mixture that is more like kerosene to a mixture that is more like diesel. Due to capital constraints and lack of sophisticated process control equipment, the average chain length of PDFO blends produced via appropriate technology is difficult to ascertain. This presentation will focus on research conducted to experimentally determine the impact of temperature, reaction time, and plastic type on the composition of PDFO produced under appropriate technology conditions. Thermogravimetric analysis (TGA) and gas chromatography-mass spectrometry (GC-MS) analysis have been used to determine the volatility and chain lengths of PDFO produced under various operating conditions and for different plastic types and blends. The results of this research can then be used to optimize production of PDFO from plastic waste using slow pyrolysis in developing regions.

Copyright © American Institute of Chemical Engineers. All rights reserved.

621841 Civic Environmentalism and Waste Recycling: The Role of Community Social Capital in Promoting Eco-Friendly Behaviors

Abstract:

Resident participation in waste valorization is an excellent way to build social health capital in poor under-resourced communities. Local civic organizations and grassroots environmental health groups have the greatest pull in influencing resident attitudes and behavior about eco-friendly practices. Residents who are civically engaged in their environments benefit their community's health. This also encourages eco-friendly health behaviors from community members. Resident attitudes about the importance of civic environmentalism requires investment by community leaders and local stakeholders engaged in environmental eco-friendly health promotion. Such involvement is highlighted in the paper with respect to promoting resident's health behavior knowledge about environmental health issues. We include several practical source reduction strategies to encourage resident civic participation in socio-environmental waste valorization activities such as educating community members about the cost to them for not recycling (i.e. increased prices for goods and services), the benefits of sustainable reusable materials in protecting the environment and human health as well as home composting and similar type energy efficient methods to reduce local waste cumulative build up.

Copyright © American Institute of Chemical Engineers. All rights reserved.

625648 Automatic Anaerobic Digestion Microplant for Decentralised Management of Waste

Abstract:

The performance of a small-scale prototype digestion plant (7.2 m^3 working volume) intended for decentralised operation was evaluated considering energy efficiency and technical suitability for biogas valorisation in producing electrical and thermal energy. The digester operated in recirculation mode to enhance organic matter conversion and improve volatile solid degradation. An energy assessment of the process assumed the incorporation of a combined heat and power (CHP) unit. The coefficient of overall performance of the plant for electrical energy (COP_{el}) was 0.95 — this values was estimated at an electrical efficiency of 22.5% and represents the ratio between energy production and consumption — for a methane yield of 360 L/kg VS and an organic loading rate (OLR) of 1.06 g VS/L d. This parameter was slightly lower than the unit thus indicating that the micro-plant was close to attaining self-sufficiency regarding electrical energy use. The temperature increase of the feed to process conditions supposed a significant amount of thermal energy which highly compromised the efficiency when operating at low organic load, thus accounting for more than 80% of the total energy demand of the installation. When the energy assessment of the process was performed at higher OLR of 2.7 g VS/ L d, the resulting COP_{el}value was1.68, demonstrating the feasibility of this configuration for decentralised digestion. [draft]

Copyright © American Institute of Chemical Engineers. All rights reserved.

630301 Anaerobic Digestion of Different Type of Wastepaper and Cardboard: Effect of Enzymatic Pretreatment

Abstract:

Anaerobic digestion (AD) is a biological waste management technology able to treatment and valorise different organic material. The main product of the process is biogas that can be used for energy generation. Boosted by new regulations, the raise in the market share of renewable energy has created the need for identifying alternative feedstock for the AD process and aiming to replace more traditional materials such as energy crops. The list with the potential feedstock for AD systems includes the by-products from the paper industry, however, the biomethanation of these materials is usually low. The AD process can be divided into four different stages including hydrolysis, acidogenesis, acetogenesis and finally methanogenesis. For many substrates, the rate-limiting step of AD is the first one, hydrolysis. This step seems to be the major controller of the whole process. During hydrolysis, complex macromolecules are breaking down into their monomers by extracellular enzymes which are excreted by the hydrolytic microorganisms.

In the present study, six different types of wastepaper products, including toilet paper, office paper, cardboard, glossy magazine, newspaper, and kitchen towel were chosen to be examined for their biomethane potential with and without a prior enzymatic pretreatment stage. Obviously, paper is mainly cellulose and due to that, the enzymatic pretreatment was conducted with the application of a commercially available product of cellulase (A868 Novozymes, Denmark). Subsequently, the biomethanation of the feedstock was examined in a biomethane potential test (BMP) experiment. Based on the results of this experiment, the application of enzymes offers the potential to improve the biodegradability of the examined paper products. The highest improvement in gas yield was found when the enzymes treated the glossy magazine paper (47%) while the lowest enhancement was reported for newspapers (6%). It is also worth mentioning that the enzymatic pretreatment not only seemed to increase the cumulative methane production but it can also effect the rate of the overall process. According to the results of this experiment, for most of the examined substrates, the enzymatic pretreatment seemed to minimize the lag phase and to increase the gas production during the first days of the experiment. This is an interesting outcome as it shows that the enzymatic pretreatment can be used not only to enhance the final gas yield but also to decrease the required digestion time, which would be significantly important, especially for difficult to digest substrates that usually require very long retention times in order to be efficiently digested.

Copyright © American Institute of Chemical Engineers. All rights reserved.

630480 Aviation Fuel from Solar Energy: A Spain Supply Chain Network.

Abstract:

The aviation industry is responsible for approximately 2% of total anthropogenic greenhouse gas emissions. Consuming 350 Mt of jet fuel per year, the sector releases around 910 Mt of life cycle CO_2 (IATA, 2019). Having an expected four to six fold growth in consumption by 2050, increased attention has been placed to reduce its carbon footprint (Sustainable Aviation, 2020; Transport & Environment, 2018). In this context, solar energy is the most abundant type of energy available to mankind and has the potential to be transformed into liquid fuels (Nocera, 2017). This presents an opportunity for the large-scale production of liquid fuels from solar energy to cover future demand at relatively low CO_2 emissions. However, realising such a pathway implies the deployment of a range of technologies at each step along the whole production process. As a result, the identification of the most sustainable production route still remains a challenge. Aiming to address this problem and inform decision making, we present a model-based approach to identify optimal routes for the production of jet fuel from a cost and CO_2 emissions perspective.

The model presented relies on an MILP superstructure that accounts for resources and technologies whose availability is constrained to a particular region. The routes considered for the production of jet fuel include the Fischer-Tropsch (FT) (Klerk, 2011) and Methanol to Fuels (MtF) (Tabak & Yurchak, 1990) processes. The optimization of the network provides the temporal and spatial interconnectivity design of imported resources and intermidiate/final products along with the material transportation network and related process technologies. The methodology presented is applied using Spain as the location of the case study, adjusting the regional demand and solar energy availability according to daily and seasonal time scales.

The results of the network, which could vary for a different location, show current production costs per kg of liquid fuel from 3.5 € (MtF) to 4.4 € (FT) using solar radiation as unique source of energy. From these, ≈90% come from capital costs of solar PV and elecrolysis technologies. These costs are nearly ten-fold current production costs from fossil fuels estimated at 0.47 €/kg fuel. In terms of emissions, both MtF and FT processes cut CO_2 life cycle emissions by ≈25% compared to their fossil-based counterpart, releasing 2.5-2.7 kg CO_2eq/kg liquid fuel on a cradle-to-grave basis. In this case, the CO_2 embodied in the installation of solar PV, and particularly solar panels, is the main contributor to the impact. Projections toward 2050 show potential cost reductions up to 2.5 € for MtF and 3.0 € for FT with lifecycle emissions below 1.00 kg CO_2eq/kg liquid fuel. The performance of the network is also analyzed at different conditions and fuel demand, including the import of electricity and varying costs for electricity and hydrogen storage. Ultimately, the use of solar energy to supply jet fuel would represent an increase of costs in a flying ticket by two- to three-fold for a competitive route at distances ~1,000 miles (London-Madrid).

References

IATA. (2019). *IATA Fact Sheet - June 2020.* https://www.eesi.org/papers/view/fact-sheet-the-growth-in-greenhouse-gas-emissions-from-commercial-aviation

Klerk, A. De. (2011). Fischer-Tropsch fuels refinery design. In *Energy and Environmental Science*

(Vol. 4, Issue 4, pp. 1177–1205). https://doi.org/10.1039/c0ee00692k

Nocera, D. G. (2017). Solar fuels and solar chemicals industry. In *Accounts of Chemical Research* (Vol. 50, Issue 3, pp. 616–619). https://doi.org/10.1021/acs.accounts.6b00615

Sustainable Aviation. (2020). *Decarbonisation Road-Map: A Path to Net Zero.*

Tabak, S. A., & Yurchak, S. (1990). Conversion of methanol over ZSM-5 to fuels and chemicals. *Catalysis Today*, *6*(3), 307–327. https://doi.org/10.1016/0920-5861(90)85007-B

Transport & Environment. (2018). Roadmap to decarbonising European aviation. In *Transport & Environment*. www.transportenvironment.org

Copyright © American Institute of Chemical Engineers. All rights reserved.

625513 Production of Syngas Via Gasification Using Aspen Hysys Simulation

Abstract:

Gasification is the thermochemical process of converting a biomass into valuable gas fuel. Several parameters affect the gasification processes efficiency, thereby affecting syn gas production and quality. In this study, the first objective was to develop and validate a downdraft gasifier simulation model by using Aspen Hysys software. The second objective was to use the validated Aspen Hysys model to investigate the effect of equivalence ratio on the syngas production of downdraft gasifier. Five different values of equivalence ratios were chosen for simulation conduction. Waste wood was used as a feed stock for all simulation cases. The Peng-Robinson equation of state with Boston-Mathias was used to estimate the physical properties in the simulation model. Hydrodynamic parameters and reaction kinetics modeling were described by this model. The developed model was validated by comparing the predicted results of syngas composition against available experimental data for different equivalence ratios. The comparison showed a good agreement between predicted and experimental results. Also, the obtained results showed that the equivalence ratio had a significant effect on the gasification process efficiency.

Copyright © American Institute of Chemical Engineers. All rights reserved.

632227 Feedstock Impact on Hydrothermal Liquification (HTL)

Abstract:

Enough sewage sludge, food waste, Fats Oils and Grease, and manure are generated every year to produce over 5.5 billion gallons of fuel per year via hydrothermal liquefaction. Waste feedstock type and composition plays a significant role not just in the hydrothermal liquefaction (HTL) process but also in upgrading HTL biocrudes to fuels via hydrotreating. This presentation will compare the hydrothermal liquification and catalytic hydrotreating for a variety of wet waste feedstocks, including sewage sludge, pig manure, dairy manure, biosolids, and other feedstocks, to ultimately produce fuel blendstocks.

Copyright © American Institute of Chemical Engineers. All rights reserved.

*** Formatting Issue - Best Available Paper/Graphic ***

623621 Pathways for UK Power and Industry to Net-Zero

Abstract:

Power and industry are the largest contributors to point source CO_2 emissions in the UK. Their decarbonisation is critical for the UK to reach carbon targets and minimise climate change and the associated social and economic loss. The two sectors vary greatly with regard to their pathways to net-zero. While decarbonisation of the power sector is well-analysed, industrial emissions are considered notoriously difficult to abate. Many options for low- and zero-carbon power (renewables, carbon capture and storage (CCS), storage) exist, and the provision of negative emissions through bioenergy with CCS (BECCS) can lead to carbon-negative electricity. In industry, no clear pathways forward exist, many technologies rely on CCS, and substantial residual emissions are to be expected. In addition to residual and negative emissions, power and industry are connected through the additional power demand arising from abatement in industry. Much of the important work on industrial decarbonisation is currently focused on individual technologies. However, there is interest in analysing the whole of industry and how it might reach net-zero together with the power sector, which is where this study is aimed at.

We build our model on ESO[1], a power systems model, extend it to include industry, and apply it to the UK. The cement, steel and refinery sectors and options for abatement in these sectors are modelled explicitly. Remaining industrial sectors, such as chemical industry, paper, natural gas processing, are included as fixed emissions requiring offsets. Abatement for these small point sources is difficult to achieve and likely uneconomical unless incentives are particularly high. The additional power demand caused by abatement in industry is added to the baseline power demand. We model import and export of industrial products explicitly and in the first part of the results analyse four scenarios with regard to abatement and trade. In the "BAU & offset" scenario industry is restricted to conventional technologies, and BECCS is required to provide offsets for the entire sector. For "abate & offset", abatement is made available. In the "import & offshore" scenario, all industrial products are imported and therefore the emissions offshored. Lastly, the "abate & export" scenario assumes an increase in UK production of low-carbon products and subsequent export. A net-zero carbon target on the total emissions with a linear trajectory is enforced for all scenarios. For the second part of the results, we investigate how trajectories to net-zero might be incentivised using a carbon price (CP) and negative emissions credit (NEC).

Figure 1 shows the results for industrial production for the four scenarios. When abatement is available, oxy-combustion is the optimal choice for cement, production with biochar and CCS is determined preferable for steelmaking, and a high amount of post-combustion capture (PCC) is optimal for refineries. Notably, all optimal abatement technologies require CCS. Abatement is deployed in the abate & offset scenario – indicating it is preferred over BECCS offsets. Examination of the emissions trajectories and emissions for transport and storage in figure 2 reveals that residual and negative emissions are highest (54 Mt-CO_2/yr) when all industrial emissions are offset (BAU & offset). CCS in industry reduces the emissions by 90%, resulting in only 3 Mt-CO_2/yr in residual emissions from cement, steel and petrochemicals. When industrial emissions from the three sectors are offshored, the negative emissions requirement is only marginally lower. This is one indicator that the economic damage from offshoring industrial emissions might not be worth the reduction in system cost. The impact of an abate & export scenario on the power sector is minor as residual emissions from cement, steel and refining are small. Consequently, strategies for exporting low-carbon British products can be pursued without major adjustments in power. The CO_2 emissions requiring transport and storage range between 65 and 130 Mt-CO_2/yr, with the abate & offset scenario capturing 85 Mt-CO_2/yr. With only carbon budgets to drive decarbonisation, CCS is deployed first in the industrial sector in 2035, then CCGT-CCS and BECCS are added in the power sector.

The decarbonisation of power and industry can be incentivised using a carbon price (CP), *i.e.*, a penalty paid by all emitters in the system, and a negative emissions credit (NEC), *i.e.*, a reward per t-CO_2 removed

*** Formatting Issue - Best Available Paper/Graphic ***

by negative emissions technologies (in this case, BECCS). For this study, ranges of 0-300 £/t-CO_2 for the CP and 0-200 £/t-CO_2 for the NEC are simulated. The results are detailed in figure 3. Many combinations of CP and NEC are able to deliver net-zero in our model (thin solid contour line). Higher combinations of both lead to cumulative net-zero (thick solid contour line). NEC values between 90-135 £/t-CO2 across the range of CP values achieve net-zero in 2050. A NEC alone allows to reach net-zero at the minimum cost to the private sector. In this scenario, no abatement in industry is deployed and the burden of decarbonisation is placed entirely on BECCS. Adding the CP reduces the cost to the public sector. It also decreases cumulative emissions (not shown). Indicated in the graph is the break-even line for the public sector. For every NEC granted to the negative emissions technologies, there is a minimum CP necessary to offset the cost to the public. An even higher CP allows the public sector to use the penalties on emitters to support other decarbonisation efforts, such as funding R&D. The highest possible CP and the lowest required NEC constitute the public sector optimum, combining the highest income through penalising emissions and lowest expenditure towards BECCS. In our model, this public sector optimum is located at a NEC of 90 £/t-CO_2 and a CP of 300 £/t-CO_2. This optimum is preferable to the private sector optimum from a burden-sharing perspective, as it leads to CCS in industry, renewables expansion, and therefore a lower demand for BECCS. Arguably, a CP this high may place an undue burden on emitters. A more sensible optimum might therefore be located at the CP beyond which no significant decreases in cumulative emissions are gained.

In this work, we have modelled power and industry together and evaluated trajectories to net-zero. We found that emissions in cement, steel and refining can be reduced by 90% with technologies heavily relying on CCS. Under carbon budgets, CCS is deployed first in industry and later in power. We quantified the required amount of carbon offsets for varying scenarios as well as the volume of captured CO_2. We identified combinations of CP and NEC that incentivise decarbonisation to net-zero and evaluated optimal pathways from both the private and public sector perspective. Future work may involve including more industrial sectors explicitly and modelling industrial demand side response.

References

1. Heuberger CF, Rubin ES, Staffell I, Shah N, Mac Dowell N. Power capacity expansion planning considering endogenous technology cost learning. *Appl Energy*. 2017;204:831-845.

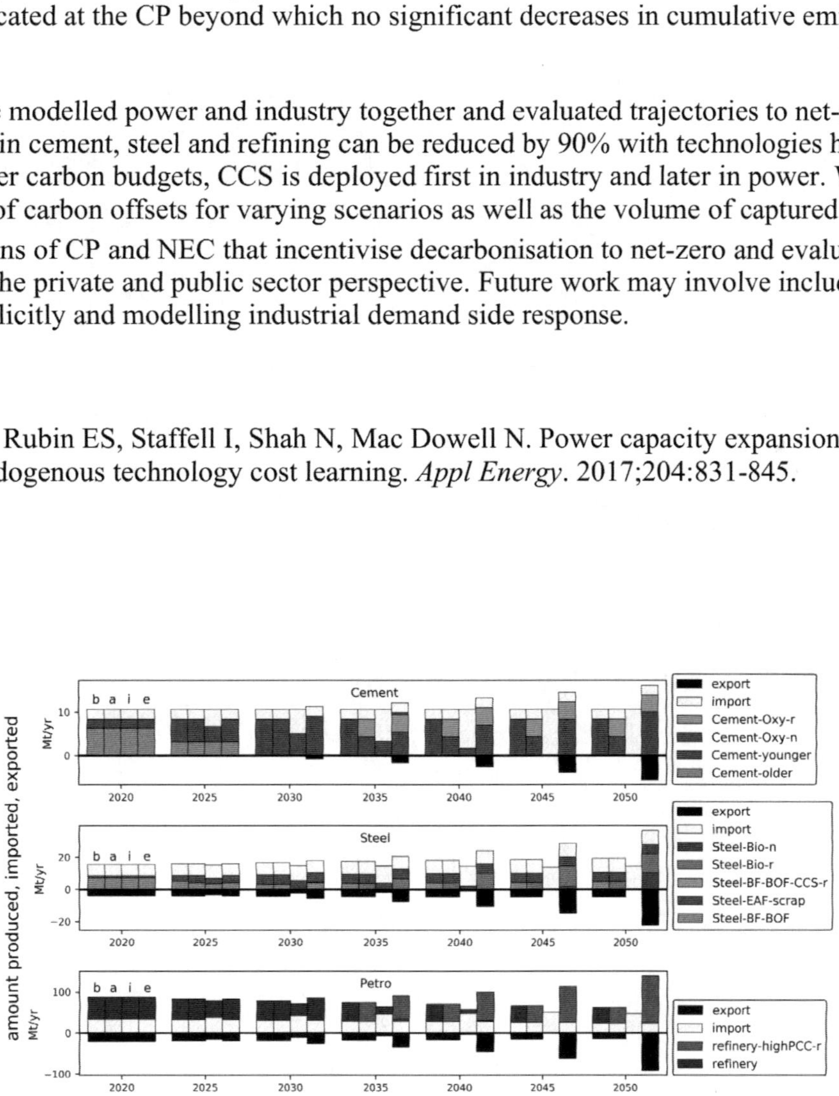

Figure 1: Industrial sector results for BAU & offset (b), abate & offset (a), import & offshore (i) and abate & export (e) scenarios.

***** Formatting Issue - Best Available Paper/Graphic *****

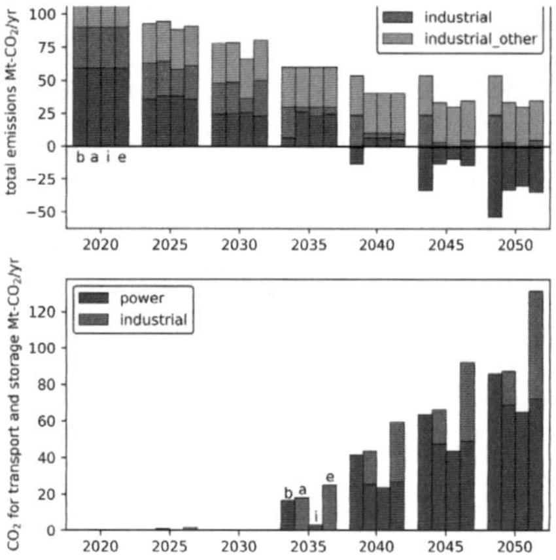

Figure 2: System emissions and emissions for transport & storage for BAU & offset (b), abate & offset (a), import & offshore (i) and abate & export (e) scenarios.

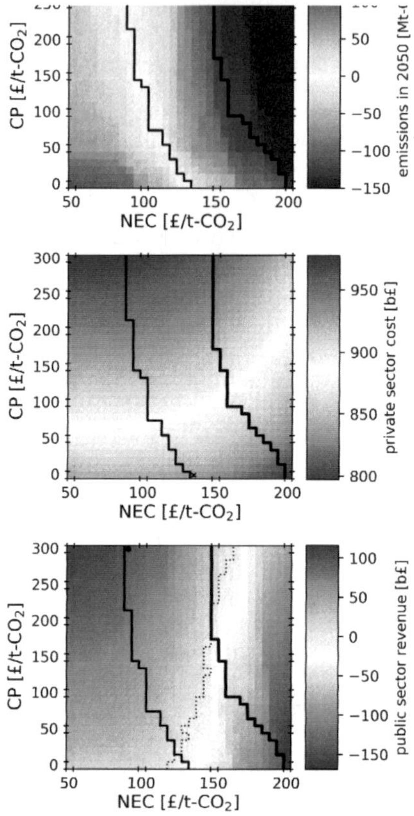

Figure 3: Emissions in 2050, private sector cost and public sector revenue for varying carbon price (CP) and negative emissions credit (NEC). Thin and thick solid contour lines represent boundaries for net-zero in 2050 and cumulative net-zero, respectively. ··· shows the break-even for the public sector. × denotes private sector optimum. • denotes public sector optimum.

Copyright © American Institute of Chemical Engineers. All rights reserved.

632238 Towards a Circular Chemical Industry: Mapping the Flow of Fossil Carbon through Chemical Manufacturing Processes

Abstract:

The rising levels of carbon dioxide in the atmosphere threaten humanity with erratic changes in global weather, better known as global climate change. Researchers have drawn many parallels between the ongoing Covid 19 crisis and the incumbent consequences of climate change.[1, 2] This naturally necessitates the need for urgent and sustainable action to reduce greenhouse gas emissions. The global chemicals industry has shown considerable decarbonization potential, accounting for 7% of the world's emissions.[3] This fraction greatly increases when the cement or iron and steel industries are lumped in with chemicals manufacturing.

Various sources of literature identify hotspots of greenhouse gas emission in the path to deep decarbonization. In this pursuit, modelling of carbon flow is an important step to pinpoint the sectors and processes with the greatest carbon footprint. Existing diagrams, such as the ones built by the Lawrence Livermore National Lab or the Stanford Global Climate and Energy Project (GCEP) suffer from aggregation of sectors and processes and thus do not offer complete insight into the problem.

In this work, we develop a steady-state material flow model of the chemical industry. This model will trace the flow of carbon from fossil, biomass, and other feedstocks and fuel sources, to intermediates, and finally valuable products, byproducts and emissions. To develop this model, we rely on process data and models of the chemical and petrochemical industries obtained from existing studies about the chemical industry,[4, 5] the economy,[6, 7] government databases,[8] life cycle inventory databases,[9, 10] and greenhouse gas emissions inventories.[11, 12] We use fundamental material and energy balance models with stoichiometric information to estimate missing data and reconcile available data. Results are represented in the form of a Sankey diagram to facilitate a graphical view of the flows and opportunities for improvement.

We demonstrate how this study aids the identification of specific processes for adoption of alternatives such as electrification, hydrogen based manufacturing and carbon capture, utilization, and storage. This model lays the foundation for further work towards the vision of a carbon neutral and circular chemical industry that is also economically, ecologically, and socially sustainable.

References

[1] Rubén D. Manzanedo and Peter Manning. COVID-19: Lessons for the climate change emergency. Science of the Total Environment, 742:140563, nov 2020.

[2] David Klenert, Franziska Funke, Linus Mattauch, and Brian O'Callaghan. Five Lessons from COVID-19 for Advancing Climate Change Mitigation. Environmental and Resource Economics, 76(4):751–778, aug 2020.

[3] Maria van der Van, Yoshimitsu Kobayashi, and Rainer Diercks. Technology Roadmap: Energy and GHG Reductions in the Chemical Industry via Catalytic Processes. Energy Technology Perspective, pages 1–60, 2013.

[4] Peter G. Levi and Jonathan M. Cullen. Mapping Global Flows of Chemicals: From Fossil Fuel Feedstocks to Chemical Products. Environmental Science and Technology, 52(4):1725–1734, feb 2018.

[5] Sean E. Derosa and David T. Allen. Impact of natural gas and natural gas liquids supplies on the united states chemical manufacturing industry: Production cost effects and identification of bottleneck intermediates. ACS Sustainable Chemistry and Engineering, 3(3):451–459, mar 2015.

[6] Stanford University - The Global Climate and Energy Project). https://gcep.stanford.edu/.

[7] Shweta Singh and Bhavik R. Bakshi. Accounting for emissions and sinks from the biogeochemical cycle of carbon in the U.S. economic input-output model. Journal of Industrial Ecology, 18(6):818–828, dec 2014.

[8] Lawrence Livermore National Laboratory. https://www.llnl.gov/.

[9] Ecoinvent version 2.2. https://www.ecoinvent.ch.

[10] United States Life Cycle Inventory (USLCI). https://www.nrel.gov/lci/.

[11] United States Environmental Protection Agency. https://www.epa.gov/ghgemissions/inventory-us-greenhouse-gas-emissions-and-sinks.

[12] IPCC - The Intergovernmental Panel on Climate Change. https://www.ipcc.ch/.

Copyright © American Institute of Chemical Engineers. All rights reserved.

632757 The Role and Value of Disruptive CCS Technologies for Electricity Systems

Abstract:

It has been recognized that carbon capture and sequestration (CCS) is a key technology in energy systems decarbonization, including in the electricity system. However, despite the readiness of conventional CCGT with post-combustion CCS (CCGT-PostCCS), the incumbent technology is constantly perceived as a "pre-commercial technology". Consequently, instead of focusing on the deployment of the technology, many proposed projects were frequently delayed, and the current "action" is mainly waiting for the appearance of a unicorn technology, *i.e.,* a very efficient and flexible, yet cheap technology. While the appearance of such technology is highly uncertain and rare, deep decarbonization of the electricity system is critical to mitigating climate change. While various studies have shown that decarbonization without CCS will be less likely or very costly, the relative value of the unicorn technology compared to the incumbent is rarely investigated. Accordingly, the objective of this research is to quantify the role and value of such a disruptive technology for the electricity system.

In this study, we used Electricity System Optimization (ESO) framework to understand how the unicorn features might affect the electricity system transition from 2020 to 2050. ESO is an MILP unit commitment model for electricity systems capacity expansion planning to minimize the total system cost considering system-wide and technology-wise constraints. To quantify the role and value of the unicorn in the electricity system, numerous combinations of technology cost, thermal efficiency, and flexibility (*i.e.,* minimum stable generation) are implemented in ESO. Moreover, to understand the impact of different characteristics a system may have, we include the UK, Poland, and ERCOT and PacifiCorp East (PACE) systems in the USA as cases.

Our results show that increasing flexibility is the least valuable improvement compared to cost reduction and efficiency improvement. Accordingly, any levels of flexibility improvement are disadvantageous if such an improvement compromises the technology cost or efficiency. However, if the value between cost and efficiency improvements are critical, flexibility (*i.e.,* min. stable generation of 20% of rated capacity) can increase the share of the unicorn in total CCS capacity by 40%-points compared to that of inflexible CCS (*i.e.,* min. stable generation of 70% of rated capacity).

In the UK, owing to expensive natural gas prices, thermal efficiency and technology cost are equally critical if the efficiency is greater than 50% and the capital cost is lower than 4000 £/kW. Accordingly, the drawback of having a higher capital cost can be compensated by having a higher efficiency cost which reduces the cost of fuel and carbon tax. However, if the capital cost is greater than £ 4000/kW, the technology becomes less competitive than nuclear power, and thus, is not deployed. Interestingly, the unicorn remains competitive although the efficiency is between 40% and 50% (i.e., lower than the incumbent CCGT-CCS) as long as the capital cost is the same as the incumbent's capex.

The effect of natural gas prices on the significance of efficiency improvement is observed to be even stronger in Poland than that in the UK as the price of natural gas in Poland is more expensive. As the result, although the capital cost is around 6000 £/kW, a 60% efficient unicorn remains competitive in the system. These results suggest that systems with expensive natural gas prices, such as UK and Poland, should prioritize efficiency improvement.

Unsurprisingly, owing to the cheap natural gas price in ERCOT, the impact of efficiency on unicorn competitiveness against the incumbent CCGT-CCS is observed to be weak whilst the impact of capital cost reduction remains strong. However, whilst the magnitude of efficiency improvement impact is dictated by the price of natural gas, we found that load profile and the quality of renewable sources also play an important role. In PACE system, for instance, efficiency improvement in unicorn not only increases the share of the unicorn in total CCS capacity but also significantly increases the capacity and capacity factor of CCS in the system owing to the combination of flat load profile, poor renewable sources quality, and cheap natural gas in the system.

Without unicorn technology, CCS capacity in 2050 ranges from 5 – 15 % of peak load for the UK, ERCOT, and Poland and up to 38% in PACE. A minor cost or efficiency improvement, however, can significantly increase the optimal capacity of CCS from such a range to 20 - 57 % of peak load in all cases. Moreover, such an improvement can also increase the capacity factor of CCS by 2000 full load hours in all cases.

From the system perspective, the value of each improvement, *i.e.,* its ability to reduce the total system cost, depends on other improvements. For instance, although a 60% efficient unicorn can be materialized, the improvement doesn't have value if the capital cost is too expensive (*i.e.,* more than £ 2000/kW for ERCOT, £ 4000/kW for UK and PACE, and £ 6000/kW for Poland). Similarly, even if the capital cost is very low, the technology is less valuable if the efficiency drops, particularly for cases with expensive natural gas such as Poland. This is because the combination of capital cost, efficiency, and flexibility on a technology dictate the level of optimal deployment of the technology in the system which, in turn, proportionally determines the value the technology can provide to the system. Although technology improvements are undoubtedly beneficial for the technology competitiveness, we found that any technical improvement of a unicorn can only decrease the total system cost by between 2 – 4 % of the initial cost in UK, Poland, ERCOT, and up to 8% on PACE.

Finally, we conclude that Incumbent CCS technology is already cheap and efficient and thus cost-competitive to be deployed in the system. Accordingly, improving the technology slightly may significantly improve the competitiveness of the technology and increase its level of deployment in the system. Interestingly, in some cases, no improvement is needed to increase the total capacity of CCS. This indicates that quick deployment can be equally or even more important than chasing the unicorn. However, it is important to note that while such improvements can be extremely valuable from the technology point of view, the value it can offer to the system tends to be negligible in all cases.

References:

1. IPCC (2014). Climate Change 2014: Synthesis Report. Contribution of Working Groups I, II and III to the Fifth Assessment Report of the Intergovernmental Panel on Climate Change Core Writing Team, R. K. Pachauri, and L. A. Meyer, eds. (Intergovernmental Panel on Climate Change (IPCC))

2. Heuberger, C.F. et al. Impact of myopic decision-making and disruptive events in power systems planning. Nature Energy 3 (2018) 634–640

3. Heuberger, C.F. et al. Power capacity expansion planning considering endogenous technology cost learning. Applied Energy 204 (2017) 831–845

Copyright © American Institute of Chemical Engineers. All rights reserved.

632985 H_2 Roadmap Alternatives for Decarbonizing Singapore through 2050

Abstract:

In the early 2020, the National Climate Change Secretariat of Singapore published Singapore's Long Term Low Emissions Development Strategy (LEDS)[1]. Singapore plans to reduce emissions by half from its peak by 2050, and reach net-zero emissions as soon as is viable in the second half of the century. Industry, power, and transport are three major CO_2 emissions sectors, whose shares are 46%, 39%, and 13% in 2017 respectively. And around 40% of emissions from power sector are indirect emissions from transport and industry sectors' power usage. As hydrogen is a versatile energy carrier that can achieve deep decarbonization, Singapore is exploring the potential of hydrogen in the energy mix to decarbonize high-CO_2-emission sectors. In the past decades, researchers have taken many studies on hydrogen roadmap, including the techno-economic analysis of hydrogen production[2], transport[3] or utilization, and the design and optimization of hydrogen supply chain[4]. We also carried out a techno-enviro-economic analyses of hydrogen supply chains, providing insights on the landed cost of hydrogen, carbon avoidance, and carbon avoidance costs. However, there is no comprehensive hydrogen roadmap studies simultaneously considering both local and regional hydrogen supply chains, multiple demand sectors, multiple periods, and comparisons with other decarbonization alternatives.

In light of the aforementioned concerns, this research was undertaken to understand the evolution and development needs of a hydrogen economy in Singapore through 2050. Singapore, as a city-state with only 700 square kilometers of land, relies on importing energy resources. Natural gas, the cleanest form of fossil fuel, has been the dominant fuel for Singapore, where over 96% of electricity is generated using natural gas. As Singapore is exploring the potential of hydrogen in the energy mix to reduce her carbon footprint, importing hydrogen from overseas and producing hydrogen locally are two alternatives under consideration. Thus, the possibility of producing hydrogen in Singapore is thoroughly compared with importing hydrogen from neighbors or distant countries to achieve deep decarbonization of transport, power, marine, and industrial sectors. Both renewable hydrogen production technologies and fossil fuel-based hydrogen production technologies with carbon capture and sequestration (CCS) are included to cover the energy source diversity in the ASEAN region. Various hydrogen transportation pathways are considered for the regional hydrogen supply chain. The variation of hydrogen demand over a long-term planning horizon and technological progress, in the form of lower costs, are considered. What's more, techno-economic analysis of natural gas utilization (power generation and industry heating) with CCS is also investigated, and compared with hydrogen economy as well as other decarbonization options (e.g. importing electricity via regional grid or subsea cables). Several what-if scenarios are explored to derive insights on the landed cost of hydrogen, levelized cost of electricity (LCOE), carbon avoidance, and carbon avoidance costs associated with various hydrogen roadmaps for a hydrogen economy and other decarbonization methods in Singapore.

1. National Climate Change Secretariat, 2020, Singapore's Long-Term Low-Emissions Development Strategy.
2. Nikolaidis, P., Poullikkas, A., 2017, A comparative overview of hydrogen production processes. *Renewable and Sustainable Energy Reviews*, 67, 597-611.
3. Niermann, M., Drünert, S., Kaltschmitt, M., Bonhoff, K., 2019, Liquid organic hydrogen

carriers (LOHCs) – techno-economic analysis of LOHCs in a defined process chain. *Energy & Environmental Science*, 12 (1), 290-307.

4. Sunny, N., Mac Dowell, N., Shah, N., 2020, What is needed to deliver carbon-neutral heat using hydrogen and CCS? *Energy & Environmental Science*, 13 (11), 4204-4224.

Copyright © American Institute of Chemical Engineers. All rights reserved.

628394 Packed Bed Column Adsorption and Parametric Analysis of CO2 Onto Palm Kernel Shell Activated Carbon Using Computational Modelling Approach.

Abstract:

Increment in Carbon emission to the atmosphere has become an issue of great concern all over the world due to its remarkable contribution to climate change. One of the major recognized approaches to prevent carbon dioxide from entering the atmosphere is carbon capture and storage. They have been several CO_2 removal technologies that have been reported, including absorption, adsorption, membrane separation, and Bio-energy with Carbon Capture and Storage (BECCS).

In this study, there was an investigation of the performance of a locally sourced adsorbent, Palm kernel shell (PKS) activated carbon in removing CO_2 from a continuous gas stream in a fixed bed adsorption column using Computational Fluid Dynamics (CFD) study. The adsorption properties are defined through the use of user defined function (UDF). The investigation of the influence of flow rate and bed height on the CO_2 removal efficiency and adsorption capacity by varying the inlet feed velocity and bed packing heights was done using CFD code ANSYS R20 academics.

The results of the investigation showed that the CO_2 removal efficiency increases with the bed heights as it was observed when the gas flowed at a rate of 0.00008kg/s to the adsorption column filled with the PKS activated carbon adsorbent of 10 cm, 15cm and 25cm respectively in height.

The parametric study involves the investigation of influence of two operating factors namely inlet and outlet mass weighted average and Area weighted Average on the adsorption process and on the carbon dioxide removal efficiency of the process.

Copyright © American Institute of Chemical Engineers. All rights reserved.

630500 Dynamic Modeling and Soft Sensor Based Predictive Control of a Moving Bed Process for CO_2 Capture Using a Micro-Encapsulated Solvent

Abstract:

Microencapsulation of carbon sorbents (MECS) is a novel technique where a solvent for CO_2 capture is encapsulated inside a CO_2-permeable polymer material [1]. MECS can enable use of solvents that have excellent CO_2 capture properties but suffer from undesirable characteristics like high viscosity, or solids formation upon CO_2 capture, that make transport difficult and use of conventional adsorption towers impractical. Using microcapsule sizes ranging from 100 – 600 , MECS can provide very high specific surface area thus offering excellent mass and heat transfer. One of the key difficulties of using the MECS is water management. Water is also known to permeate through polymers that can be used for encapsulation. Ineffective water management can lead to undesirable solvent concentrations, excess energy penalties, and varying capsule sizes, all of which penalize performance. Therefore, efficient water management is one of the critical requirements for commercial use of the MECS technology. While there are a few studies that have investigated different reactor configurations for MECS [2-3], no studies have been conducted on water management for MECS. While control of water content within the MECS capsule is required to ensure optimal performance, it is not practical to measure water concentration inside the microcapsules. In this work, we propose a computationally based soft sensor for estimating water concentration in the solvent.

Moving bed (MB) reactors have been reported to exhibit strong potential for solid-sorbent based CO_2 capture [4-5]. Therefore, in this work, the MB technology is evaluated for MECS. A detailed first-principles dynamic model of the MB absorber and regenerator is developed using Na_2CO_3 as the encapsulated solvent. The capsule model used in the moving bed reactor is validated with experimental data from our previous study [3]. The soft sensor for estimating water concentration in the solvent is developed to infer difficult to measure water content by using the data of the easily measured variables [6]. State space models for both absorber and desorber are used to estimate the capsule H_2O concentration.

Finally, linear and nonlinear model predictive controllers for maintaining CO_2 capture, desorber temperature and minimizing water loss are developed and evaluated by simulating a number of disturbances. The performance of the predictive controllers are compared with conventional PID controllers. It is observed that while model predictive controllers without the soft sensor exhibit excellent tracking and disturbance rejection performance, they can lead to high energy loss. Predictive controllers with the soft sensor for estimating core concentration at the outlet of the absorber and desorber results in much lower energy usage mainly due to reduced water evaporation in the desorber.

References

1. Stolaroff, J. K. et al., Microencapsulation of advanced solvents for carbon capture. Faraday Discussions. DOI: 10.1039/c6fd00049e (2016).
2. Hornbostel, et al, Packed and fluidized bed absorber modeling for carbon capture with micro-encapsulated sodium carbonate solution, Applied Energy 1192-1204, 2019.

3. Kotamreddy, et al., Process Modeling and Techno-Economic Analysis of a CO2 Capture Process Using Fixed Bed Reactors with a Microencapsulated Solvent, Energy & Fuels 2019 33 (8), 7534-7549 DOI: 10.1021/acs.energyfuels.9b01255.
4. Knaebel, K.S., Temperature swing adsorption system. 2009, Google Patents.
5. Mondino, Giorgia et al. "Effect of Gas Recycling on the Performance of a Moving Bed Temperature-Swing (MBTSA) Process for CO_2 Capture in a Coal Fired Power Plant Context." Energies, 10(6), 745. 2017.
6. Fortuna, L. et al., Soft sensors for monitoring and control of industrial processes. Springer. 2007.

Copyright © American Institute of Chemical Engineers. All rights reserved.

631367 Evaluation of Real Time Optimization – Model Predictive Control and Economic Model Predictive Control Strategies for Post Combustion CCS Operations within Refineries

Abstract:

Carbon Capture and Storage (CCS) technologies has become key in abating post combustion CO_2 emissions from large point sources, including powerplants, petroleum refineries, and chemical production facilities. Although, CCS technology has seen an abundance of scientific efforts being invested in capturing CO_2 emissions from power plants, lately, this technology is seeing an increased interest within the petroleum refinery sector as a method to decrease carbon dioxide emissions. According to available data, ~ 65 % of the CO_2 emissions from refineries are due to stationary combustion of natural and fuel gases. Post combustion CO_2 capture, therefore, is vital for the overall CO_2 emissions reduction within refineries.

Captured CO_2 will most likely be transported over long distances via pipelines. CCS operations, therefore, can be considered to consist of three energy intensive sub-operations: CO_2 capture, CO_2 compression, and CO_2 dehydration to meet pipeline specifications. A steady state simulation of a CO_2 capture process for a 300,000 bbl/day refinery shows that ~150 MW of energy is required to regenerate CO_2 from the amine regeneration unit to capture 90 % of the available post combustion CO_2 emissions. Additionally, ~ 14 MW of energy is required to compress and dehydrate the CO_2 to pipeline conditions. Hence, the energy intensive nature of the three CCS sub-operations requires one to consider them as a combined system. Additionally, due to the inherently dynamic refinery processes, the CCS operations rarely reach steady state. Hence, the CCS operation must be studied in a transient state for better economic optimization and control. In this study, a dynamic optimization and control strategy is developed for a CCS system in context of a standard mid-sized refining facility.

To begin, steady state simulation of the CCS operation was performed using an aqueous solution of 30 wt % monoethanolamine (MEA) to capture CO_2 and a 98 wt % triethylene glycol (TEG)/water to dehydrate the captured CO_2 gas. These absorbents were chosen due to current industry practices for CO_2 capture and dehydration technologies. A 7-stage CO_2 compression unit was used to compress the gas to pipeline conditions (15.17 MPa). For a refinery of this size, results indicated that at least two absorber units, one CO_2 regeneration unit, one dehydration, and one glycol regeneration unit would be required. Two design limits were imposed on the simulation while determining the number of columns: Height to Diameter Ratio (H/D) > 2 and Diameter of Column (D) < 12 m. A minimum threshold of 90 % CO_2 recovery was maintained. The steady state layer was then optimized using a cost function. This layer was then linked to the Model Predictive Control layer, thereby forming a Real Time Optimization – Model Predictive Control (RTO-MPC) optimization and control infrastructure for the CCS operation. The RTO-MPC infrastructure is considered as an industry standard while implementing optimization and control processes. An important assumption of the RTO-MPC infrastructure is that the process eventually reaches steady state. However, this assumption is not likely to occur for CCS capture operations within refineries. Therefore, the result of the RTO–

MPC layer may not be an accurate representation of a real time process control operation with respect to the process economic performance.

To overcome this problem, a novel Economic Model Predictive Control (EMPC) strategy was developed in which the process cost function was optimized in a dynamic and transient manner with respect to the process constraints in a single layer. Unlike the RTO-MPC process, EMPC does not require the process to be optimized at steady state conditions. In addition, the EMPC strategy uses a single layer to optimize and control the cost function. The hypothesis is that such a control strategy will be able to reduce the response time to optimize the cost function, thereby achieving a better economic optimization of the entire CCS process. This study compared the economic performance of the two mentioned control strategies for CCS operation within the refinery.

Once the base case (using MEA and TEG absorbents) was completed, the study was expanded to replace MEA and TEG with ionic liquids, Trihexyl(tetradecyl)phosphonium 2-cyanopyrrolide ([P_{66614}][2-CNPyr]) and 1-Ethyl-3-methylimidazolium methanesulf onate [EMIM][$MeSO_3$], respectively. According to the literature, [P_{66614}][2-CNPyr] has shown to possess high CO_2 absorption capacity, low regeneration energy, and negligible solvent loss during regeneration. Additionally, [P_{66614}][2-CNPyr] has also shown to be less sensitive to H_2O on the ionic liquid due to the anion $[2\text{-CNPyr}]^-$. [EMIM][$MeSO_3$] has shown to be highly hygroscopic while also possessing the benefits of low regeneration energy and negligible solvent losses and was studied as a replacement for TEG.

Copyright © American Institute of Chemical Engineers. All rights reserved.

624558 Physical Property Characterization of Ionic Liquids in HFC-32 and HFC-125 Using a High-Pressure View Cell

Abstract:

The current hydrofluorocarbon (HFC) refrigerants used in automobile air conditioning, home appliances and air-conditioners, and retail refrigerant systems are being phased out due to high global warming potentials, resulting in millions of kilograms of refrigerant needing to be recycled and repurposed. Many of these refrigerants are HFC mixtures, and if separated, the components may be repurposed or reused to prevent release of greenhouse gases into the atmosphere. Since most HFC mixtures are azeotropic, separation is difficult and energy intensive and thus no reasonable technologies exist today. Ionic liquids (ILs) offer a sustainable way to separate HFC refrigerants if an IL can be found with a large solubility difference between the components of an HFC mixture. This research is part of Project EARTH (Environmentally Applied Research Toward Hydrofluorocarbons) that is working to create an efficient and sustainable process for the separation of HFC refrigerants. The group has measured the solubility of HFC-125 (CHF_2CF_3) and HFC-32 (CH_2F_2), which make up the refrigerant R-410A (50-50 wt%), in selected ILs using a microbalance. The microbalance does not allow for visually observing what phase the IL might exist in when pressurized with the refrigerant, therefore this project utilizes a high-pressure view cell with quartz windows to visually study the solid-liquid phase change of imidazolium halide based ILs. In some cases, the ILs are solid at room temperature, but in the presence of HFC-125 and HFC-32 as a function of pressure melt and undergo a phase change. Understanding the phase behavior of an IL during HFC sorption allows for the management of heat in separation processes as heat of fusion can help reduce the heat of sorption. Other property measurements based on visual data, including melting point, the presence of metastable states, and occurrence of foaming, are taken in addition to phase change data. This presentation will be discussing how the physical properties of imidazolium halide based ILs change with the sorption of HFC-32 and HFC-125 to indicate the viability of the ILs for separation of the refrigerant R-410A.

Copyright © American Institute of Chemical Engineers. All rights reserved.

629168 Material Selection and Design of Ionic Liquid-Based Extractive Distillation for Hydrofluorocarbons Separation

Abstract:

Hydrofluorocarbons (HFCs) and their mixtures are used as refrigerants for commercial and domestic cooling systems. Because of high global warming potential (GWP) of some of these mixtures, the Kyoto Protocol has recommended their gradual phase-out. There is an economic incentive to separate these HFC mixtures into their constituent components rather than venting or incinerating. However, many of these HFC mixtures are azeotropic in nature, and there is no commercial technology available to perform the separation. Extractive distillation, first proposed in the 1930s [1], is a technology that can separate close boiling azeotropic mixtures by using a higher boiling point entrainer. Ionic liquids (ILs) are attracting attention as entrainer for the extractive distillation of HFCs because of their negligible vapor pressure and chemical and thermal stability [2-3]. Due to significantly large number of possible ions combinations, theoretically, it is possible to have millions of ILs [4]. Selecting an appropriate IL for the separation of a particular HFC mixture is important for effective and economic distillation [5]. We have developed a multiscale process design, synthesis, and material selection framework called SPICE_ED for the synthesis of optimal extractive distillation process for separating HFC mixtures. SPICE_ED stands for Synthesis and Process Intensification of Chemical Enterprises involving Extractive Distillation. The framework is based on a building block-based novel representation of chemical phenomena, tasks, and operations in the chemical process industry [6-8]. From a candidate set, the framework can automatically select the most optimal IL that offers minimum energy requirement and/or has minimum processing cost. We generate process flowsheets and equipment designs considering the appropriate refrigerant and IL inlet trays, IL flow rates, reflux ratio, column operating conditions, and IL regeneration processes involving separate distillation, stripping, and flash operations. Finally, the framework can be utilized to map out the thermophysical properties of hypothetical ILs, thus directing the efforts towards discovering novel ILs with desired process economics. When searching for an optimal process for the separation of R-410A (50 wt% R-125 (C_2HF_5) and 50 wt% R-32 (CH_2F_2)), we obtained a heat-integrated optimized process that consumes 65% less energy and emits 72% less indirect CO_2 emission compared to previously reported designs with an existing IL [9]. In this presentation, we will also discuss how the framework can be employed for large-scale computational screening of ILs and other solvents for the separation of mixers with optimized process designs.

References:

[1] Randall, M. and Webb, W.A., 1939. Separation process. *Industrial & Engineering Chemistry, 31*(2), pp.227-230.

[2] Morais, A.R.C., Harders, A.N., Baca, K.R., Olsen, G.M., Befort, B.J., Dowling, A.W., Maginn, E.J. and Shiflett, M.B., 2020. Phase Equilibria, Diffusivities, and Equation of State Modeling of HFC-32 and HFC-125 in Imidazolium-Based Ionic Liquids for the Separation of R-410A. *Industrial & Engineering Chemistry Research*, *59*(40), pp.18222-18235.

[3] Shiflett, M.B. and Maginn, E.J., 2017. The solubility of gases in ionic liquids. *AIChE Journal*, *63*(11), pp. 4722-4737.

[4] Shiflett, M.B. and Yokozeki, A., 2006. Solubility and diffusivity of hydrofluorocarbons in

room-temperature ionic liquids. *AIChE Journal*, 52(3), pp.1205-1219.

[5] Lei, Z., Dai, C., Zhu, J. and Chen, B., 2014. Extractive distillation with ionic liquids: a review. *AIChE Journal*, *60*(9), pp.3312-3329.

[6] Demirel, S.E., Li, J. and Hasan, M.M.F., 2017. Systematic process intensification using building blocks. *Computers & Chemical Engineering*, *105*, pp.2-38.

[7] Demirel, S.E., Li, J. and Hasan, M.M.F., 2019. A General Framework for Process Synthesis, Integration, and Intensification. *Industrial & Engineering Chemistry Research*, *58*(15), pp.5950–5967.

[8] Demirel, S.E., Li, J., El-Halwagi, M.M. and Hasan, M.M.F., 2020. Sustainable Process Intensification using Building Blocks. *ACS Sustainable Chemistry & Engineering*, *8*(48), pp.17664–17679.

[9] Shiflett, M.B. and Yokozeki, A., 2006. Separation of difluoromethane and pentafluoroethane by extractive distillation using ionic liquid. *Chimica oggi*, *24*(2), pp.28-30.

Copyright © American Institute of Chemical Engineers. All rights reserved.

631872 The Effect of Alkyl Chain Length on Imidazolium-Based Ionic Liquids and Choice of Halide Anions (Cl, Br, I) on Phase Equilibria for HFC-32, HFC-125, and Binary Mixtures (R-410A)

Abstract:

The current generation of refrigerants, hydrofluorocarbons (HFCs), are being phased out due to their high global warming potential (GWP) while the next generation of more environmentally friendly refrigerants, hydrofluoroolefins (HFOs), are developed and commercialized. Consequently, millions of tons of HFC mixtures will need to be recycled. Project EARTH (Environmentally Applied Research Toward Hydrofluorocarbons) is focused on developing efficient and selective separation processes for HFC mixtures. One promising approach is to use ionic liquids (ILs) as an entrainer to selectively separate azeotropic HFC mixtures like R-410A, which is composed of 50 wt.% HFC-125 (CHF_2CF_3) and 50 wt.% HFC-32 (CH_2F_2). The HFC-32 can be blended with HFOs to produce new products such as R-454B (68.9 wt.% HFC-32 and 31.1 wt.% HFO-1234yf (CH_2=$CFCF_3$)) and HFC-125 can be repurposed as a high-value fluorinated feedstock for production of new materials with a low-GWP.

This presentation will compare the effects of alkyl chain length on imidazolium-based ILs and anion substitution (Cl, Br, I) on the phase equilibria data for HFC-32 and HFC-125 and mixtures (e.g., R-410A). This presentation will also discuss ILs that are both solid and liquid at 25 °C. Phase equilibria data for HFC-32 and HFC-125 were measured using a Hiden Isochema IGA gravimetric balance and mixtures of HFC-32 and HFC-125 were measured using the first Hiden Isochema XEMIS Integral Mass Balance (IMB) method. The impact of alkyl chain length and anion on diffusion and selectivity will also be discussed for both pure HFC+IL and binary HFC+IL mixture measurements.

Copyright © American Institute of Chemical Engineers. All rights reserved.

631738 Liquid-Liquid Equilibria of Ternary Systems of Thermally Robust Ionic Liquids and Hydrocarbon Mixtures

Abstract:

Hydrocarbon (HC) separations is an energy intensive process due to similar boiling points and formation of azeotropes. Here, we test different novel thermally robust perarylphosphonium and perarylsulfonium ionic liquids (ILs) for separating aromatic compounds from HC mixtures. Molecular dynamics (MD) simulations are used to provide a molecular-level understanding of these systems. Simulations of binary mixtures of toluene/ILs and *n*-heptane/ILs are performed to validate and study the miscibility behavior. Ternary mixtures consisting of IL/toluene/*n*-heptane at different concentrations of toluene in *n*-heptane are carried out to determine the amount of toluene absorbed into IL phase. Further, the fundamental interactions between aromatics and IL leading to this behavior are assessed through radial distribution functions and nearest neighbor analysis. Selectivity and distribution coefficients are determined to evaluate the affinity of toluene between IL and HC phase. The free energy of solvation ($\Delta G_{solvation}$) and transfer free energy ($\Delta G_{transfer}$) of toluene in different solvents are calculated using thermodynamic integration (TI) method. The results show that $\Delta G_{solvation}$ of toluene is more favorable in all the ILs compared to *n*-heptane. Overall the simulation results provide a molecular-level understanding of the solvent environment provided by these ILs in separating aromatics from hydrocarbon mixtures.

Copyright © American Institute of Chemical Engineers. All rights reserved.

629325 The Deterministic Role of Hydrogen Bonding on Water Diffusivity in Methylimidazolium Halide Ionic Liquids and Ionogels

Abstract:

Polymeric ionogels, crosslinked gels swollen by ionic liquids (ILs), are useful vehicles for the release and storage of molecular solutes in next generation separation and delivery applications. Although rapid solute diffusion is often critical for performance, it remains challenging to predict diffusivities across multi-dimensional composition spaces. Recently, we showed that water (a model neutral solute and ubiquitous contaminant) diffuses through alkyl-methylimidazolium halide ILs by hopping between hydrogen bonding sites on relatively immobile cations. Here, we expand on this activated hopping mechanism in two significant ways. First, we demonstrate that water diffuses through poly(ethylene glycol) diacrylate ionogels via the same mechanism at a reduced rate. Second, we hypothesize that the activation energy barrier can be determined from relatively simple ^{1}H NMR chemical shift measurements of the proton responsible for H-bonding. This relationship enables water's diffusivity in ionogels of this class to be predicted quantitatively, requiring only (1) the composition-dependent diffusivity and Arrhenius behavior of a single IL, and (2) ^{1}H NMR spectra of the ionogels of interest. High-throughput microfluidic Fabry-Perot interferometry measurements verify prediction accuracy across a broad formulation space (4 ILs, $0 \leq x_{H2O} \leq 0.7$, $0 \leq \varphi_{PEGDA} \leq 0.66$). The predictive model may expedite IL material screening. Moreover, it intimates a powerful connection between solute mobility and hydrogen bonding, and suggests targets for rational design.

Copyright © American Institute of Chemical Engineers. All rights reserved.

628524 Detection of Disinfection By-Products and Oil Spill Pollutants in Water Utilizing Colorimetric Reactions for a Portable, Inexpensive Nanofiber-Based Sensor

Abstract:

Aquatic ecosystems are a critical part of maintaining life on the planet. There are, however, many sources of contamination within the oceans—a large source of which being pollutants from oil spills and disinfection by-products (DBPs). Common classes of such pollutants are polycyclic aromatic hydrocarbons (PAHs) and trihalomethanes (THMs), many of which are suspected of causing cancer and are regulated by the Environmental Protection Agency (EPA). Here, a polymer-based sensor is used for the detection of these contaminants at concentrations similar to those in the environment and relevant to regulations. The sensor utilizes a nanofiber mat and a colorimetric reaction chemistry. To produce the nanofibers, electrospinning was employed. To provide a theoretical connection, computer modeling of the system was done with finite element analysis. This on-site and novel sensor provides an alternative to the traditional, expensive analytical equipment (HPLC, GC-MS), and allows for enhanced water monitoring of carcinogenic pollutants.

Copyright © American Institute of Chemical Engineers. All rights reserved.

630402 Investigating the Sensitivity of CO_2 Reduction Costs

Abstract:

The problem of global warming has induced a spreading awareness about the need to reduce the production of greenhouse gases, especially CO_2. The emissions can be reduced through implementing various options such as enhancing efficiency of existing processes, implementing renewable energy, and applying CO_2 capture, utilization, and sequestration (CCUS). The implementation of such projects is governed by various social and economic factors, and the cost of CO_2 reduction has been considered a key parameter for assessing the options. The costs may vary significantly depending on the applied technologies, their performance, and other factors such as the demand and accessibility to resources. Consequently, many studies surveying such pathways report the costs within wide ranges due to the varying assumptions. This work studies the effect of such factors on the cost of different CO_2 reduction pathways and presents a comparison between the different options based on their economic and environmental performances. The focus is on the impact of power options, resource availability, and demand for power on the cost and CO_2 reduction potential of renewable energy and CCUS options. The integrated systems associated with the different scenarios analyzed are represented on Minimum Marginal Abatement Cost (Mini-MAC) curves. Mini-MAC is a recently developed analysis tool that allows the illustration of key insights related to the cost of CO_2 reduction while considering the integration between the different components of the system. The results identify the future opportunities of economic CO_2 reduction to assist planners and policy makers when designing CO_2 reduction pathways.

Copyright © American Institute of Chemical Engineers. All rights reserved.

624489 Planktonic and Biofilm Cultures of *Shewanella Oneidensis* MR-1 to Remediate Iodate Contamination

Abstract:

Radioiodine-129 (I^{129}) from leaking nuclear waste storage tanks is one of the major contaminants of groundwater at the Hanford Site in Washington State. Speciation of I^{129} in the contaminated groundwater is predominantly as iodate, organo-iodide and iodide.

Shewanella oneidensis MR-1 is a strain of bacteria known to reduce silver and uranium and has been validated for its potential to serve as an ex situ bioremediation agent against iodate. A proposed bioremediation solution is to use microbial dissimilatory reduction of iodate to iodide. This study hypothesizes that planktonic and biofilm cultures of *S. oneidensis* can effectively reduce iodate to chemical species that are more readily remediated through chemical or biological approaches. For the purpose of this study, *S. oneidensis* planktonic cultures were grown in minimal media (GM9) in the presence and absence of iodate under both, aerobic and anaerobic conditions. Iodate concentrations were monitored by colorimetric assays with UV Spectrophotometry. Data suggests that planktonic cultures of *S. oneidensis* can reduce iodate by 28% under aerobic conditions, and no reduction occurs under anaerobic conditions.

Further assays are underway to analyze the effect of biofilm cultures of *S. oneidensis* on the bioremediation of iodate. Biofilm cultures will be grown under static conditions in GM9 in the presence or absence of iodate, and iodate concentrations will be monitored by colorimetric assays.

Success of this project will result in the validation of using planktonic and/or biofilm cultures of *S. oneidensis* for possible contamination cleanup of nuclear waste. This may lead to the development and design of novel microbial bioreactors for the bioremediation of groundwater, soils and sediments.

Funder Acknowledgement(s): This study was supported by an NSF DEB Award#1832140 to Dr. Patrick Ymele-Leki.

Copyright © American Institute of Chemical Engineers. All rights reserved.

632717 Game Theory Applications to Pollution Trading

Abstract:

Pollution trading, also known as a Cap & Trade strategy, is both a market-based and an environmental strategy which adds flexibility to the pollution abatement decision making. This strategy aims to satisfy environmental regulations at a cost lower than that required by treatment technology implementations. The trading is based on the idea that some of the pollutant sources (firms or organizations) have enough financial capacity and infrastructure to reduce their emissions below the required limits through treatment technology implementation; hence, if the pollutant concentration of their discharges is below the allowable limits, such organizations obtain credits, which can be sold. Simultaneously, organizations which are not able reduce its emissions due to the lack of infrastructure can meet environmental regulations by purchasing credits.

This work is concerned with pollution trading in watersheds. Several mathematical programming approaches have been developed to optimize the decision making in the pollution trading strategy. In general, such approaches intend to minimize the global cost of technology implementation (the summation of technology implementation costs for every pollutant source). However, these modeling approaches are not realistic, since most of the firms are not concerned with the overall costs and benefits of the trading; they are concerned with their own costs and benefits. Therefore, the trading is really a market strategy where every pollutant source can cooperate or compete with any other source. That is, the approach should be represented as a mathematical game, where every source could be a player.

This work proposes optimization models that help pollutant sources to make optimal decisions in a polluting trading strategy. The mathematical models are formulated based on game theory. In particular, two approached have been analyzed. The first model considers a bilevel Stackelberg game; the second formulation involves a Nash equilibrium game represented as a Mixed Complementary Problem (MCP) with equilibrium constraints. Both of the formulations are solved through the GAMS modeling environment. The results include the decisions for each source of pollutants in order to satisfy global environmental regulations. To evaluate the performance of the optimization models, a mercury trading problem is considered. In particular, this presentation will discuss recent findings with respect to the impact on the solution of the numbers of leaders and followers used in the game, as well as the impact of grouping the pollutant sources in terms of the geographical location or the amount of polllutant discharges.

Copyright © American Institute of Chemical Engineers. All rights reserved.

627707 Responsive Porous Materials for Selective SO2 Capture from Flue Gas

Abstract:

Sulfur dioxide (SO_2) is harmful to human and environment, mainly produced by burning low-grade fossil fuels. The efficient removal of SO_2, especially for flue gas desulfurization (FGD) is of great significance and has attracted immense attention. Currently, the FGD processes are mainly accomplished by wet scrubbing using limestone slurry, but the method is energy-intensive and incapable to realize efficient removal of trace SO_2 (about 2000 ppm). In addition, the similar molecular size of SO_2 to other gases in flue gas, like N_2, CO_2, makes the SO_2 capture from flue gas be a great challenge. Herein, we report the discriminatively responsive behavior of SO_2 to other gases (e.g. CO_2, N_2) in the material with designed flexible pore structure. The material excludes the adsorption of N_2 and CH_4, and exhibits high SO_2/CO_2 (10/90) IAST selectivity, about 62.3 under ambient conditions. The binding domains for adsorbed SO_2 molecules and the binding energy are determined by single crystal X-ray diffraction experiments along with first-principles density functional theory calculations. The high-density host-guest interactions and guest-guest interactions between SO_2 molecules are responsible for the sensitive response of the material towards trace SO_2 (2000 ppm), and its SO_2 capture ability is further demonstrated by the breakthrough experiments. Moreover, the material is cheap and easy to scale up, making it be promising for trace SO_2 removal.

Copyright © American Institute of Chemical Engineers. All rights reserved.

633017 Oxidative Stress Response in Human Pulmonary Cells Following Titanium Dioxide Particulate Exposure

Abstract:

Fine particulate matter (PM2.5) is a major health concern, impacting the respiratory system through impaired lung function, infection, cancer, and contributing to 4.2 million deaths globally in 2016 [1]. Exposure to particulate matter causes oxidative stress through the direct introduction of exogenous ROS and compounds that drive free radical reactions, or indirectly through the recruitment and activation of inflammatory cells which release free radicals. Manufactured nanoparticles contribute to PM2.5 and are a cause for concern of occupational exposure. One such component, titanium dioxide nanoparticles are widely used in industrial and consumer products including paints, plastics, pharmaceuticals, cosmetics, and food products. TiO_2 nanoparticle toxicity has been studied extensively in *in vitro* and *in vivo* dermal models and acute pulmonary exposures, however, there is a need for further characterization of oxidative stress response in respiratory cells following acute and chronic exposure.

The goal of this work was to characterize oxidative stress response of human pulmonary cell line A549 following isolated and repeated exposures to titanium dioxide nanoparticles in a dynamic *in vitro* model. Size and charge of TiO_2 nanoparticles were characterized using DLS, zeta potential, and SEM prior to exposure. Post-exposure investigation included cell viability by MTT assay and physiological response parameters including IL-8 secretion, apoptosis execution determined by caspase 3 activation by western blotting, ratio of reduced and oxidized glutathione quantified by HPLC, and immunohistochemical characterization of surfactant protein expression.

[1] https://www.who.int/gho/phe/outdoor_air_pollution/en/

Copyright © American Institute of Chemical Engineers. All rights reserved.

627292 Electrocoagulation-Electroflotation for Primary Treatment of Animal Rendering Wastewater to Enable Recovery of Fats

Abstract:

Rendering wastewater (RW) contains high levels of fats collected from material processed during rendering. Primary treatment of RW to remove floatable fats is necessary for to avoid issues in downstream treatment. Recovery of waste fats that can be sent back to the rendering process can be of value to the rendering enterprise. Dissolved air flotation (DAF) is commonly used for primary treatment of RW but requires polymer coagulants that must be minimized to keep their concentrations low in the recovered fat stream. Oxidation of fats caused by dissolved air is another undesirable effect of DAF. In this work, RW was treated by electrocoagulation-electroflotation (EC-EF) in a novel horizontally-placed electrode pair design as an alternative to DAF. EC-EF performed better in COD removal with increasing current densities. In situ flotation within the electrochemical cell was able to concentrate and preserve long chain fatty acids (LCFA). Treatment at higher current densities within the electrochemical cell generated surplus cation release and allowed further coagulation in a subsequent step. Longer mixing duration in the post-electrochemical treatment step resulted in increased overall COD removal. The overall COD removal for EC-EF with applied current densities of 2 mA/cm^2 reached $69.49 \pm 4.83\%$ after 15 minutes of mixing at 100 rpm followed by 15 minutes of sedimentation. EC-EF thus not only provides competitive treatment performance for RW, but also recovery of well-preserved rendering products.

Copyright © American Institute of Chemical Engineers. All rights reserved.

626400 Optimization of Waste Plastics to Fuel Oil Plants' Deployment Using Mixed Integer Programming. (MIP)

Abstract:

Mixed Integer Programming (MIP) is an approach that involves the optimization of a range of decision variables in order to minimize or maximize a particular objective function. The main objective of this study was to apply the MIP approach to optimize the deployment of waste plastic to fuel oil processing plants in Uganda. The processing plants are meant to reduce plastic pollution by pyrolyzing the waste plastic into a cleaner fuel that can be used to power diesel/ paraffin engines, so as (1) to reduce the negative environmental impacts associated with plastic pollution and also (2) to curb down the energy gap by utilizing the fuel oil. A programming model was established and tested in two case study applications that are, small-scale applications in rural towns and large-scale deployment across major cities in the country. In order to design the supply chain, optimal decisions on the types of waste plastic to be processed, size, location and number of plants, and downstream fuel applications were concurrently made based on the payback period, investor requirements for capital cost and production cost of fuel and electricity.

The model comprises qualitative data gathered from waste plastic pickers at landfills and potential investors, and quantitative data obtained from primary research. It was found out from the study that a distributed system is suitable for small rural towns whereas a decentralized system is only suitable for big cities. Small towns of Kalagi, Mukono, Ishaka, and Jinja were found to be the ideal locations for the deployment of distributed processing systems whereas Kampala, Mbarara, and Gulu cities were found to be the ideal locations initially utilize the decentralized pyrolysis technology system. We conclude that the model findings will be most important to investors, engineers, plant developers, and municipalities interested in waste plastic to fuel processing in Uganda and elsewhere in developing economies.

Copyright © American Institute of Chemical Engineers. All rights reserved.

631121 Triboelectric Separation of Plastic Waste: How to Increase the Selectivity?

Abstract:

The triboelectric separation is a promising method in plastic waste recycling, as it is a mechanical method to separate different kinds or grades of grinded plastic waste, which can be utilized in the following recycling processes. This method consists of three main steps. In the first step, plastic waste is grinded into small pieces. In the second step, the charge is generated on their surface due to tribocharging and in the third step, the pieces are separated during the free fall between electrodes in the electrostatic separator. The charge is 'generated' on the surface of the pieces during their contact and friction and after their subsequent separation. The amount and direction of the transferred charge is generally determined by the properties of both contacted materials, especially by the work function, which is well-defined only for conductive materials. In the setup of the triboelectric separator, two contacting groups of material are involved, the first is represented by the grinded plastic waste mixture and the second group is represented by the charging adaptors.

Although the triboelectric separation is a promising method with the potential to become an important member of the recycling chain for a number of reasons (low costs, automatization potential, possible separation of black materials...), there are several challenges, which need to be addressed to ensure the high selectivity and yield of this method. Especially challenging is the selection of an adaptor, as it plays a key role in the controlling of the charge obtained by individual plastic pieces. Currently, the tribocharging of dielectrics is not well understood yet even to simply determine a suitable counter-material, although some theoretical or empirical concepts are corresponding well with experimental results. Thus, there is not sufficient knowledge to predict the charging behavior of such similar materials as polymers, when distinguishing between their grades (for example, HDPE and LDPE), and that is why most of the work needs to be done experimentally. Furthermore, the nature of the materials is not the only parameter influencing the direction and magnitude of the transferred charge, but a role plays also the type of contact counter-material, the properties of the surrounding atmosphere as well as the dimensions and the shape of the grinded pieces.

We use self-constructed charging apparatus based on different contacting approaches: contact during rotational motion, enhanced by vibrations or contact during sliding motion. To evaluate the charge on particles, we use two different methods: we can evaluate the total charge in a sample in a Faraday pail connected to the electrometer (used for experiments with pure materials, not mixtures), or we are using our self-constructed free-fall separator to evaluate the charge distribution in the sample. After the separation, plastic pieces are collected in boxes based on their trajectory in the electric field and the charge of the fraction of pieces in individual boxes is measured in the Faraday pail. Then, the representation of different plastic materials in collectors is evaluated.

Our work is focused on improving electrostatic separation by finding proper counter materials to effectively separate multiple-component plastic mixtures by separating one material per cycle and recharging the remaining mixture with different counter material during each cycle. We compared the most promising tribocharging mechanisms for automatization (rotational and vibrational), and we combined these methods with a corona discharge to decrease the time needed to achieve sufficient, i.e., saturation charge on individual pieces. We performed a study addressing the charging dependent on material composition with 8 adaptors with different composition and 9 waste plastic materials, and we showed that the magnitude and, in some cases, even polarity can be controlled by the choice of the adaptor and the mode of charging. Additionally, we implemented an external electric field during the vibrational motion of plastic materials in the tribocharging apparatus to fine-tune the transferred charge, to enhance the sensitivity of separation and to decrease the requirements on the variability of the used adaptors during the whole process. These results can be used as guidance to choose a suitable surface for the efficient separation of various plastic mixtures. Then, we used these results to perform the separation of 3-component mixtures, aiming for the successful separation of halogenated material – PVC from the mixture, as PVC is inconvenient for incineration. We began with a one-stage separation with very good purity of the two materials but in a low yield, and then we introduced the repeating cycle in two different ways. One is to reach higher purity by separating only one material in one stage and another one aims to increase the yield. Finally, we reached the purity of separated PVC up to 91 % at 88% recovery, the purity of PET up to 95% with 97% recovery, and PP with 93% purity and 70% recovery.

Copyright © American Institute of Chemical Engineers. All rights reserved.

ADSORPTIVE REMOVAL OF LEAD FROM AQUEOUS SOLUTION USING BIOCHAR

Ahnaf Hossain, Auburn High School, Auburn, AL
Shamim Begum, Tuskegee University, Tuskegee, AL
Zhihua Jiang, Auburn University, Auburn, AL

Biography

Ahnaf Hossain is an IB Diploma Program Candidate. He is an Auburn High School Ambassador and also Auburn Senior Class Vice President.
Shamim Ara Begum, Ph.D., P.E. is an Associate Professor in Chemical Engineering Department at Tuskegee University. Begum's research focuses on physicochemical treatment processes for water and wastewater, and also biological methods for wastewater treatment.
Zhihua Jiang, Ph.D., is an Auburn Pulp and Paper Foundation Assistant Professor and also Director of the Alabama Center for Paper and Bioresource Engineering. His research focuses on bioresource engineering, added-value products from lignin and nanocellulose, pulping and bleaching, pulp and paper, and wood chemistry.

ABSTRACT

This study focuses on the removal of lead from aqueous solution by batch adsorption experiments using biochar derived from wood chips (spruce, pine, and fir) at 700 °C. Adsorption characteristics of lead onto biochar were evaluated as a function of biochar dosages using batch adsorption experiments. At the end of each batch tests, samples were collected and analyzed to determine the residual lead concentration using an Inductively Coupled Plasma Mass Spectrometry. Results demonstrated that the percentages of lead removal increased with the increase of biochar dosages. The percentages of lead removal were 14.53%, 38.10%, 82.07% and 99.10% with biochar dosages of 1 g/L, 2 g/L, 4 g/L and 6 g/L, respectively for an initial lead concentration of 100 mg/L in 24 hours. The pH values of the solution were not controlled during the tests. It was observed that the pH values were 4.6, 4.8, 5.9 and 6.7 for 14.53%, 38.10%, 82.07% and 99.10% of lead removal, respectively at the end of the test. Lead adsorption capacity of biochar was about 17 mg/g. This study indicated that biochar can remove lead from aqueous solution very effectively. Further experiments are in progress to investigate the optimum parameters (dose of biochar, initial lead concentration, pH and contact time), adsorption kinetics, adsorption isotherm, and removal mechanisms of lead onto biochar.

INTRODUCTION

Heavy metal pollution in aqueous solution is a worldwide problem because of its toxicity and non-biodegradable properties. Lead is one of the major heavy metal contaminants, which can cause disastrous consequences on human health, such as serious blood disease, infant diarrhea, influenza, and scarlet fever, as well as impair brain development and interfere with neurotransmitter systems [1, 2]. In addition, lead is harmful to the environment with its toxic characteristics. Lead can enter into aqueous solution from process industries like mining, smelting, printing, metal plating, dying and explosives manufacturing. Allowable limit of lead in potable drinking water is up to 0.015 mg/L according to the United States Environmental Protection Agency [3]. Due to stringent environmental regulations of potable water, it is imperative to develop cost-effective and environmental-friendly technologies for removing lead from aqueous solution.

Trying to find ways to reduce lead contamination to purify water, it was found that an environmental-friendly method was adsorption using biochar. Adsorption is one of the most cost-effective methods to remove heavy metals from aqueous media. In recent years, biochar as an adsorbent has received significant attention because of its distinctive characteristics [4]. These include high adsorption capacity and specific surface area, microporosity, and ion exchange capacity [5, 6]. Biochar is a charcoal-like substance, which is made from a process known as pyrolysis of organic material seen in Figure 1. In the process of pyrolysis, organic materials are essentially burned in an enclosed container with little amount of oxygen to reduce the amount of fumes released [7]. Different types of waste biomass have been used to produce biochar, which can be used as an adsorbent to remove contaminants from water [8]. Adsorption capacities of biochars depend on its physical and chemical properties on the basis of type and pyrolysis temperature of feedstock materials [9, 10].

Figure 1: Transformation of organic material such as wood chips, into powdered biochar

The main process of the investigation is the adsorption process in which the biochar derived from wood chips (spruce, pine, and fir) at 700 °C, the adsorbent, will bind with the lead in the contaminated water with the use of molecular attractions. According to Helmenstine [11], adsorption is when particles or molecules bind to the surface of the adsorbent material, while the absorption process involves the filling of pores or holes in the entire volume of the absorbent material.

In the investigation, Lead chloride and biochar derived from wood chips were used for the experiment. While the amount of biochar was varied to evalute its effect on the amount of lead removed, the initial amount of lead concentration of solution, time, and initial pH of the solution were constant. Along with that, the data were processed by calculating the percent removal of lead from the water and the adsorption capacity of the biochar. It was expected that the percent removal of lead will have a direct relationship with the amount of biochar used in the experiment because as the amount of biochar increases, the surface area increases as well, demonstrating that more lead should be adsorbed on the surface of the biochar during the experiment.

MATERIALS AND METHODS

Preparation and characterization of biochar

The biochar used in this study was purchased from Phoenix Energy, San Francisco, California, USA. The analytical specifications of the bio-char were provided by the company as carbon content, ash, bulk density, total surface area and particle size of 90.8%, 3.9%, 0.11 g/cm^3, 186 m^2/g and 19-150 mm, respectively. It is a carbonaceous granular material prepared by pyrolysis of wood chips at 700^{0}C. The biochar was crushed and sieved to get the particle size between 53 and 250 μm. The surface area and pore volume of the reduced size of biochar were determined using a NOVA 1200 surface area analyzer (Quantachrome Instruments, version 2.0, Boynton Beach, FL) with the Bruanuer-Emmett-Teller nitrogen adsorption method at 77°K.

Adsorption procedure

Analytical grade reagents were used during batch experiments. The batch adsorption experiment was conducted to identify the effects of bio-char dosage on the removal of lead using biochar with an initial lead concentration of 100 mg/L and a solution volume of 50 mL. Lead Chloride salt was used to prepare 100 mg/L lead solution. The biochar dosages used were 0g (control), 0.05g, 0.10g, 0.20g, and 0.30g to obtain its concentrations of 0 g/L, 1 g/L, 2 g/L, 4 g/L and 6 g/L, respectively. The pH of the solution containing lead and biochar was not controlled. The solution containing lead and biochar was placed onto the platform shaker (New Brunswick Scientific, model: innova 2300) at 150 rpm at room temperature and allow the adsorption process to take place for 24 hours. Then, about 5 mL samples were collected and filtered through a syringe filter to remove the biochar. The filtered solution was used to determine the residual lead concentration of each sample using the Inductively Coupled Plasma Mass Spectrometry (ICP-MS). The ICP-MS directly gave the final lead concentration in mg/L. There were no calculations in finding the final lead concentration. Removal efficiency and adsorption capacity of lead were obtained using equations 1 and 2, respectively [12].

$$R = \frac{(C_0 - C_f)}{C_0} x\ 100 \quad (1)$$

$$q_t = \left(\frac{C_0 - C_t}{m} \right) V \quad (2)$$

where, C_0, C_f and C_t are the initial, final and at time t solution-phase concentration (mg/L) of lead, respectively. m and V are the mass of biochar (g) and volume of the solution (L), respectively. The pH values of the solutions before and after adding biochar were monitored using a pH meter. All the biochar that were used had the same adsorption capacity.

All batch tests were performed in triplicate. The experimental blanks were run in parallel. All working solutions were prepared freshly prior to use in the batch tests. The glassware were presoaked in 5% HNO_3 solution overnight and then washed with tap water, rinsed three times with deionized water and dried in air before use.

RESULTS AND DISCUSSIONS

Textural properties of bio-char

After reducing the particle size between 53 and 250 μm, the surface area (BET), pore volume, and average pore diameter were found as 285 m^2/g, 0.1507 cm^3/g and 2.117 nm, respectively. The surface area of the bio-char increased by over 50% from a value of 186 m^2/g to 285 m^2/g due to the size reduction as expected.

Effects of biochar dosages on the Adsorption Process

To evaluate the effects of biochar dosages, 0g (control), 0.05g, 0.10g, 0.20g, and 0.30g of biochar dosages were used separately during the tests. All the tests were conducted with an initial solution pH of 4.5 with 100 mg/L of lead concentration before the addition of biochar. However, pH values of the solutions were 4.6, 4.8, 5.9 and 6.7 after the addition of 0.05g, 0.10g, 0.20g, and 0.30g of biochar, respectively.

Different amounts of biochar have distinct effects in the removal of lead from the water sample, and it has been shown throughout Tables 1-5. It can be seen that the more biochar used in the lead sample, the higher the lead percentage that is removed from the sample. Throughout the investigation, the highest recorded % lead removal was 99.10 using 0.30 grams (i.e. 6 g/L) of biochar. Lead adsorption capacity of biochar was about 17 mg/g at 99.10% removal of lead. Cairns et al., [13] observed that the maximum adsorption capacity of lead was 7.8 ± 0.1 mg/g using biochar derived from European larch (Larix decidua Karst.) wood chips. However, Amin et. al., [12] obtained that the maximum adsorption capacity of lead was 98.9 mg/g using date palm derived bichar dosage of 10 g/L. The variation of results may be due to the various types of biochar and experimental conditions.

The uncertainty value for the percent removal of lead for the value shown in the Tables 1 to 5 was calculated like so:

$(0.08 + 0.08) = 0.16 \div 17.40 \times 100 = 0.92\%$ (Percent Uncertainty of Numerator)
$0.08 \div 100.00 \times 100 = 0.08\%$ (Percent Uncertainty of Denominator)
$= 0.92\% + 0.08\% = 1.00\%$ (Percent Uncertainty of Value)
$= 1.00 \div 100 \times 17.40 = 0.17$ (Absolute Uncertainty of Value)

This calculation was done for all different amounts of biochar and is seen in the Percent Lead Removal column in all tables.

Table 1: 0 grams of Biochar vs. Percent Removal of Lead

Mass of Biochar, g ±0.0001	**Initial Lead Concentration, C_0, mg/L ±0.08**	**Final Lead Concentration, C_f, mg/L ±0.08**	**Volume of Solution, mL ±0.08**	**% Lead Removal ±0.00**	**Average % Lead Removal**
0.0000	100.00	100.00	50.00	0.00	0.00
0.0000	100.00	100.00	49.89	0.00	
0.0000	100.00	100.00	49.99	0.00	

Table 2: 0.05 grams of Biochar vs. Percent Removal of Lead

Mass of Biochar, g ±0.0001	**Initial Lead Concentration, C_0, mg/L ±0.08**	**Final Lead Concentration, C_f, mg/L ±0.08**	**Volume of Solution, mL ±0.08**	**% Lead Removal ±0.17**	**Average % Lead Removal**
0.0517	100.00	82.60	49.96	17.40	14.53
0.0508	100.00	86.60	50.00	13.40	
0.0513	100.00	87.20	49.89	12.80	

Table 3: 0.10 grams of Biochar vs. Percent Removal of Lead

Mass of Biochar, g ±0.0001	**Initial Lead Concentration, C_0, mg/L ±0.08**	**Final Lead Concentration, C_f, mg/L ±0.08**	**Volume of Solution, mL ±0.08**	**% Lead Removal ±0.19**	**Average % Lead Removal**
0.1019	100.00	62.60	49.96	37.40	38.10
0.1012	100.00	59.90	49.93	40.10	
0.1027	100.00	63.20	49.98	36.80	

Table 4: 0.20 grams of Biochar vs. Percent Removal of Lead

Mass of Biochar, g ±0.0001	**Initial Lead Concentration, C_0, mg/L ±0.08**	**Final Lead Concentration, C_f, mg/L ±0.08**	**Volume of Solution, mL ±0.08**	**% Lead Removal ±0.22**	**Average % Lead Removal**
0.2032	100.00	26.50	49.96	73.50	82.07
0.2046	100.00	12.70	50.00	87.30	
0.2024	100.00	14.60	49.98	85.40	

Table 5: 0.30 grams of Biochar vs. Percent Lead Removal

Mass of Biochar, g ±0.0001	**Initial Lead Concentration, C_0, mg/L ±0.08**	**Final Lead Concentration, C_f, mg/L ±0.08**	**Volume of Solution, mL ±0.08**	**% Lead Removal ±0.24**	**Average % Lead Removal**
0.3035	100.00	0.20	50.00	99.80	99.10
0.3021	100.00	1.50	49.99	98.50	
0.3029	100.00	1.00	49.95	99.00	

Additionally, seen in Figure 2, as the mass of biochar was increased in the sample, the average percent removal was seen to increase proportionally as well demonstrating the effectiveness of using biochar to remove lead as other significant variables were controlled to avoid confounding variables. This signifies the importance of using increased amount of biochar as numbers of functional groups and active adsorption sites will be increased due to the presence of greater surface area [14].

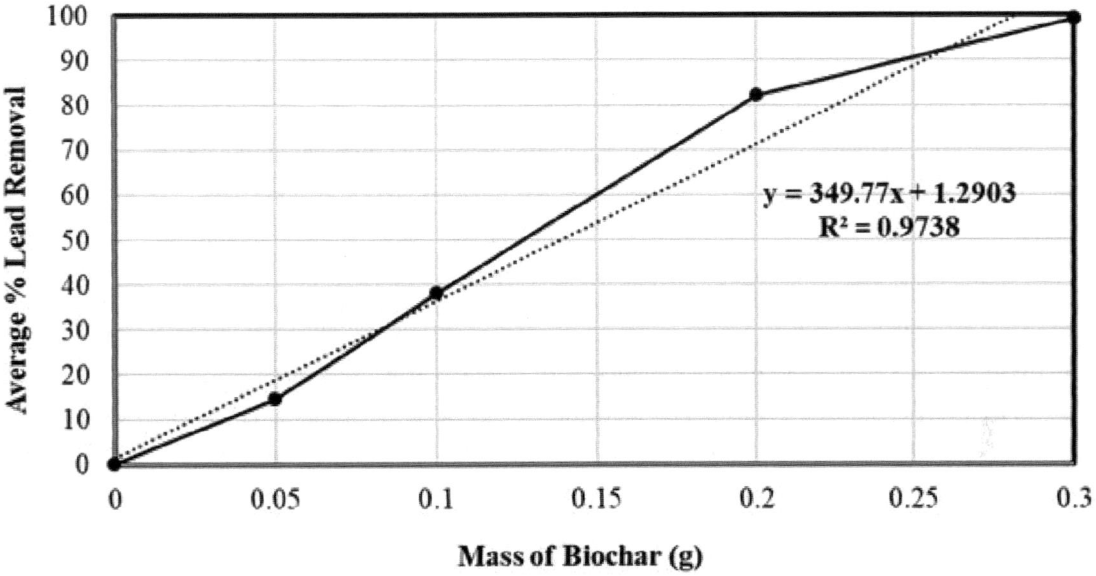

Figure 2: Mass of biochar vs average % lead removal

CONCLUSIONS

After completing the investigation, many conclusions were determined based on the results in the experiment. It was found that the implementation of biochar into a lead-based water sample led to a distinct adsorption amount of lead from the water sample. It was found that as more biochar was used to remove lead, the greater the percent lead removal was from the water sample. This is true due to the constant adsorption capacity of biochar. Since the biochar had the same adsorption capacity, the greater quantity of biochar added into the sample, the greater the removal of lead because there is more surface area for the adsorption process to occur. This was proven from the investigation as the lowest percent removal was seen when using 0 grams of biochar (0% lead removal), while the greatest percent lead removal was seen when using 0.30 grams of biochar (99.10% lead removal). It can be concluded that based on the $R^2 = 0.9738$ there is a linear relationship and proportion between the amount of biochar used and the percent lead removal from the water sample.

EVALUATION

In the conclusion, the main type of error seen in the investigation was random error. There was no systematic error as all the data that were produced yielded accurate data values. However, the data was not as precise as expected. The limitation in the current materials and human error led to sporadic values in measuring accurate volumes of water samples leading to different concentration values. Even so, the random error was minimalized by completing three different trials for each mass of biochar to find an accurate average result.

Additionally, the wood chip-derived biochar may not have been grinded to a complete powdered state, which may have resulted in lower percent lead removal values. Besides human error in measuring volumes of water and grams of lead chloride, there was not much error found within the materials themselves to support systematic error. Seen in Figure 2 above, it can be determined that there was a fluctuation in the values and resulted in a deviation of values from the regression line. This further demonstrates the idea that random error existed throughout the experiment.

For future study, this investigation could be extended to using other forms of biochar to figure out the most optimal form of biochar to adsorb the most lead from the water sample. It can also digress into the effects that biochar has on the water and what changes occur to the pH value of the water. The main hope was to purify contaminated water, but this investigation only covered a portion of the study. The investigation might continue to see what other methods are required to completely filtrate the water sample for human consumption. It also may be taken into consideration if the amount of time affects the total lead removal from the water sample. Even though 0.30 grams of biochar yielded the best results, the investigation could have even greater amounts of biochar to evaluate if there is a limit to how much biochar can be added until the biochar becomes ineffective.

REFERENCES

1. Chen, L. F., Liang, H. W., Lu, Y., Cui, C. H., and Yu, S. H. (2011), "Synthesis of an attapulgite clay@ carbon nanocomposite adsorbent by a hydrothermal carbonization process and their application in the removal of toxic metal ions from water," Langmuir, 27(14), pp 8998-9004.
2. Luconi, M. O., Silva, M. F., Olsina, R. A., and Fernández, L. P. (2000), "Cloud point extraction of lead in saliva via use of nonionic PONPE 7.5 without added chelating agents," Talanta, 51(1), pp 123-129.
3. Ji, M., Hook, G., and Fumiyo, F., (2019). In the Advances in Empirical Translation Studies: Developing Translation Resources and Technologies, Ji, M., Oakes, M., ed, Cambridge University Press, Cambridge, pp. 77-93.
4. Mohan, D., Rajput, S., Singh, V. K., Steele, P. H., and Pittman, C. U., (2011), " Modeling and Evaluation of Chromium Remediation from Water Using Low Cost Bio-char, A Green Sorbent," Journal of Hazardous Materials, 188, pp 319-333.
5. Ahmad, M., Rajapaksha, A. U., Lim, J. E., Zhang, M., Bolan, N., Mohan, D. and Ok, Y. S. (2014), "Biochar as a sorbent for contaminant management in soil and water: a review," Chemosphere, 99, 19-33.
6. Mohanty, P., Nanda, S., Pant, K. K., Naik, S., Kozinski, J. A., and Dalai, A. K. (2013), "Evaluation of the physiochemical development of biochars obtained from pyrolysis of wheat straw, timothy grass and pinewood: effects of heating rate," Journal of analytical and applied pyrolysis, 104, pp 485-493.

7. Spears, S., “What is Biochar?,” Regeneration International, https://regenerationinternational.org/2018/05/16/what-is-biochar/ (accessed January 12, 2021).
8. Ippolito, J.A., Laird, D.A., and Busscher, W.J., (2012), “Environmental benefits of biochar,” Journal of Environmental Quality, 41, pp 967–972.
9. Mukome, F.N.D., Zhang, X., Silva, L.C.R, Six, J., and Parikh, S.J. (2013), “Use of chemical and physical characteristics to investigate trends in biochar feedstocks,” Journal of Agricultural and Food Chemistry, 61(9), pp 2196-2204.
10. Li, S., Barreto, V., Li, R., Chen, G., and Hsieh, Y.P. (2018), “Nitrogen retention of biochar derived from different feedstocks at variable pyrolysis temperatures,” Journal of Analytical and Applied Pyrolysis, 133, pp 136-146.
11. Helmenstine, A.M., (2020), "What Adsorption Means in Chemistry," ThoughtCo. https://www.thoughtco.com/definition-of-adsorption-605820 (accessed November 1, 2020).
12. Amin, M.T., Alazba, A.A. and Shafiq, M. (2019), “Application of biochar derived from date palm biomass for removal of lead and copper ions in a batch reactor: Kinetics and isotherm scrutiny,” Chemical Physics Letters, 722, pp 64-73.
13. Cairns, S., Chaudhuri, S., Sigmund, G., Robertson, I., Hawkins, N., Dunlop, T., and Hofmann, T. (2021), “Wood ash amended biochar for the removal of lead, copper, zinc and cadmium from aqueous solution,” Environmental Technology & Innovation, 24, pp 101961
14. Uzunoğlu, D., Gürel, N., Özkaya, N, and Özer, A., (2014), “The single batch biosorption of copper (II) ions on Sargassum acinarum,” Desalination and Water Treatment, 52, pp 1514–1523.

629185 A Simulation Study of Mineral Carbonation through Electrolysis of Rejected Brine

Abstract:

Greenhouse gas emissions owing to the relentless burning of fossil fuels have negatively impacted the atmospheric environment. Global warming has caused abnormal phenomena such as ocean acidification, rise in sea levels and extreme climate changes. Among the six major greenhouse gases: carbon dioxide (CO_2), methane(CH_4), nitrous oxide(N_2O), hydrofluorocarbons (HFCs), sulfur hexafluoride(SF_6), perfluorocarbons(PFCs), carbon dioxide has the lowest global warming potential but has the greatest contribution to global warming due to its excessive production.

To build resilience against climate change, Carbon Capture and Utilization (CCU) technology has been employed as a viable solution. Our study proposes the simultaneous treatment process for both wastewater and air pollutant using desalinated rejected brine and their performances are evaluated within the simulated data. The high content of magnesium and calcium ions allows rejected brine to become a stable feedstock in the application of CCU technology.

The electrolysis of desalinated seawater is done to produce sodium hydroxide and is later applied in the carbon dioxide capture and ion separation processes. Throughout the process, above 90% of CO_2 is captured with the alkali absorbent in a form of ionic carbon dioxide such as carbonate and bicarbonate. Metal cations present in seawater rejected brine are extracted via electrolysis and then utilized to regenerate the CO_2-loaded absorbent and produce metal carbonates. Simulations for electrolysis were run in MATLAB2020a while the carbonation process was modeled in ASPEN PLUS simulation tool.

The mineral carbonation process is developed in favor of high metal ion content in the desalinated rejected brine. Using the different solubilities of magnesium and calcium ions at different pH, sodium hydroxide produced via electrolysis is applied in the metal ion separation process via the pH swing method. While magnesium ions precipitate into magnesium hydroxides at pH 11, the calcium ions precipitate into calcium hydroxide at pH 13. The results showed that 99% of calcium and 60% of magnesium recovered from seawater were converted to calcium carbonate and magnesium carbonate. The metal ions are selectively separated in the form of metal hydroxides for a faster reaction rate and higher yield of the final product. While calcium undergoes spontaneous reaction, magnesium ions have a lower conversion yield due to kinetic hindrance present caused by the hydrophilicity of the magnesium ion.

The proposed design process connects our previous studies based on mineral carbonation of carbon dioxide using metal ions present in desalinated rejected brine into the commercialization of the conceptual design. Our research team has previously studied a series of works based on mineral carbonation using desalinated rejected brine to enhance its efficiency with metal ion separation and purification of the final product through speculation study of its morphology. The simulation study has helped in identifying the optimal operating conditions for further scale-up studies. It is expected that our designed process serves as a cornerstone for a step forward in the commercialization of mineral carbonation of carbon dioxide and contributes to mitigation in global warming issues for sustainable development.

Copyright © American Institute of Chemical Engineers. All rights reserved.

629909 Fuel Recovery from the Waste Plastic: An Environmentally Friendly Approach

Abstract:

In the present era of commercialization, industrialization and upgradation, plastic becomes a very important constituent of the day to day activity. Be it the automobile industry, be it the medical industry or be it the household utilization, plastic is utilized in a numerous activities and numerous ways. Due to the upgradation, it is a well known fact that there is a generation of more than 100 million tones of plastic waste globally on an annual basis.

The present status of the reuse of the plastic waste is not satisfactory. The most common use of the plastic waste is in the landfills but it is not as per the requirement. Thus, to utilize this waste to reduce the environmental impact is always a question for research. The research is also towards the development of the bio-degradable plastic. Another opportunity is the conversion of the waste plastic to the value added fuels.

The present work is focused on the utilization and conversion of the waste plastic to the value added fuel which can be helpful in the reduction of the environmental impact of the plastic waste. This method is helpful in dual way as troubles of waste plastic and problems of fuel scarcity are being tackled all together. In the present work, more focus is given to the utilization of Low Density Poly-Ethylene (LDPE) for the production of different fuels through pyrolysis process at about 300°C. The process focuses on the de-polymerization, pyrolysis, thermal cracking and distillation to attain diverse value added fuels such as petrol, kerosene and diesel, lube oil, etc.

Transform the waste plastics into fuel cling to an immense assurance mutually to the environmental and economic scenarios. Therefore, the course of converting plastics to fuel has now curved the problems into a prospect to build prosperity from waste.

Copyright © American Institute of Chemical Engineers. All rights reserved.

627691 Plastic Waste Recycling: A Parametric Study of Dissolution Recycling Method

Abstract:

One of the current environmental issues is products service life made of polymeric materials and their subsequent processing. The materials service life very often exceeds the time for which we designed them. Therefore, plastics recycling is becoming a major scientific problem and economically and ecologically advantageous process needs to be developed, at its end a recycled polymer comparable to a freshly produced polymer will appear.

At present, two basic methods predominate: (i) mechanical (physical), and (ii) chemical recycling. Neither one is universal nor has significant advantages. The mechanical method is problematic in processing of wide range of materials with different densities, colours, additives, etc. The chemical recycling based on polymer pyrolysis is often accompanied by process difficulties. An approach in between these two methods is plastics recycling by dissolution.

This alternative is based on a polymer dissolution into a suitable selective solvent to create a polymer solution. The solution is further refined to get rid of added pigments and additives. This approach requires finding suitable solvents for both for original material dissolution and for the selective extraction of additives. The toxicity, impact on the environment and cost shall be considered.

Polyolefins, mainly semicrystalline polyethylene (PE) and polypropylene (PP) are the polymers with the highest annual production. Therefore, our work focuses on PE and PP dissolution in organic solvents. We have performed a parametric study of various parameters: temperature (up to 200 °C), pressure (up to 10 MPa) and solvent type. The solvents were chosen according to the affinity of the solvent to the studied material.

The experiments have been conducted in a 100 cm^3 mixing vessel with a glass window for monitoring the visual changes in polymer dissolution. Subsequently, an extraction reagent was added to get rid of the potential additives in the dissolving material. Then the refined and dried polymer has been analysed by Differential Scanning Calorimetry (DSC) and Time-Domain Nuclear Magnetic Resonance (TD-NMR). We have studied the polymer crystallinity changes of the original (waste) and refined material. Promising preliminary results show good polymer characteristics recovery. The up-to-date challenge is the removal of the additives from the original material.

The understanding of interaction of the crystalline phase with the solvents will be based on our wide database of polyolefin thermodynamic data. Polymer crystallinity decreases with concentration of gaseous penetrants as measured by TD-NMR. And we observed by gel permeation chromatography (GPC) that short polymer chains are washed out in the presence of liquid penetrants that decrease polymer crystallinity. The obtained data broaden the understanding of diffusion and transport phenomena in semicrystalline polyolefins and allow the optimization of the dissolution process.

Copyright © American Institute of Chemical Engineers. All rights reserved.

623032 Electrospun Ca_2CuO_3 Fibrous Sorbent for Post-Combustion CO_2 Capture

Abstract:

CaO is the well-known sorbent for post-combustion CO_2 capture. High capacity, abundancy and low-cost makes this sorbent attractive. Despite the advantage, it suffers from sintering during carbonation and regeneration process at high temperature. Structural modification can be the key for the stable long-term operation of the sorbent. One of the effective modification methods is incorporation of an inert phase to hinder sintering. We synthesized Ca_2CuO_3 that spontaneously decomposes into CaO and copper under reducing condition. The bulk oxide powder was made by sol-gel synthesis from calcium nitrate, copper acetate, and citric acid complexant. The same strategy was used to make fibrous Ca_2CuO_3 by electrospinning with PVP as the polymeric structure directing agent. In all cases, 900 °C calcination in air removed all organic component and resulted in the targeted oxide phase. Both the bulk powder Ca_2CuO_3 and the fibrous Ca_2CuO_3 were decomposed into CaO and Cu by hydrogen reduction. The endogenous copper phase acts as a sintering barrier for CaO in the calcium looping process. The conversion kinetics and stability of these sorbents were investigated in the carbonation and regeneration process at high temperature. The results were compared to pure CaO of the two different physical forms. We observed higher stability and faster kinetics for the electrospun sample. The faster kinetics is due to high porosity and large specific area of electrospun nanofibers that gives them a unique potential for high catalytic activity. The presence of copper phase enhanced the stability of the sorbent by sintering suppression

Copyright © American Institute of Chemical Engineers. All rights reserved.

626672 Steady State Flow Model for Impure CO2 Transportation in Industrial Processes

Abstract:

Carbon capture, utilization and storage, or CCUS, has been recognized as an accepted technology in the United Nations Framework Convention on Climate Change (UNFCCC) to achieve the Paris agreement commitments [1]. Within the CO_2 capture and utilization trend, the flow must be transported. The presence of impurities in the CO_2 stream affects thermo-dynamical fluid properties (density, critical point, triple point, viscosity, J-T coefficient, ect.), causing corrosion and modifying the CO_2 behavior during operational activities such as pipe depressurizations [2].

In order to prevent these negative effects, this paper describes one-dimensional steady-state flow model for impure CO_2 implemented in a rigorous process simulator, CHEMCAD 7.0. Focusing on the thermodynamical model selected in the process simulator, a benchmarking exercise was developed comparing Peng-Robinson equation of state (PR EoS) and Soave-Redlich-Kwong (SRK EoS). The model was validated with experimental data that were obtained in a demo transport plant (TRL 7) with a total length of 3.000 m (10 coils), 2" diameter and operation conditions that can be set within a range of 10-31 ºC and 80-110 bar(a); the facility includes a doping system to modify the composition of the processed CO_2, being the impurities selected SO2, O2 and moisture with a maximum composition of 500 ppmv that could be classified as low contaminant levels and water content below the solubility content.

The validation of the model for both thermodynamic models was really closed. The experimental results were matched with a maximum difference of 6,6 mbarg per linear meter of pipe in the pressure drop. Regarding the pressure drop behaviour, it is well known that one of the main parameters is the velocity (or volume flow rate) for a given diameter: the higher the velocity, the higher the pressure drop. Nevertheless, velocity is not the final analysis, but the CO_2 contaminants and the composition have also a crucial role in the pressure drop. To study this effect, a set of tests were done with constant flow rate (in the range of 14.600 l/h), obtaining that for the case of CO2 with 100 ppmv of SO2, an important increase in the pressure drop is produced whereas for CO2 with 500 ppmv of O2, a diminution of pressure drop was obtained.

References

[1]| IEA- International Energy Agency, «Global Energy & CO2 Status Report,» [En línea]. Available: https://www.iea.org/geco/electricity/. [Último acceso: 30 06 2019]

[2] S. Tollak, M. Hammer y S. Lovseth, «CO2 transport: Data and models - A review,» *Applied Energy,* pp. 499-523, 2016.

Copyright © American Institute of Chemical Engineers. All rights reserved.

630610 CO_2 Capture Using Phase-Changing Bis-Iminoguanidines (BIGs) with Amino Acids

Abstract:

CO_2 capture using phase-changing bis-iminoguanidines (BIGs) with aqueous amino acid (sarcosine and glycine) is investigated in this study. This work seeks to understand the phase-changing bicarbonate crystallization of glyoxal-bis-iminoguanidine (GBIG). CO_2 capture processes involving BIGs are of major interest due to their comparatively lower regeneration energy requirements. Unlike in monoethanolamine (MEA), which does not change phase, only the insoluble bicarbonate precipitate undergoes thermal regeneration instead of the bulk solution, thus saving energy. A critical step in the proposed CO_2 capture mechanisms is the regeneration of the deprotonated amino acid. This regeneration was modelled and studied kinetically and thermodynamically through the crystallization of $GBIGH_2^{2+}(HCO_3^-)_2(H_2O)_2$. A theoretical model simulating glycine regeneration was developed to determine the rate limiting step of the system. A combined analysis using experimental data and modeling results indicates that the GBIG-bicarbonate crystallization step provides most of the thermodynamic driving force for the deprotonated glycine regeneration and that the protonation of GBIG prior to crystallization is the rate-limiting step. The CO_2 loading and amino acid regeneration steps were intensified into a single-step process using a bubble column reactor. The CO_2 loading capacity of the intensified GBIG process was experimentally determined at 1.36 mol CO_2 per mol GBIG. These results provide the fundamental basis for developing an effective carbon capture technology with phase-changing amino acid/guanidine absorbents that can be used to effectively capture both point-source emissions and atmospheric CO_2.

Copyright © American Institute of Chemical Engineers. All rights reserved.

630200 Resilience in the Face of COVID-19: Managing a User Program at the Molecular Foundry

Abstract:

Introduction

The Molecular Foundry is a Nanoscale Science Research Center, located at Lawrence Berkeley National Laboratory (Berkeley Lab) in Berkeley, California, sponsored by the Department of Energy (DOE). The Molecular Foundry is a knowledge-based User Facility, where leading-edge resources in instrumentation, analytical capabilities, and scientific expertise in nanoscale science are made available via a peer-reviewed proposal process. Roughly 1000 researchers or "users" come to the Foundry from around the world to conduct their work each year. The COVID-19 pandemic presented challenges and opportunities in conducting collaborative on site research. This presentation focuses on the operation of the user program amid the pandemic, with an emphasis on strategies for maintaining research productivity and user access. We also present an outlook on ways that gains in virtual collaboration and science during the pandemic can provide opportunities in a post-pandemic world.

The Molecular Foundry provides scientific expertise and capabilities distributed among seven facilities: Inorganic, Organic, and Biological facilities for synthesis, preparation, and assembly; Nanofabrication, for processing and integration; the National Center for Electron Microscopy and Imaging and Manipulation, for characterization; and Theory, for understanding and predicting material properties.

The Molecular Foundry maintains a highly collaborative research model. Staff scientists and technical staff at the Molecular Foundry interact with users through instrument training, consultation during experiments, support with data interpretation, and collaborative writing of journal articles. For each 12 month project, users stay in the local area for an average of three months while they work on site.

Spring 2020 Shelter-in-Place

In spring 2020, the scale and urgency of the COVID-19 pandemic in the United States became apparent, and the city of Berkeley enacted a Shelter-in-Place order effective March 17, 2020. The Molecular Foundry responded by pausing research operations on site while keeping research facilities safe and ready for reopening. This directly impacted every one of the hundreds of research projects that were active at the time, and presented a challenge to the Molecular Foundry's commitment to safety and user access.

The Foundry staff collaborated to respond to this challenge with a holistic approach to keep science moving forward safely. All user projects received a 12-week extension, to account for this pause on work. During this period, the Foundry developed protocols for safe work on site, aligned with requirements for Berkeley Lab and applicable regional, state, and national guidelines. These included the use of face coverings, hand-washing, distancing/density control, and surface disinfection.

While maintaining a very low density of people on site, a few high-priority projects were allowed to proceed, including research on COVID-19 and SARS-CoV-2 detection, treatment, and transmission mitigation. These early research activities served as a testing ground and model for the further expansion of access in the months to follow. Staff from across all seven facilities began to explore investment in virtual modes for training, user support, and instrument access amid the pandemic. Berkeley Lab developed new training modules for safe work on site during the COVID-19 pandemic, launched a lab density study to determine appropriate safe occupancy limits for all laboratories across the site, and developed a Health Pledge,symptom check system, and a specific COVID19 training module that was required for all on site personnel.

Summer 2020 and Beyond

The months that followed the onset of the pandemic were characterized by continual adaptation to the emerging pandemic. As of the writing of this abstract, this phase continues to evolve as vaccination campaigns are gaining momentum nationally. These are some of the strategies employed to manage the user program during this phase of the pandemic:

Communication

Communication with the user community and the staff of the Molecular Foundry was a central pillar of maintaining safe work at the Molecular Foundry. This included email messages to all users generally, frequent discussions among staff across all roles, and communication with Berkeley Laby leadership.

Shift Work to Manage Density

Upon emerging from the full shelter-in-place order, moderate numbers of researchers were allowed to work on site at the Molecular Foundry. The total number of persons accessing Berkeley Lab was managed centrally by the Laboratory, and the Molecular Foundry worked with an allotment of "slots" for site access throughout the duration of the pandemic restrictions. A three-shift model was deployed. Slots were distributed among the seven facilities, and researchers requested a slot ahead of time, subject to approval by Molecular Foundry leadership and compliance with the Health Pledge, weekly symptom check, and a specific COVID19 training requirements.

Prioritizing Research Projects

Facing limited access compared to normal operations, it was necessary to consider carefully which staff and user work could be accomplished on site. The Molecular Foundry values equity of access for users, held in balance with safety precautions. We aimed to carefully consider such factors as the potential to impact the fight against COVID-19, DOE and Berkeley Lab research priorities, inherently safe activities, experimental complexity, travel considerations, and user training.

Badged Entry for Approved Work

During the pandemic, the Molecular Foundry implemented badge-reader entry for all buildings. Coupled with weekly calendars and daily posting of all persons with prior approval to enter the buildings, this provided a layer of accountability around access to the spaces.

Regular Safety Meetings

Berkeley Lab implemented weekly virtual meetings focused on COVID-19 safe operations and other safety and space planning topics. These meetings require attendance by at least one staff representative from each of the seven facilities, and are open to all Molecular Foundry staff and users. Each facility hosted a daily safety check-in for work planning.

Carefully Considered Training

Training presented a particular challenge during the pandemic, as many users receive training on complex tasks at the Molecular Foundry, traditionally accomplished side-by-side. After several months, the Foundry formalized guidelines to consider for new user training, considering such factors as lab occupancy, the potential to develop tools to support distanced training, inherent hazards in the proposed work, and robustness of the instruments.

Development of New Remote and Virtual Training and Access

Molecular Foundry staff rallied around the call to support user access virtually. Specific responses are adapted to the needs and opportunities of each facility. A sampling of tools include: training videos, holographic augmented reality for training, audio headsets to support distancing, remote operation of select equipment, remote login for data processing, virtual presence via video conferencing during experiments, telepresence robots, and a variety of informal communication and collaboration tools.

Calls for Proposals Scope

Two of the standard calls for proposals have been fully within the pandemic. While we kept the calls open for the most exciting research across nanoscience, we balanced the practical considerations with some emphasis for these calls. We encouraged users to submit proposals that included COVID-19 research, theoretical and computational work, remote access to instruments, or data collection for remote analysis, and proposals requesting samples of materials developed at the Foundry.

Expanded Definition of Remote Users

The Molecular Foundry is one of five Nanoscience Research Centers sponsored by the DOE, and therefore shares certain reporting parameters for our user programs. Effective for FY 2020 reporting, the centers collectively adopted an expanded definition of a remote user, to recognize the increasing role of remote access for science at these knowledge-based user facilities.

Virtual Meetings and Conferences

Meetings have shifted online, supported by zoom video conferencing. This includes smaller team meetings, larger seminars spanning the Molecular Foundry, and our annual User Meeting--a nanoscale science research conference which drew 900 registrants in 2020.

Outcomes and Future Outlook

The pandemic presented challenges to maintaining a strong nanoscience research user program. Nevertheless, overall metrics reflect resilience in the face of these challenges. The User Program remains strong, serving 643 on-site and 97 remote users in FY 2020, according to formal reporting definitions, with a total of over 1500 named researchers on active projects during FY 2020. Demand for the user program remains strong, with around 225 standard proposals received during the most recent spring 2021 call for proposals, roughly two thirds our pre-COVID maximum. Research productivity as measured by journal articles has remained robust when comparing 2020 to 2019 and 2018.

The shift to increased virtual access presents opportunities to expand the geography of our user base to include more researchers outside our local area, due to reduced barriers in travel once post-pandemic travel resumes. We have seen an increase in sample-only proposals, enabling new ways of working together across institutions. We also have potential to include more people in the active experimental data collection phase, as undergraduate researchers and principal investigators increasingly log in remotely to participate in experiments that previously were completed by just one member of the research team. Furthermore, shifting our annual User Meeting online has facilitated a tripling of conference registrations, with potential to reach new communities with our science.

As we look ahead to 2022 with optimism, we hope to learn from the unique challenges presented by the pandemic. We will have new facets of this experience to manage, including a safe return to work for our users. The Molecular Foundry remains committed to advancing user science to bring solutions to the world.

Copyright © American Institute of Chemical Engineers. All rights reserved.

629703 Global Supply Chain Optimization for COVID-19 Vaccines Under Covax Facility

Abstract:

With the pandemic in action, over 130 million people across 219 countries have been infected by COVID-19. The number of mortalities due to COVID-19 has climbed to around 3 million [1]. Though around 80% of the infected patients have recovered, several patients are becoming reinfected recently appeared. As a result, efforts for COVID-19 vaccines have accelerated around the world. The world's top pharmaceutical companies and research organizations have developed multiple vaccines. It is expected that the transmission can be slowed considerably if a sufficient population is vaccinated. Though a typical vaccine takes around ten years for approval, several COVID-19 vaccines have already been developed and approved in record times. As of 31 March 2021, around 750 million people worldwide have been vaccinated [2]. With world governments aggressively supporting vaccine developments and procurements, the vaccination drive is progressing rapidly.

However, to end the pandemic, we not only need COVID-19 vaccines, but the development of vaccines must be followed by a fair, fast, and efficient global distribution. The current challenge is the timely delivery of the vaccines. The expected volume of the vaccines per month adds up to around 850 tons and is more than the UNICEF suppliers' capacity combined [3]. Two main concerns regarding the global COVID-19 vaccine distribution are 1) Limited supply of resources for the delivery of the vaccines, e.g., cold chain equipment, and 2) Vaccine nationalism, i.e., governments signing agreements with companies to procure surplus vaccines. To overcome these challenges, the COVAX facility was launched by WHO for multiple vaccines to prevent vaccine nationalism. It aims to ensure fair access to COVID-19 vaccines for all countries, irrespective of their financial status. It allows countries to pool resources for effective vaccine development and distribution. Under COVAX, deals for vaccine procurement have already been made. However, vaccine distribution is still under development. It faces various challenges due to (1) the variety of vaccines, (2) person-specific doses, 2) unknown country-specific demands required for herd immunity threshold, especially low-middle income countries, (5) low-temperature storage protocols for most vaccines, and (6) unanticipated disruptions in both production and distribution.

Addressing these challenges requires a multi-prong approach that must go beyond straightforward pharmacological measures. The two primary goals are to (1) estimate the herd immunity threshold and (2) effect an efficient global vaccine distribution. In this work, we explore the time-varying nature of reproduction number during the first wave of the

COVID-19 pandemic using confirmed cases, recovered cases, fatalities, and timings of lockdown phases. We propose a mean-field epidemiological model for COVID-19 that extends the classical SEIR model [4]. We describe NEIASCFR, an 8-compartmental model with variable transmissibility over time, by considering the time-varying nature of parameter values over different lockdown phases. We estimate our model's parameters with public data on confirmed cases, recovered cases, and fatalities for each country. The parameter values are used to determine reproduction number and herd immunity threshold. The model is applied to study the herd immunity threshold of 9 countries (India, Nigeria, Iran, Pakistan, Thailand, Mexico, Democratic Republic of Congo, Egypt, Thailand), constituting 50% of demand under the COVAX facility.

Using this within the framework of COVAX, we develop a global supply chain model to optimize the production and allocation of vaccines to participating countries from various production centers. The supply chain was optimized considering potential disruptions, vaccine approval dynamics, and vaccine candidates' approval with different risk mitigation strategies. Given: 1) Network configuration consisting of production facilities and different regions. 2) Initial vaccine procurement quantities. 3) Capacity of different storage containers. 4) Costs of procurement, storage, transportation, and risk mitigation per unit. 5) Different scenarios, we determine 1) Production quantities of vaccines at different centers and vaccine allocation to each demand region. 2) Extra procurement of vaccines at each center. 3) Cost of procurement and transportation.

References:

1. Worldometer. COVID-19 CORONAVIRUS PANDEMIC. 2021; Available from: https://www.worldometers.info/coronavirus/.

2. Bloomberg. Covid-19 Tracker. 2021; Available from: https://www.bloomberg.com/graphics/covid-vaccine-tracker-global-distribution/.

3. UNICEF, UNICEF outlining plans to transport up to 850 tonnes of COVID-19 vaccines per month on behalf of COVAX. 2020.

4. Li, M.Y. and J.S. Muldowney, Global stability for the SEIR model in epidemiology. Mathematical biosciences, 1995. 125(2): p. 155-164.

Copyright © American Institute of Chemical Engineers. All rights reserved.

632079 Sustainability and Resiliency Assessment for Industrial Synergies between Renewable Energy and Chemical Production

Abstract:

Photovoltaic (PV) is the fastest growing energy technology and is expected to maintain this growth in the coming years. Due to the durability of PV modules, there is a lengthy time gap between installation and decommissioning of the modules, and it is expected to create large scale waste worldwide in the near future. On the other hand, chemical manufacturing systems are undergoing significant transition towards bio-based production to meet the goal of sustainable low carbon feedstock. This work presents a potential synergy between renewable energy wastes and bio-based polymer production through extending a previously proposed system to circularize the PV lifecycle by diverting organic byproducts from PV recycling to agricultural production. We identify and introduce additional synergies as links in the network to close the loop between PV module disposal and manufacturing. The chemical outputs from the PV recycling process are used as fertilizers for crops that are potential feedstock for biopolymer production. These biopolymers are used in the manufacturing of PV backsheets, replacing the current petroleum-based products. We quantify the total life cycle environmental and economic benefits at scale of implementing this synergy along with total reduction in GHG emissions. To evaluate the design, we also use a network-based approach to understand the structure and topology of the proposed system by exploring various network measures as well as information theory-based measures, to assess resilience and long-term sustainability. We show quantitatively that network measures like diameter and density are ineffective at scoring the network resilience due to the nature of these measures in original network science literature, while centrality measures are useful for finding the relative importance of certain nodes, which may highlight network vulnerabilities that factor into the overall resilience. Likewise, ascendency and reserve values for the network appear useful in determining the overall resilience and sustainability of the network, however, based on their application to comparing two alternative network scenarios, these metric interpretations should be more rigorously defined for industrial networks before being incorporated for industrial network design.

Copyright © American Institute of Chemical Engineers. All rights reserved.

633441 A Community-Level Energy Planning Model for the Dynamic Optimization of Waste-to-Gas and Waste-to-Power Anaerobic Digestion Supply Chains.

Abstract:

Society has growing concerns over the long-term environmental impacts of climate change and the accumulation of organic waste in local communities. Community energy planning is increasingly focused on the use of novel technologies to provide sustainable and cost-effective heat and power. Furthermore, changes in energy demand patterns support the development of dynamic optimization models of energy dispatch. This study focuses on the use of anaerobic digestion (AD) technology for converting human, agricultural, and industrial waste into renewable natural gas (RNG) and power. We develop a dynamic mixed integer linear programming (MILP) in Calliope to optimize the production and dispatch of RNG and power.
Anaerobic Digestion (AD) is a microbiological process that is used in processing organic, biodegradable wastes and side streams into methane-rich biogas and nutrient-rich digestate. Previous techno-economic (TEA) and lifecycle (LCA) analyses indicate that AD can be a commercially competitive and sustainable technology for communities to generate clean energy. In this study, we evaluate the use of AD in Ames, Iowa to convert human and animal waste, biomass, and wastewater into RNG and power. RNG and power are consumed locally to provide industrial, commercial, and residential heat and power demand. The AD process generates by-product digestate that is returned to Iowa farms. The model covers simulates 1 year of energy demand in hourly and monthly increments.

The MILP identified 13 locations in the vicinity of Ames, Iowa capable of producing power or RNG. The electricity price ranged between \$0.065 - \$0.18/kWh and the RNG price is estimated at \$0.026/kWh (\$7.62/mmbtu). Electricity production varied between 30 and 50 million kWh per month. Gas production varied between 15 and 40 million kWh per month. All the electricity demand in Ames, Iowa could be supplied from local waste resources. Summer gas demand can be supplied locally. However, the gas supply will need to be supplemented with gas from the external grid to meet winter demand. Technological improvements and public incentives could increase the adoption of AD in community energy planning systems. Further research will evaluate the impacts of variabilities in energy supply and demand to address grid disruptions.

Copyright © American Institute of Chemical Engineers. All rights reserved.

625575 Asset Integrity Management (AIM) for Hydrogen: Major Incident Risk Reduction By Integrating a Risk-Based Safety Approach

Abstract:

Asset Integrity Management (AIM) is a well-understood process for optimizing the management of assets over their whole life-cycle. If applied correctly, AIM can offer asset owners and operators the ability to manage risk and enable the assertion, with confidence on an evidential basis, that their assets are safe and reliable. For AIM to be most effective, the assets' design must consider the whole life-cycle of the assets' operation. Adopting a risk based approach to safety and operations as early as possible in the design phase of a facility is demonstrated to improve long-term reliability, availability and maintainability. Combining a risk based approach at the design stage with AIM during operations has the potential to simplify and align processes. Additional benefits include the improvement of long-term reliability, availability and maintainability. By integrating a risk based approach with AIM the realization of common goals can be facilitated: the reduction of major accident hazard risks and consistent performance of assets. For a sector on which many emergent national decarbonization policies are reliant upon to deliver "net zero", an all-encompassing risk based safety and AIM approach offers asset owners and operators in the beginnings of the hydrogen economy boom a means by which to deliver business objectives profitably without major incident.

Copyright © American Institute of Chemical Engineers. All rights reserved.

625565 More Haste, Less Speed: Proportionate Supply Chain Risk Management for the Accelerating Hydrogen Economy

Abstract:

The long awaited acceleration of the hydrogen economy is dawning. It is viewed as a necessity in the race to decarbonize industry, heat and transport, amongst other key sectors. Existing hydrogen economy supply chains serving the current markets face the challenge of ramping up to meet future demand. Aside from hydrogen growth, many organizations are increasingly concerned about supply chain risk, not only the traditional supplier issues of late delivery and poor quality but also other risks with security, environmental, social and ethical impacts. Furthermore, hydrogen as an energy vector requires significant safety management to mitigate its potential to cause major accidents. The confluence of increasing demand, major accident hazards associated with hydrogen and concerns of supply chain risks present a significant challenge for those organizations in the hydrogen supply chain. When viewed through the lens of rapid demand acceleration, supply chain risks are magnified by the potential for short-cuts and risk taking. Such risks are exacerbated by the potential set-back in the global decarbonization goal, should a major accident be realized. Formal Supply Chain Risk Management (SCRM) techniques adopting a risk based approach are explored to help mitigate supply chain risks in a proportionate way, reducing supply chain vulnerability, ensuring continuity if risks occur and increasing value within the supply chain through improved performance. Crucial attributes of successful SCRM are considered to be: application of good governance, early supplier engagement, proportionate assurance levels based on risk, supply chain mapping, monitoring of risk events/risk event pre-cursors and risk register maintenance.

Copyright © American Institute of Chemical Engineers. All rights reserved.

HEAT OF ABSORPTION OF CO_2 IN NOVEL SOLVENTS FOR BIOGAS UPGRADING

Sai Hema Bhavya Vinjarapu[1*], *Philip Loldrup Fosbøl*[1]
[1] *Technical University of Denmark, Lyngby, Denmark*

Introduction

The increasing levels of CO_2 in the atmosphere have proven to be a cause for serious concern. Energy productions of all forms contribute to 72% of greenhouse gas emissions [1]. A majority of the world's energy needs are still met by fossil fuels. Renewable energy sources are expected to account for 34% of all energy sources by 2050. Even in this scenario, the carbon emissions are projected to be 22% higher than required, to limit the global temperature rise to 2°C [1]. Therefore, there is an increasing urgency in switching to carbon-neutral sources while ensuring their cost-efficiency and economic viability.

One such potential alternative is Biogas. Biogas is produced by anaerobic digestion of organic waste from wastewater sludge, industrial wastes, and agricultural wastes. The composition of biogas depends on the raw organic material used. However, it generally comprises 60-70% Methane, 30-40% CO_2, and minor impurities of water vapour and H_2S [1]. Owing to the high CO_2 concentration, biogas has a low calorific value, making it a less appealing replacement to energy-dense fossil fuels. To increase its energy density, this CO_2 is removed. This process of biogas purification is called biogas upgrading and the technology used for achieving this goal is carbon capture.

Of the numerous carbon capture technologies available, chemical absorption by primary amines is state-of-the-art. Amines like Monoethanolamine (MEA), which are basic, react reversibly with acidic CO_2 gas. The process involves an absorber where CO_2 from the biogas dissolves in MEA, releasing pure bio-methane. The CO_2-rich solvent then enters a desorber where it is heated to regenerate MEA and a pure CO_2 stream. The regenerated MEA is recycled to the absorber and the pure CO_2 obtained is either sequestered or utilized as a raw material. MEA has a high absorption capacity of CO_2 and a high reaction rate, which make it a suitable solvent.

Conventionally, the CO_2-rich solvent is heated to 105°C by lean-amine before being pumped to the top of the desorber, which is at 110°C. The reboiler at the bottom of the column is at 120°C. A high-energy requirement is the main limitation of this technology. The heat required for the solvent's regeneration is given as the sum of three terms [2]:

$$Q=Q_{sens}+Q_{vap,H2O}+Q_{abs} \qquad (1)$$

where, Q_{sens} is the sensible heat required to raise the temperature of the solvent to that of the reboiler's, $Q_{vap,\,H2O}$ is the heat of evaporation required to produce the stripping steam in the reboiler, and $Q_{abs,\,CO2}$ is the heat of absorption of CO_2 into the solvent. The equivalent amount of heat released by the exothermic absorption of CO_2 should be supplied to desorb it. The energy required for regeneration accounts for 70-80% of the whole process. Additionally, the high-temperature conditions result in degradation of the solvent leading to corrosion of the equipment. All of these result in significant operating costs. These disadvantages inhibit the wide-scale implementation of the process.

To optimize the regeneration of the solvent, new additives, known as vapour reduction additives (VRAs) are being investigated. Typically, a 30 wt% aqueous MEA solvent is used for the capture process. The high water content implies a high heat of vaporization, increasing the heat of regeneration and solvent losses. The VRAs are designed to replace water, thereby reducing the solvent's vapour pressure. This abates the amount of water evaporated along the column, thereby decreasing the energy requirements at the condenser situated at the top of the desorber. Further, the size requirements of the desorber are expected to decrease resulting in a decline in both capital and operating costs.

The heat required for regeneration of the solvent, as mention in equation (1), is dependent on the heat of absorption of CO_2 in the solvent. It is, therefore, important to determine the heat of absorption of CO_2 in the presence of the new additives and establish that they do not significantly increase the $Q_{abs,\ CO2}$. Experimentally determining the heat of absorption for these solvents is also expected to aid in the development of accurate thermodynamic models like the Extended UNIQUAC model for representing the new solvent systems.

Experimental setup and operating conditions

The heat of absorption of CO_2 in aqueous solutions of MEA and VRAs was measured using a flow mixing-cell adapted to a Setaram C-80 Calorimeter. The experiments were conducted at 40°C and 120°C, which is the temperature range of the carbon capture process. The pressure in the system was maintained at 250 psi by means of a backpressure regulator. The flow rate of the solvent was varied in the range of 0.02-0.3 ml/min while the flow of CO_2 was maintained constantly, either at 0.4 or 0.6 g/h. The inlet pressure of CO_2 was maintained in the range of 32-35 bar. However, the inlet pressure of CO_2 does not have any impact on the measurements. Experiments were done for loadings in the range of 0.1-0.5 mol CO_2/mol MEA. The loading of the solvent was determined by the ratio of the molar flow rates of CO_2 and MEA, given as:

$$\alpha = n_{CO2}/n_{MEA} \qquad (2)$$

Both the fluids were pumped into the system by use of ISCO pumps. The molar flow rate of the solvent was determined from the volumetric flow rate, density, and concentration of the solvent. The density of the solvent was measured at 298K and atmospheric pressure. The flow of CO_2 was controlled by the use of a Coriolis mass flow controller. The molar flow rate of CO_2 was determined directly from this mass flow rate, eliminating the need for density estimation. Experiments were conducted for 30wt% MEA with and without the VRAs.

The system consists of three sub-units: the pre-heater, the mixing cell, and the calorimetric block. The pre-heater heats the fluids to the measurement temperature before they enter the mixing cell. The mixing cell consists of two parallel tubes for the flow of the solvent and CO_2. The tubes are coiled together at the lower end, where the mixing of the two fluids takes place. The outlet tube leaves the calorimeter and the pre-heater parallel to the inlet tubes. There are two mixing cells, one used for the mixing of the fluids, and the other as a reference cell.

The mixing cells are situated in a calorimetric block. The heat emitted by the absorption of CO_2 in the MEA solvent is transmitted to the calorimetric block through the wall of the vessel in an isothermal mode and is detected by thermopiles.

Results and Discussion

Validation of the results

The experimental results were validated by determining the heat of absorption of CO_2in 30 wt% MEA at 40 and 120°C and comparing them to data found in the literature. The results for 40°C are presented in Figure 1. From Figure 1', it can be observed that the experimental data are fairly close to the data found in Kim et al [3]. However, a maximum in the heat if absorption is observed in our work. Theoretically, the heat of absorption is expected to be a constant value until saturation. This increasing trend can be considered as an experimental error as it lies within the uncertainty range of ±3%. The difference between the experimental values and the data from literature was also observed to lie with a range of ±5%. The validation of the method was also carried out for 30 wt% MEA at 120°C.

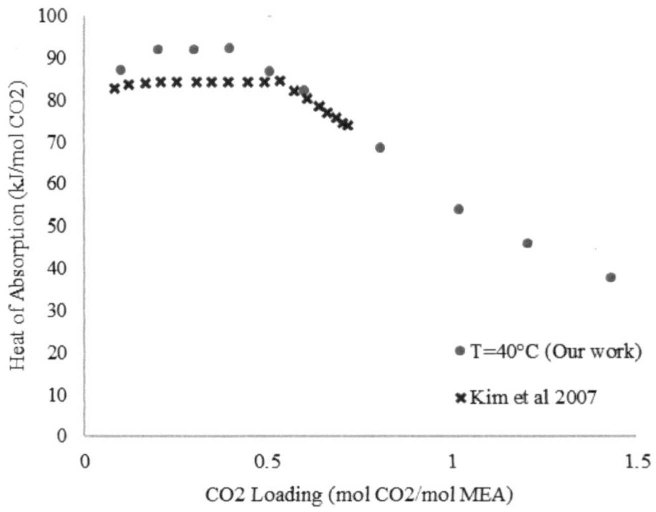

Figure 1: ΔH_{abs} of CO_2 in 30 wt% MEA at 40°C

Results for 30 wt% MEA and VRA1

Figure 2 represents the heat of absorption of CO_2 in 30wt% MEA and 7.5, 30 wt% VRA1 at 40°C. It can be observed that the heat of absorption in the presence of VRA1 is slightly higher than the base case scenario. The heat of absorption in the presence of 7.5 wt% VRA1 is higher by 4.4% than 30 wt% MEA and 30 wt% VRA1 is higher by 8.3%. Owing to the presence of uncertainty, these values are not significantly higher than 30 wt% to impact the $Q_{abs,CO2}$ for the regeneration of the solvent. Experiments were also carried out for the remaining operating conditions.

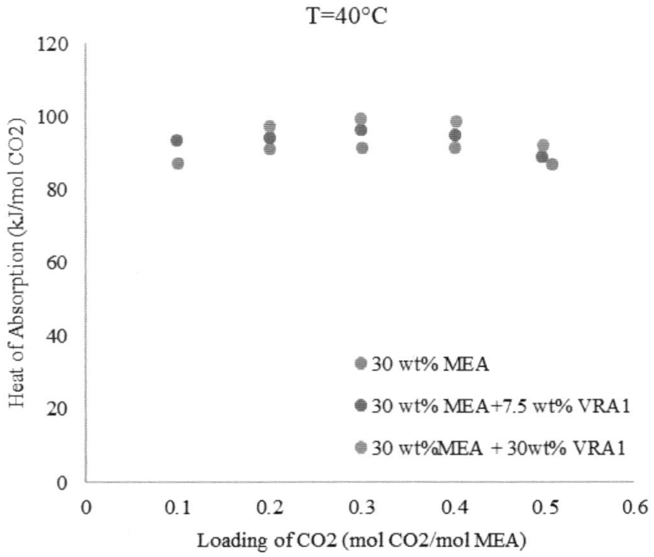

Figure 2: ΔH_{abs} of CO_2 in 30 wt% MEA+VRA1 at 40°C

Conclusion

Energy reduction of the solvent regeneration process plays a crucial role in the extensive establishment of biogas upgrading plants. The new solvent technology has the potential to meet these requirements. The vapour reduction additives are capable of reducing the energy requirements. Through this work, it has been established that the heat of absorption of CO_2 in MEA based solvents is not significantly higher in the presence of these additives. As a result, these additives can aid in reducing the energy consumed for solvent regeneration.

Acknowledgment

This project is co-funded by the Danish Government through the EUDP (Energy Technology Development Program) Agency.

References

1. Global Energy Perspective 2019. (n.d.). Retrieved July 23, 2019, from https://www.mckinsey.com/industries/oil-and-gas/our-insights/global-energy-perspective-2019
2. Oexmann, J., & Kather, A. (2010). Minimising the regeneration heat duty of post-combustion CO2 capture by wet chemical absorption: The misguided focus on low heat of absorption solvents. International Journal of Greenhouse Gas Control, 4 (1), 36-43.

628617 Design of Carbon Neutral Industrial Clusters over a Time Horizon

Abstract:

Climate change concerns have led to a diverse range of efforts to ultimately mitigate the rising carbon dioxide emissions and reduce fossil fuel dependence. Advances taken in this direction comprise of different emission reduction pathways, which include improvement of process efficiency, adoption of carbon capture, utilization and storage systems, implementation of renewable energy options, and more recently, development of negative emission technologies and carbon neutral fuels. Policies are frequently coupled with these advances to bring about a systematic change towards sustainable development using carbon reduction targets. Industrial ecology and symbiosis are evolving sustainable concepts that aim to tackle emission footprints and resource conservation within the system of interest. Industrial networks, which adopt both industrial ecology and symbiosis, allow for the simultaneous incorporation of these cleaner reduction pathways to achieve carbon reduction targets and other economic, environmental, and social objectives. The fundamentals of process integration, circular economy, and economic diversification are integral to these networks. The concurrent integration of multiple resources to design these multi-objective networks considers overall balances, capital and operating costs, and resource prices. However, the operation of these networks varies with fluctuations over time, some of which have the potential to alter their design. Improvements in existing technologies, emerging technologies, network expansions, and changing market conditions, are some forms of these fluctuations that can cause variations over time. Furthermore, policies are designed to achieve a goal by a future date, making the development of planning strategies for their realization critical. To accommodate and plan for these variations, it is therefore essential to incorporate multi-period analysis into the network design. The approach introduced in this work allows develops a carbon neutral industrial network that considers fluctuations over time. A mixed-integer linear programming model will determine the optimal network configurations that can meet set objectives over multiple time periods through optimization. The multi-period model will also assess the network's economic and environmental benefits to allow stakeholders to evaluate and implement the designs. A case study will demonstrate the approach developed using a carbon reduction policy and assess the design obtained in terms of economic and environmental indicators.

Copyright © American Institute of Chemical Engineers. All rights reserved.

633170 Multiple Greenhouse Gases Mitigation (MGM): Process Intensification to Mitigate Non-CO_2 Gases and CO_2 from Air

Abstract:

Background and Motivation

The Paris Agreement is an important milestone that brought together 196 countries to agree upon climate targets through emission reduction. Although the main focus has been on mitigating CO_2 in the IPCC's 1.5 °C report [1], mitigating non-CO_2 greenhouse gases like CH_4 and N_2O will have a significantly higher climate impact. Majority of these non-CO_2 GHGs are from the agricultural and farming sector, accounting for 39% and 72% of CH_4 and N_2O emissions worldwide in 2017 [2, 3]. One of the main challenges with these emissions is that they are very diverse and dilute. However, the confidence gained through development of Direct Air Capture (DAC) technology for CO_2 capture from air [4] presents opportunities to mitigate CH_4 and N_2O emissions. The scope of this paper is to provide an insight into an intensified process that can mitigate multiple GHGs (CH_4, N_2O and CO_2) from air in a single system. The proposed process concept is termed as Multiple Greenhouse Gas Mitigation (MGM) and can accelerate achieving negative emissions. The envisioned process technology can be first applied to mitigate GHGs from air closer to animal ventilation stables, that contains CH_4 (2-300 ppm).

Two types of approaches have been studied to mitigate CH_4 and/or N_2O from air ***i)*** gas separation, capture and regeneration, and ***ii)*** catalytic conversion/decomposition. In the first approach, CH_4 and N_2O are generally adsorbed over porous zeolites and metal-organic frameworks (MOFs) [5, 6]. However, these processes have challenges with presence of CO_2 and moisture in air [7]. The second approach is to convert CH_4 and N_2O over a catalyst surface (thermal or photocatalyst). Thermal catalyst can convert CH_4 and N_2O only at temperatures above 300 °C [8-10]. It is worth noting that the studies didn't include results for simultaneous CH_4 and N_2O mitigation. However, photocatalyst have potential to convert both CH_4 and N_2O at room temperatures [11, 12]. Based on this concept, a solar chimney power plant (SCPP) [11] was proposed that can convert CH_4 and N_2O while generating electricity from solar panels. However, there is still a gap in science with respect to mitigating all the three major GHGs (CH_4, N_2O, CO_2) in one single system.

This paper presents a first-of-its-kind process intensification concept to mitigate all the three major GHGs (CH_4, N_2O, CO_2) from air. A simple schematic of the process is shown in Figure 1. Photocatalytic surface is used to convert the CH_4 into CO_2 and H_2O, and N_2O into N_2 and O_2, followed by DAC of CO_2. The captured CO_2 can be then compressed, transported and utilized or stored. The proposed concept can advance the development of DAC technologies and can have greater impact by mitigating CH_4 and N_2O that have higher global warming potential (28 and 265 respectively, following fifth assessment report AR5). In this paper, we propose the intensified process and present analysis with respect to the energy consumed per ton of CO_2-equivalent mitigated.

Methods

A 0D model for the photocatalytic converter is modelled using MATLAB. The 0D model incorporates the thermodynamic and kinetic data, and radiation transport equation to predict the conversion of CH_4 and N_2O in air. 0D model does not consider the shape of the catalytic converter nor the flow profiles. The remainder of the process is modelled using commercial simulation software tools like Aspen Plus. The energy consumed in the process for blowers and regeneration of sorbent is estimated. The energy consumed in the DAC of CO_2 is considered as the reference. The typical amine-impregnated oxide-based DAC for CO_2 [4] technology is considered in this study.

The key performance indicator in this study is the energy consumed per ton of CO_2-equivalent mitigated. The tons of CO_2-equivalent includes the tons of CO_2 and tons of CH_4 and N_2O multiplied by their global warming potential.

Results

The results from this study mainly compares the estimation of the conversion, material and energy requirements in the process against the DAC for CO_2. We will also make a comparison of the results against the process where each gas would been separately mitigated. The amount of greenhouse gases mitigated and the energy consumed per ton of CO_2-equivalent mitigated will be calculated. The DAC for CO_2 capture consumes 12 GJ/ton-CO_2 for a technology that uses amine impregnated oxide-based DAC. Preliminary results show that the energy consumed per ton of CO_2-equivalent mitigated will be 4 GJ in the MGM concept. A detailed process analysis, with material (catalyst and sorbent) and energy consumed, quantum yield (as we consider photocatalysts) and amount of greenhouse gases mitigated will be presented in the final paper.

Acknowledgements

The work is part of the project "Energieffektiv negativa utsläpp från jordbrukssektorn" (project number 50340-1) funded by Energimyndigheten (Swedish Energy Agency). The authors would like to thank the collaborators in the project from KTH Royal Institute of Technology (Stefan Grönkvist), Uppsala University and Swedish University of Agricultural Sciences for their contributions through discussions within the project. This work is also financially supported by the Swedish governmental initiative StandUp for Energy.

References

1. IPCC, *Summary for Policymakers. In: Global Warming of 1.5°C. An IPCC Special Report on the impacts of global warming of 1.5°C above pre-industrial levels and related global greenhouse gas emission pathways, in the context of strengthening the global response to the threat of climate change, sustainable development, and efforts to eradicate poverty [Masson-Delmotte, V., P. Zhai, H.-O. Pörtner, D. Roberts, J. Skea, P.R. Shukla, A. Pirani, W. Moufouma-Okia, C. Péan, R. Pidcock, S. Connors, J.B.R. Matthews, Y. Chen, X. Zhou, M.I. Gomis, E. Lonnoy, T. Maycock, M. Tignor, and T. Waterfield (eds.)]. World Meteorological Organization, Geneva, Switzerland, 32 pp.* 2018.

2. © *FAO. FAOSTAT Emissions database. http://www.fao.org/faostat/en/#data/GT.*

3. Gütschow, J., et al., *The PRIMAP-hist national historical emissions time series (1850-2017).* 2019, GFZ Data Services.

4. Sanz-Pérez, E.S., et al., *Direct Capture of CO2 from Ambient Air.* Chemical Reviews, 2016. **116**(19): p. 11840-11876.

5. Kim, J., et al., *New materials for methane capture from dilute and medium-concentration sources.* Nature Communications, 2013. **4**(1): p. 1694.

6. Yang, J., et al., *MIL-100Cr with open Cr sites for a record N2O capture.* Chemical Communications, 2018. **54**(100): p. 14061-14064.

7. Jackson, R.B., et al., *Methane removal and atmospheric restoration.* Nature Sustainability, 2019. **2**(6): p. 436-438.

8. Anderson, R.B., et al., *Catalytic Oxidation of Methane.* Industrial & Engineering Chemistry, 1961. **53**(10): p. 809-812.

9. Kapteijn, F., J. Rodriguez-Mirasol, and J.A. Moulijn, *Heterogeneous catalytic decomposition of nitrous oxide.* Applied Catalysis B: Environmental, 1996. **9**(1): p. 25-64.

10. Gélin, P. and M. Primet, *Complete oxidation of methane at low temperature over noble metal based catalysts: a review.* Applied Catalysis B: Environmental, 2002. **39**(1): p. 1-37.

11. de_Richter, R., et al., *Removal of non-CO2 greenhouse gases by large-scale atmospheric solar photocatalysis.* Progress in Energy and Combustion Science, 2017. **60**: p. 68-96.

12. Ming, T., et al., *Fighting global warming by greenhouse gas removal: destroying atmospheric nitrous oxide thanks to synergies between two breakthrough technologies.* Environmental Science and Pollution Research, 2016. **23**(7): p. 6119-6138.

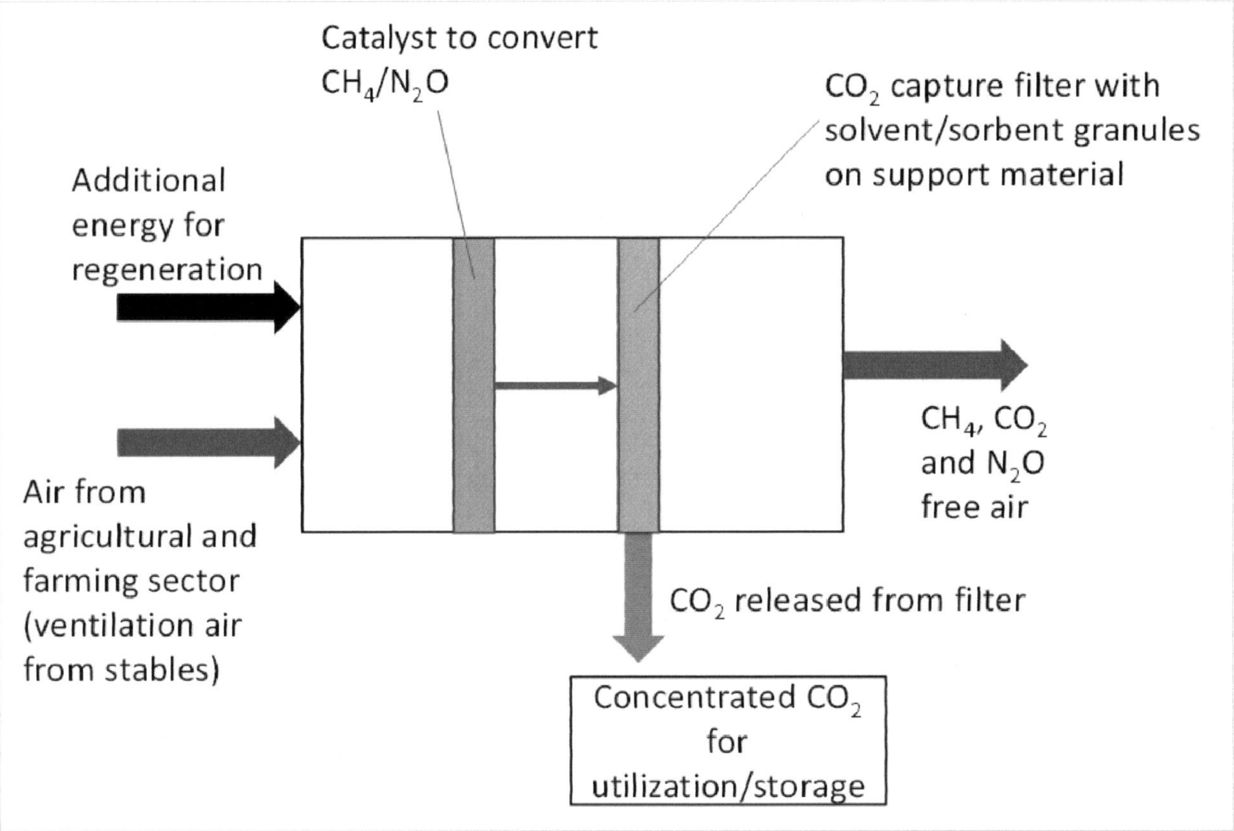

Figure 1: Schematic of multiple greenhouse gases mitigation (MGM) process concept

Copyright © American Institute of Chemical Engineers. All rights reserved.

631001 Rate-Based Dynamic Modeling and Optimization of an Amine-Based Carbon Capture Unit for Flexible Operation

Abstract:

Mitigating carbon emissions is one of the greatest challenges for human society. Modeling and simulation of advanced multiscale process units have great potential in accelerating the commercialization of carbon capture technologies[1]. Amine-based CO_2 capture technology is the most commercially mature technology for retrofitting existing power stations. The integration of amine-based post-combustion CO_2 capture technology with power plants to reduce greenhouse gas emissions incurs a high energy penalty decreasing a power plant's efficiency by about 23 percentage points[2]. Understanding the capture plant dynamics plays an important role in its technical and economic performance. Capital and operating costs of capture plants strongly depend on the configuration[3]. For a given configuration, operating conditions play a vital role in reducing the operating costs[4]. Flowrate and composition of the flue gas change as the host power plant ramps up and down to meet required power output. Hence, various strategies can be considered for improving the economics of the capture unit. For example, loaded solvent can be stored when the price of electricity is high and regenerated when the price drops. Thus, the capture unit must be operated optimally under dynamic, off-design conditions. For dynamic optimization, an accurate dynamic model that captures off-design operation of the amine-scrubbing unit is required. Rate-based dynamic models of amine scrubbing units are limited in the open literature compared with steady-state models. Most existing dynamic rate-based models have been developed assuming linear driving force across the liquid and gas films[5-8]. This assumption can lead to considerable inaccuracy under off-design conditions due to nonlinearity in the transport variables in the film region, especially under off-design dynamic operation. Furthermore, existing rate-based dynamic models consider simple thermodynamic models and constant heat of reaction that do not adequately represent the complex thermodynamics of electrolyte systems.

In this study, a dynamic model of a packed tower is developed for a monoethanolamine(MEA)-based capture unit considering simultaneous mass transfer and chemical reactions in the films. Multi-component transport of molecular and ionic species are modeled using an extended Maxwell-Stefan (MS) transport equation that can capture thermodynamic non-idealities and the effect of electrostatic forces for reactive absorption process. Modeling the transport profile within the film leads to a multi-scale problem as the film thickness is in the order of millimeters while the packing heights are in the order of meters. For primary amines exhibiting fast reaction kinetics, there is significant steepness and nonlinearity in the transport profile especially close to the liquid-gas interface. In addition, the chemical equilibrium in the adjacent liquid bulk region, introduces a high-index problem in the system of equations. These lead to considerable convergence issues especially when good initial guesses are difficult to generate.

To tackle the aforementioned issues, several approaches are developed including null-space projection, model reformulation, scaling, and development of new dimensionless quantities. The dynamic model is validated using transient data from the National Carbon Capture Center in Alabama, USA. For dynamic optimization, the simultaneous approach is adopted over the sequential approach by fully discretizing space and time[9]. While the fully discretized approach offers many advantages, including direct Jacobian and Hessian calculation within the optimizer and efficient decomposition strategies that can exploit structure and sparsity of the system of equations[9], this approach leads to a large-scale optimization problem. The problem is solved using the flexible, open-source Institute for the Design of Advanced Energy Systems Integrated Platform (IDAES)[10] which provides access to efficient large-scale NLP solvers. Several approaches are developed for generating good initial guesses for future state and algebraic variables. In addition, capabilities in IDAES for activating and deactivating constraints are exploited to develop a sequential initialization strategy. Our results show that dynamic optimization not only reduced the energy usage, but also reduces mass transfer limitations thus improving the economics of capture processes under fast, part-load operations.

References

1. Miller, D. C.; Siirola, J. D.; Agarwal, D.; Burgard, A. P.; Lee, A.; Eslick, J. C.; Nicholson, B.; Laird, C.; Biegler, L. T.; Bhattacharyya, D., Next Generation Multi-Scale Process Systems Engineering Framework. In *Computer Aided Chemical Engineering*, Elsevier: 2018; Vol. 44, pp 2209-2214.
2. Supekar, S. D.; Skerlos, S. J., Reassessing the efficiency penalty from carbon capture in coal-fired power plants. *Environmental science & technology* **2015,** *49* (20), 12576-12584.
3. Bhattacharyya, D.; Miller, D., Post-combustion CO 2 capture technologies — a review of processes for solvent-based and sorbent-based CO 2 capture. *Current Opinion in Chemical Engineering* **2017,** *17*, 78-92.
4. Akula, P.; Eslick, J.; Bhattacharyya, D.; Miller, D. C., Model Development, Validation, and Optimization of an MEA-Based Post-Combustion CO2 Capture Process under Part-Load and Variable Capture Operations. *Industrial & Engineering Chemistry Research* **2021**.
5. Montañés, R. M.; Flø, N. E.; Nord, L. O., Dynamic process model validation and control of the amine plant at CO2 Technology Centre Mongstad. *Energies* **2017,** *10* (10), 1527.
6. Ziaii, S.; Rochelle, G. T.; Edgar, T. F., Dynamic Modeling to Minimize Energy Use for CO2 Capture in Power Plants by Aqueous Monoethanolamine. *Industrial & Engineering Chemistry Research* **2009,** *48* (13), 6105-6111.
7. Harun, N.; Nittaya, T.; Douglas, P. L.; Croiset, E.; Ricardez-Sandoval, L. A., Dynamic simulation of MEA absorption process for CO 2 capture from power plants. *International Journal of Greenhouse Gas Control* **2012,** *10*, 295-309.
8. Nittaya, T.; Douglas, P. L.; Croiset, E.; Ricardez-Sandoval, L. A., Dynamic modeling and evaluation of an industrial-scale CO2 capture plant using monoethanolamine absorption processes. *Industrial & Engineering Chemistry Research* **2014,** *53* (28), 11411-11426.
9. Biegler, L. T.; Zavala, V. M., Large-scale nonlinear programming using IPOPT: An integrating framework for enterprise-wide dynamic optimization. *Computers & Chemical Engineering* **2009,** *33* (3), 575-582.
10. Institute for the Design of Advanced Energy Systems (IDAES) https://idaes-pse.readthedocs.io/en/stable (accessed 4/1/2021).

Copyright © American Institute of Chemical Engineers. All rights reserved.

632159 Low-Temperature Chemical Looping Driven NO_X Decomposition with Natural Gas Utilization

Abstract:

Nitrogen oxide (NOx) refers to the family of nitrogen and oxygen compounds including NO, NO2, N2O, NO3, and N2O5, where the emission of NO and NO2 from several combustion processes remain a major concern. These compounds are extremely harmful environmental pollutants and contribute significantly to acid rain and the destruction of the polar ozone layer. Current catalytic and non-catalytic NOx decomposition processes exhibit drawbacks such as the thermal instability, poor selectivity, poisoning by acidic gases, and the requirement of high temperatures. Chemical looping NOx decomposition (CLND) is an advanced process scheme that can address these concerns. This scheme is based on the second law of thermodynamics which translates to the reduction of process irreversibility thereby enhancing the process efficiency. Moreover, it provides an additional degree of freedom in process optimization of the operating conditions for NOx decomposition reaction. The scheme employs a low temperature isothermal operation which encompasses reduction of the carrier material utilizing the natural gas such as methane at 400°C (reducer) and subsequent oxidation of the reduced carrier material by NOx at 400°C in the subsequent step (oxidizer). This culminates in a NOx free pure N2 stream and methane combustion products (majorly CO2)
at the oxidizer and reducer, respectively. Novelty of the proposed process manifests in the form of selective
separation of NOx from a flue gas stream consisting of CO2, NOx, H2O as the presence of CO2 and H2O in
the stream is found to not affect the NOx uptake of the carrier, thereby enabling the purification of the flue gas stream. Additionally, the process utilizes methane that is a greenhouse gas for regeneration and produces a sequestration ready pure CO2 stream. The overall reaction is exothermic, and heat can be extracted from the process to improve the process efficiency. A metal oxide, NiO is employed as the active carrier material, wherein the material exists in the form of a crystalline phase. Furthermore, to increase the dispersity of the active material and reduce sintering effect across redox cycles, five supports i.e., SiO2, CeO2, Al2O3, TiO2 and ZrO2 are investigated. Besides, the best performing support is tested at different concentrations to find the optimum support concentration. The supports chosen for the study varied by the nature of active sites. Acidic support SiO2 exhibited the lowest NOx uptake owing to its weak interaction with the NOx while basic support CeO2 is found comparatively better. Interestingly, the amphoteric supports
performed the best with Al2O3 performing significantly better than others. Experimental investigation of the performance of carrier samples is conducted using a Thermogravimetric analyzer (TGA) and a fixed bed reactor setup. Results indicate a NOx uptake of 412.5μg uptake per mg of the active material (NiO). The morphological transformation and active material dispersion are investigated for all supports across multiple redox cycles, using microscopy-assisted solid characterization. Additionally, XRD and BET analyses are used to study the phase and surface area changes respectively for pre and post reaction samples.
DFT calculations are performed further to investigate the active sites and identify the possible reaction pathway. The CLND process thus offers an opportunity for a low temperature NOx decomposition alongside
natural gas utilization in an economically viable and environmentally sustainable manner.

Key words: NOx decomposition, Chemical looping, Flue gas processing, Methane utilization, Oxygen carrier, Environment

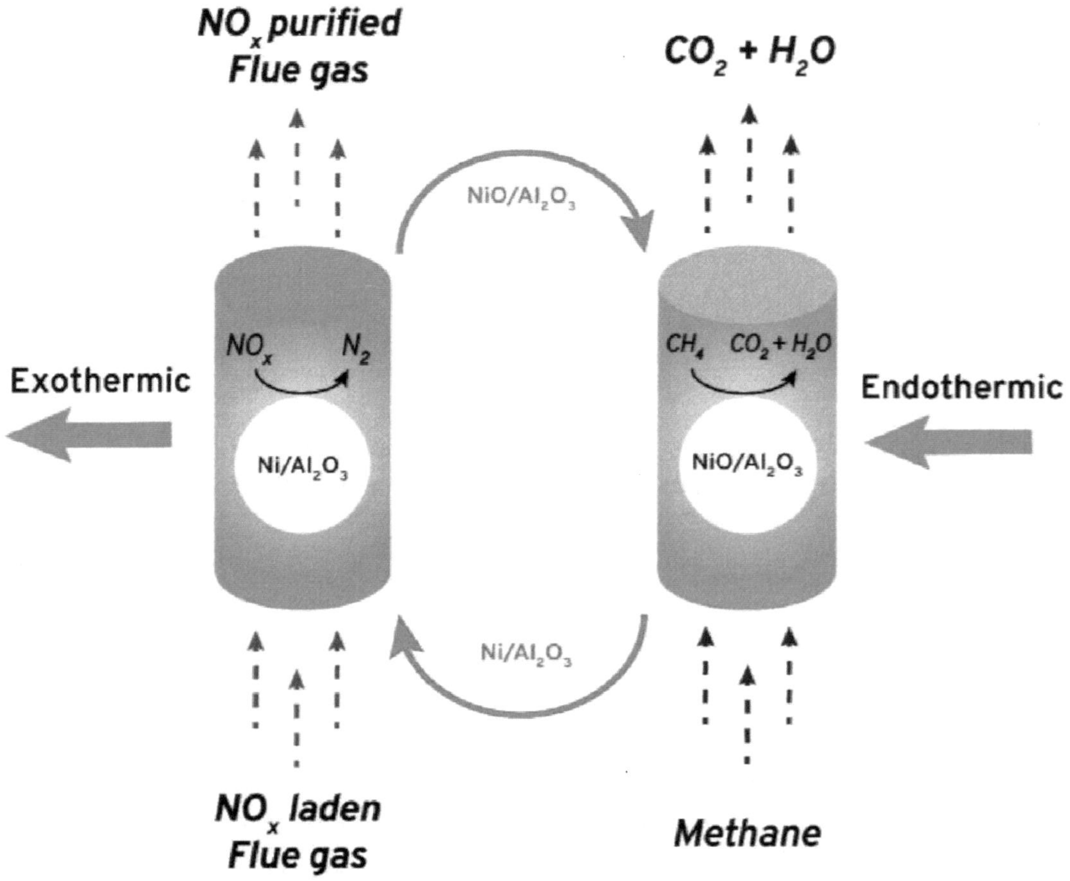

Figure 1. Chemical looping NO_x Decomposition (CLND) process schematic

Copyright © American Institute of Chemical Engineers. All rights reserved.

622570 Optimal Design of Controlled Environment Agricultural Systems Under Market Uncertainty

Abstract:

Global population growth and increased climate variability have posed major challenges to traditional agricultural production [2,6]. As these trends persist, meeting increased consumer demand under heightened uncertainty will further stress the capabilities of our existing food production infrastructure [4,5]. For this reason, alternative methods are being sought to augment conventional food production. Controlled environment agriculture (CEA) is an alternative strategy that decouples crop production from environmental conditions through indoor cultivation. CEA offers a more sustainable alternative to conventional agriculture and presents opportunities for improved food security by increasing public access to nutritious fresh produce. Startup costs for CEA systems, however, are significantly higher than those of traditional growing operations, and therefore present a larger barrier to adoption. To accelerate the adoption of this potentially disruptive technology for next generation food production, a better understanding of the long-term economic viability of CEA technology is desired. To provide this valuable insight, we have developed the first economics-based optimal design and planning model for CEA systems.

To date, the majority of agricultural planning models have focused on traditional outdoor food production. Many of the existing models have been tailored to highly specific operations, few have employed mathematically rigorous optimization techniques, and only a handful have accounted for uncertainty, which is desired to realistically assess the long-term economic viability of the system [1,3,7]. Since existing models are not applicable to CEA systems, we have developed an optimal design and planning model that can be used to assess the long-term economic viability of a broad range of CEA systems. Our proposed model employs robust optimization to account for uncertainty in the design and planning stages of CEA production, which enables risk-averse investment decision making for long-term economic viability.

In this work, we present a novel methodology for the simultaneous robust design and planning of CEA systems under multi-period market price uncertainty/risk. This problem is cast as a bilevel program with a nonconvex outer program pertaining to the design objective and multiple semidefinite inner programs pertaining to mean-variance risk exposure over each period considered in the planning horizon. The bilevel program is reformulated as a nonconvex semi-infinite program with several semi-infinite constraints pertaining to the multi-period risk exposure over the horizon. The problem is solved to global optimality using a custom branch-and-bound strategy constructed with EAGO.jl [10] that exploits a combination of McCormick-based envelopes and Gurobi's nonconvex solver [8], to solve the relaxed (upper-bounding in the maximization sense) subproblems, and IPOPT [9] for local solution of the lower-bounding (in the maximization sense) subproblems. A solution to the model represents a design and operating schedule that is robust to worst-case market uncertainty and therefore provides a conservative basis for CEA engineering and investment decision-making.

Our proposed model has been formulated generally and can be used to optimize large-scale CEA systems for the cultivation of any crop portfolio via single or multi-mode cultivation. The optimization objective employed is to maximize the net present value of the CEA system over its expected lifespan. Constraints limiting risk exposure and production capacity are also included to ensure that the optimal design is both robust and practical. Capital and operating expenses, including those associated with system construction, cultivation equipment, resources, labor, and distribution, are accounted for in the economic objective through sub-models that capture the location, capacity, and crop allocation cost dependencies of the specific CEA system being modeled. These sub-models can be easily adjusted for different CEA configurations without structural changes to the main optimization algorithm. This not only enables efficient determination of the optimal design for a given CEA configuration, but also enables efficient comparison between multiple candidate systems for strategic adoption of CEA technology.

To evaluate the efficacy of our approach, the economic performance of the robust design was compared against that of naïve design and operating strategies through two case studies. In each of the presented case studies, an economically feasible design was identified, even when the naïve design was economically infeasible, and the improvement in system performance attributed to the implementation of a robust design was quantified. This approach represents, to our knowledge, the first robust optimization approach to CEA systems and is demonstrated to effectively increase the robustness of CEA systems to market uncertainty, improve the long-term economics of CEA systems over naïve operating strategies, and validate the economic viability of single and multi-mode CEA production of distinct crop portfolios. This work establishes a means of evaluating the economic viability of a wide array of CEA systems and serves as a baseline for future development of CEA design and planning models to accelerate the adoption of CEA technology.

References:

[1] Ahumada, O., Villalobos, J.R., 2009. Application of planning models in the agri-food supply chain: a review. Eur. J. Oper. Res. 196 (1), 1–20. doi: 10.1016/j.ejor.2008.02.014 .

[2] Deschênes, O., Greenstone, M., 2007. The economic impacts of climate change: evidence from agricultural output and random fluctuations in weather. Am. Econ. Rev. 97 (1), 354–385. doi: 10.1257/aer.97.1.354 .

[3] Dury, J., Schaller, N., Garcia, F., Reynaud, A., Bergez, J.E., 2012. Models to support cropping plan and crop rotation decisions. A review. Agron. Sustain. Dev. 32 (2), 567–580. doi: 10.1007/s13593- 011- 0037- x .

[4] Fedoroff, N.V., 2015. Food in a future of 10 billion. Agric. Food Secur. 4 (1), 11. doi: 10.1186/s40066- 015- 0031- 7 .

[5] Field, C.B., Barros, V., Stocker, T.F., Dahe, Q., 2012. Managing the Risks of Extreme 1083 Events and Disasters to Advance Climate Change adaptation: Special Report of the Intergovernmental Panel on Climate Change (IPCC). Cambridge University Press.

[6] Food and Agriculture Organization of the United Nations (FAO), 2018. The state of agricultural commodity markets 2018: Agricultural Trade, Climate Change and Food Security. Technical Report. FAO. URL http://www.fao.org/3/I9542EN/i9542en.pdf .

[7] Glen, J.J., 1987. Mathematical models in farm planning: a survey. Oper. Res. 35 (5), 641–666. doi: 10.1287/opre.35.5.641 .

[8] Gurobi Optimization, LLC, 2020. Gurobi optimizer reference manual. URL http://www.gurobi.com .

[9] Wächter, A., Biegler, L.T., 2006. On the implementation of an interior-point filter line-search algorithm for large-scale nonlinear programming. Math. Program. 106 (1), 25–57. doi:

10.1007/s10107- 004- 0559- y.

[10] Wilhelm, M.E., Stuber, M.D., 2020. EAGO.jl: Easy advanced global optimization in Julia. Optim. Methods Softw. doi: 10.1080/10556788.2020.1786566 .

Copyright © American Institute of Chemical Engineers. All rights reserved.

632232 Reproducible and Fast Mapping of Material Flows in Industrial Networks Using Piot Hub: A Novel Cloud Based Computational Tool

Abstract:

An important goal of industrial ecology is to ensure a balance is maintained between economic and environmental sustainability. Facilitation of this balance requires an understanding of the structure of the material economy and the material flows/waste generation in industrial networks in order to minimize the environmental consequences of these industrial systems. Physical Input-Output Tables (PIOTs) are uniquely suited for this purpose by mapping physical flows between industrial sectors. In previous work, PIOT-Hub was developed to automate the generation of both Physical Supply Tables (PSTs) and Physical Use Tables (PUTs) as well as their conversion into PIOTs by integrating the strength of mechanistic models and macroeconomic Input-Output framework [1]. Hence, PIOT-Hub, a cloud-based tool can be used to generate the PST and PUT for a region using a bottom-up approach where relevant industrial processes are simulated via mechanistic models to determine steady-state material input and output requirements. Thus, the tool establishes a connection between engineering and economic model in such a way that it is insightful for transitioning to a low carbon and zero waste circular economy. In past work, PIOT-Hub was applied to 11 agro-based economic sectors in Illinois, thus capturing elemental flows across the economy, demonstrating the automation aspect. Here we evaluate the reproducibility and time aspect of PIOT-Hub to map the material flows of equivalent economic sectors in Indiana. We develop simulation models for industries in Indiana and utilize PIOT-Hub for mapping the industrial network, thus evaluating the functionality of this novel computational platform. This work not only demonstrates the reproducibility of the tool's utilization to other geographic regions, but also serves as a case study of how the fast generation of PSTs, PUTs, and PIOTs by PIOT-Hub can be used to compare the material economies of multiple regions. We perform a comparative analysis of material flows in Illinois and Indiana using our work and the PIOT generated for Illinois in previous work. We will also demonstrate the collaborative aspect of this computational tool by live demo of the tool during this talk, thus highlighting the scaling impact of collaboration through cloud-based platform in making sustainable manufacturing decisions.

References

[1] Shweta Singh; Lan Zhao; Jaewoo Shin; Venkata Sai Gargeya Vunnava (2021), "PIOT hub for Automated Generation of Regional PIOTs," *https://mygeohub.org/resources/piotpy.*

Copyright © American Institute of Chemical Engineers. All rights reserved.

625295 Solid Oxide Fuel Cell – Thermal Energy Storage (SOFC-TES) Integrated System Modeling and Energy Efficiency Optimization.

Abstract:

Concerns over greenhouse gas emissions and their associated environmental challenges continue to increase. As a result, cleaner energy generation technologies, such as fuel cells, are receiving attention as the next generation's energy sources. One of these attractive energy technologies is the solid oxide fuel cells (SOFCs) due to its several advantages such as 1) reduced greenhouse gas emissions, 2) higher efficiency, and 3) flexible fuel selection compared to other types of fuel cells. However, the temperature of flue gas from SOFCs is significantly high due to its operating temperature between 600°C and 1000°C. Therefore, it is necessary to recover this waste heat so that we can maximize the process energy efficiency.

Many studies have been conducted to utilize the high-temperature exhaust gases from SOFCs so far. However, they only focused on the direct utilization of the heat through the electricity generation, using processes such as gas turbine[1], Rankine cycle[2], and combined heat and power (CHP)[3] in order to maximize the process energy efficiency. However, the aforementioned utilization methods result in losses where the generated electricity is further converted to other forms of energy such as thermal energy for space heating. The energy losses during this process are often significant.

Motivated by this, we propose the SOFC-integrated thermal energy storage (TES) method as a more efficient utilization technique for recovering waste heat from SOFCs. We employed a 4 kW-SOFC and combined it with the thermal energy storage system for storing the excess energy from the high-temperature exhaust gas. In order to integrate both processes, SOFC and TES mathematical models were developed using MATLAB. Next, the developed models are integrated, simulated, and the overall efficiency of the SOFC-TES model is compared to the general utilization methods. To further improve the proposed model's thermal efficiency, we developed a heat exchanger network using Aspen Energy Analyzer. Through the developed heat exchanger network, the integrated model was optimized by pinch analysis.

The optimization results revealed that the process efficiency of the integrated system, that is, SOFC-TES, showed superiority compared with the traditional methods of utilizing the waste heat from SOFCs. By this, excess heat from the exhaust gases can be reliably stored during low-demand seasons and reutilized during high-demand seasons to supplement residential and industrial energy demand or deficits. We concluded that the SOFC-TES could be widely used for various purposes due to its inherent characteristics, i.e., storing heat, such as a greenhouse in the smart-farm and human life, compared with the traditional method of waste heat utilization from the SOFCs.

References

[1] S.H. Chan, H.K. Ho, Y. Tian, **Modeling of simple hybrid solid oxide fuel cell and gas turbine power plant**, Journal of Power Sources, 109 (1) (2002), pp. 111-120

[2] Masoud Rokni, **Thermodynamic analysis of an integrated solid oxide fuel cell cycle with a Rankine cycle**, Energy Conversion and Management, 51 (2010), pp. 2724-2732

[3] S.H. Chan, C.F. Low, O.L. Ding, **Energy and exergy analysis of simple solid-oxide fuel-cell power systems**, Journal of Power Sources, 103 (2002), pp. 188-200

Copyright © American Institute of Chemical Engineers. All rights reserved.

630767 Upcycling Food Waste for Hydrogen Storage: Technoeconomic Assessment

Abstract:

While annually 1.6 billion tons of food waste is estimated to be discarded globally by 2030, according to U.S. Environmental Protection Agency (EPA) the corresponding figure is projected forecast to reach approximately 66 million tons for U.S. With the monumental challenge of food waste management, researchers are seeking avenues of upcycling FW into higher value materials such as hydrogen storage medium. Therefore, the objective of this research was to evaluate the technoeconomic prospect of activated carbon production from food waste for the ultimate purpose of hydrogen storage through the application of hydrothermal carbonization of food waste followed by chemical activation of the hydrochar using KOH. The process was simulated using Aspen Plus and the process flow diagram (PFD) was constructed such that it mainly constituted of 2 rotary kilns and filter press, 1 rotary cooler, dryer, homogenizer, as well as other necessary equipment - pumps, heat exchangers, storage tanks. The facility processes 96 ton of food waste per day to be converted to activated carbon powder with a carbon yield of 20.4%. Also, it was assumed that the activated carbon manufacturing facility obtains food waste of Brevard County, Florida and the facility operates continuously 330 days a year. Several parameters were incorporated in the study including equipment sizing, capital costs and operating costs, such as labor, utilities, maintenance and equipment depreciation. Break-even cost from the economic analysis was evaluated and sensitivity analysis was carried out to reveal the effect of 7 different process variables on the break-even cost where raw materials' (KOH and food waste) and wastewater treatment cost significantly affected it. The unprecedented analysis shed light on the economic feasibility of producing activated carbon from food waste as a cost-effective route of upcycling it.

Copyright © American Institute of Chemical Engineers. All rights reserved.

625694 Economic and Environmental Impact of Biochar Addition on Anaerobic Digestion Process

Abstract:

Anaerobic digestion (AD) technology is becoming a promising pathway for producing clean and renewable natural gas from waste resources. Despite growing industrial adoption, AD technologies still suffer from inefficiency due to process instability and microbial inhibitions. Scientists have demonstrated that biochar addition has positive impacts on AD. Biochar addition increases biogas yield and methane concentration and reduces gas contaminants. This influence is attributed to several mechanisms: providing support for microbial colonization, adsorbing excess inhibitory products, and improving electron transfer between syntrophic microorganisms. However, due to the cost and energy-intensive conversion process, the additional costs for biochar addition could offset the resulting energy benefits. This study investigates an integrated biorefinery where corn stover biochar is added to the AD of animal manure to produce biogas-derived Renewable Natural Gas (RNG) for pipeline injection.

A kinetic model of AD, biogas upgrading, and pipeline injection is simulated in Aspen Plus ^{TM} for a mid-size farm with 5000 cattle heads. The process simulation model estimates a 28% increase in biogas yield for 3% (wt. % of wet mass) of biochar addition. High pressurized water scrubbing process is used for biogas upgrading into RNG. Techno-economic analysis (TEA) determines the minimum RNG selling price for a 30-year plant lifetime with a 10% Internal Rate of Return (IRR). TEA estimates the minimum RNG selling price as \$42/MMBTU and \$35/MMBTU for the without biochar and with biochar scenario, respectively. Biochar addition lowers the RNG selling price by 16% whereas adding revenue from digestate selling and renewable identification number (RIN) credits result in up to 32% reduction. Biochar addition increases the capital and operating costs by 8% and 13%, respectively. Cradle-to-gate life cycle analysis (LCA) is performed in SimaPro ^{TM} to evaluate the GHG emissions of biochar amended AD for RNG production. GHG abatement for the without biochar and with biochar scenario is 0.038 kg $CO_{2\ eq}$/MJ of RNG and 0.123 $CO_{2\ eq}$ /MJ of RNG, respectively. LCA results show biochar addition leads to 2.18 times GHGs abatement compared to the without biochar scenario. Environmental benefits mostly come from the additional carbon sequestration with biochar land use and fossil fuel displacement by syngas from pyrolysis. Sensitivity analysis shows the influence of different major processes and economic parameters on the RNG selling price. The process is cost-prohibitive for small-scale farms due to the significant economy of scale penalty for the biogas upgrading and RNG injection facility. Increased biogas yield reduces the manure input leading to less digestate production. This would decrease the environmental benefits of emissions avoidance from conventional manure management and carbon sequestration per MJ of RNG while reducing the emissions from digestate transportation and applications. Uncertainty analysis considers the variations of key technical and economic parameters within a reasonable range and gives an estimated range of fuel selling prices and GHG abatement. Minimum RNG selling prices vary between \$8.82/MMBTU and \$16.28/MMBTU and GHG abatement vary between 0.06 kg $CO_{2\ eq}$/MJ and 0.16 kg $CO_{2\ eq}$/MJ within 10-90% confidence intervals.

Copyright © American Institute of Chemical Engineers. All rights reserved.

627985 Optimizing Syngas Production Process By Downdraft Gasifier Using Response Surface Methodology

Abstract:

The gasification process of biomass to produce combustible gases has an attracted interest recently. The downdraft gasifier has an advantage over other types of gasifiers because of its small size, high quality gas production, and low tar yield. One of the challenges in the gasification process is the production of low-quality syngas (low concentration of H2 and CO) and large amount of tar. To overcome this problem, optimization of the gasifier operating conditions was employed to achieve high-quality product gases and hence high gasification process efficiency. In this work, the influence of initial gas temperature, airflow rate, and mole fraction of oxygen in primary air, on the gasification efficiency and product quality was optimized by response surface methodology. The obtained results showed that the reduction in the temperature caused a low concentration of carbon monoxide and a high amount of tar in the product, thereby there was a decline in the efficiency of the gasification process.

Copyright © American Institute of Chemical Engineers. All rights reserved.

623925 Techno-Economic Evaluation of Nano-Porous Silica Production from Rice Husk and Sand

Abstract:

Global rice production in 2019 was estimated at about 500 million tons. Rice husk is a by-product from rice, including about 20 wt% of rice and is needed to be disposed as waste. Rice husk contains approximately 14-25% of ash. The silica content in rice husk ash (RHA) is above 90%. Nanoporous silica (NPS) used in industrial fields such as reinforced rubber additives (tires), concrete mixture (cement binder), nano-fibers, nano-filters, and membranescan be produced from RHA. Rice hisk is combusted at a relatively low temperature (650-800°C) and RHA is produced, whereas the silica sand is operated at 1400-1500°C to produce silicate. In addition, compared to the conventional NPS production from sand, The NPS production from RHA has an advantage in lowering the production cost due to low material cost and energy consumption.

In this study, the techno-economic feasibility for the sodium silicate ($Na_2O{\cdot}xSiO_2$)-based nano-porous silica (SS-NPS) production from sand and rice husk was performed. The return on investment (ROI) and payback period (PBP) were evaluated according to the plant size for the two cases. The effects of raw material and product prices on ROI and PBP were analyzed and economic conditions to be economically viable were suggested.

Copyright © American Institute of Chemical Engineers. All rights reserved.

AIChE

The Global Home of Chemical Engineers

A CAPITAL AND ENERGY EFFICIENT ALTERNATIVE EMISSIONS CONTROL TECHNOLOGY FOR CONTROLLING EMISSIONS USING BEAD ACTIVATED CARBON

Kim K. Tutin, Captis Aire LLC, East Point, GA United States
Harold R. Cowles, Environmental C&C Inc., Scottsdale, AZ United States
John M. Berger P.E., Environmental C&C Inc., Fishers, IN United States

TABLE OF CONTENTS:

I. INTRODUCTION:

In manufacturing, process exhaust gas (air) streams released may contain volatile organic compounds (VOCs), Hazardous Air Pollutants (HAPs), particulate, etc. These are considered pollutants/emissions and must be removed from the air prior to release to avoid air pollution emissions into our communities. Add-on controls either collect the VOCs for recovery and reuse or destroy the VOCs. A number of control options are available including thermal oxidizers, catalytic oxidizers, and biofiltration which destroy pollutants. Other control systems including condensation, scrubbing, and adsorption systems may recover pollutants.

This paper provides an introduction to an adsorption technology, the Fluidized Bed Concentrator (FBC) with Bead Activated Carbon (BAC) media. It describes 1) the FBC technology, 2) the BAC media used in these systems, 3) the performance characteristics, 3) typical applications of this technology, and 4) a case study review of the FBC technology demonstration for controlling emissions and collecting terpenes from wood drying.

II. DESCRIPTION OF FBC TECHNOLOGY: FOUR PRIMARY PARTS

1. Adsorber:

The adsorber section has trays filled with very adsorbent BAC media. Contaminated process exhaust enters from the bottom, passes upward through the adsorption trays, and fluidizes the Bead Activated Carbon (BAC) adsorbent media. When the process air interacts with the BAC, it adsorbs (captures) the terpenes Volatile Organic Compounds (VOCs) in its pores. VOC saturated BAC flows to the bottom of the adsorber vessel and is transferred using fans to pneumatically transfer it to the desorber. Cleaned (reusable) BAC from the desorber feeds continuously into the top of the adsorber, providing counter-current VOC removal. Cleaned air exits out the top of the adsorber.

2. Desorber:

In the desorber, the BAC media is heated (regenerated), causing it to release the VOC vapors into a low volume, gas stream. This gas stream is then circulated continuously to the next stage in the process. The regenerated (cleaned) BAC is then returned continuously to top of the adsorber for reuse. Returning this regenerated BAC to the top of the adsorber provides a final "polishing" of emissions to remove residual emissions prior to release of the process exhaust air.

"All Electric" Option

Desorber Recovery System: In the recovery system, a low volume inert carrier (nitrogen) gas is introduced into the bottom of the desorber. Gas flows up through the desorber as the BAC flows downward. The system is heated via either hot oil or electric heating. The heat causes the VOCs to volatilize off the BAC. The VOC vapors are liberated resulting in a concentrated waste flue gas stream that moves from the desorber to the condenser. This system may eliminate the on-site use of natural gas entirely. Thus, it offers an "All Electric" control technology.

Desorber Oxidation System: In the oxidation system, clean hot stripping air at 100-700°F is introduced to the bottom of the desorber. It flows upward across the BAC media to both volatilize the VOCs and fluidize the BAC. The VOC vapors are liberated resulting in a concentrated waste flue gas stream that moves from the desorber to the oxidizer.

3. Condenser or Oxidizer:

The gas stream from the desorber is circulated continuously to this system for further treatment to either capture and recover VOCs or oxidize them.

"All Electric" Option

Condenser Recovery System: In the recovery system, nitrogen gas circulates from the desorber to the condenser continuously. The VOC vapors are sufficiently concentrated such that the vapors can be cooled via a tube and shell condenser converting the vapor to liquid. The liquid drains continuously into a storage system. The liquid VOCs can then be reused or sold. The nitrogen recirculates back to the desorber where it entrains the VOCs vapors again and the cycle continues.

Oxidizer Oxidation System: In the oxidation system, the vapors are instead sent to a small thermal oxidizer to treat the VOCs. While this system does not provide recovered (collected) VOCs, it nevertheless retains FBC technology benefits including significantly reducing natural gas usage. The burner component of the thermal oxidizer is selected to provide enough energy to increase the thermal oxidizer to the 1400°F operating temperature. The burner is designed to minimize the "fire rate and thus consume negligible fuel. Optionally, a "low NO_x" design can be selected to minimize production of NO and NO_2 (oxides of nitrogen). Thus, the system minimizes energy usage and GHG emissions. [1]

4. Side Stream Reactivator (SSR) [2]:

A small portion of the BAC media exiting the desorber may be transferred to an optional SSR. The SSR reactivates the BAC by heating it to 1450°F with the introduction of a small amount of water which reactivates the BAC continuously to regenerate it and thus maintain its ability to adsorb VOCs. Reactivated BAC from the SSR is transferred back to the top of the adsorber for reuse.

III. ADSORBENT MEDIA:

The BAC adsorbent media utilized in this system is an essential element of the FBC technology. Specialized Bead Activated Carbon (BAC) efficiently adsorbs VOCs (i.e. terpenes) from the process air. The BAC has high-fill (adsorption) capacity, is a small sphere-shaped particle that is less than 1 mm in diameter with a narrow particle size distribution, has low dust, has high purity, has high flowability, and has high strength and wear resistance as measured by ASTM D 3802 – 79 (2005) Standard Test Method for Ball-Pan Hardness of Activated Carbon.[3] The spherical shape enables the BAC to fluidize and flow.

The BAC media can be reactivated and reused. This is accomplished either by either continuous or batch processes.

---

[1] "VOC Removal and Destruction System and Method". U.S. Patent and Trademark Office. (USPTO) https://appft.uspto.gov/netacgi/nph-Parser?Sect1=PTO2&Sect2=HITOFF&p=1&u=%2Fnetahtml%2FPTO%2Fsearch-bool.html&r=1&f=G&l=50&co1=AND&d=PG01&s1=20190022579.PGNR.&OS=DN/20190022579&RS=DN/20190022579 (Accessed October 4, 2021).

[2] "GASTAK® Solvent Recovery and Odor Removal by Fluidized Bed Process". Kureha Ecology Management Co. , LTD website. https://www.kurekan.co.jp/en/download/gastak.pdf (Accessed September 18, 2021).

[3] "ASTM D3802-79 (2005) Standard Test Method For Ball-Pan Hardness Of Activated Carbon. https://webstore.ansi.org/standards/astm/astmd3802792005 (Accessed September 18, 2021).

Continuous Reactivation: In the continuous process, a small amount of BAC media is diverted from the BAC exiting the desorber and sent to an in-line side stream reactivator (SSR).[4] It is then returned to the top of adsorber. Returning this reactivated BAC to the top of the adsorber provides a final "polishing" of emissions to remove residual emissions prior to release of the process air.

Batch Reactivation: In the batch process, for operators that do not have an SSR, the BAC media is unloaded periodically, shipped offsite for reactivation, and then reloaded once returned. In the interim, a second load of BAC is utilized. Reactivation is typically performed by California Carbon who has been providing reactivation services since 1975.[5] This service includes a reactivation warranty *"Carbon will be reactivated to 95% of Virgin Activity per ASTM D 3467 and Carbon will lose no more than 10% of incoming volume of the specified mesh size in the reactivation process."*[6] BAC lifetime is site and application specific. Some operators have been reusing their BAC for years.

Bead Activated Carbon (BAC) Adsorbent Media

Bead Activated Carbon (BAC), the Adsorbent Media, Fluidized in Tray

IV. FBC OPTIONS: RECOVERY AND OXIDATION

The Fluidized Bed Concentrator can be utilized to either recover the VOCs or oxidize them as shown in the figures. Operators typically select the desired option based on a number of factors especially including the anticipated value and reusability of the recovered VOCs. The value of the VOCs depends on a number of factors including the anticipated quantity recovered, purity, salability, and/or reusability.

Recovery: When the value of the recovered VOCS is high and if the cost of recovery is less than the cost of purchasing new material, then recovery makes sense.

---

[4] "GASTAK® Solvent Recovery and Odor Removal by Fluidized Bed Process". Kureha Ecology Management Co. , LTD website. https://www.kurekan.co.jp/en/download/gastak.pdf (Accessed September 18, 2021).

[5] "California Carbon Co. Inc. Welcome". California Carbon website. http://www.californiacarbon.com/index.html (Accessed September 18, 2021).

[6] "Custom Spent Carbon Reactivation". California Carbon website. http://www.californiacarbon.com/pdfs/Custom%20Reactivation.pdf (Accessed September 18, 2021).

Oxidation: If the VOC has low or no recovery value, as, for example, when it is a mixture, or if it poses disposal concerns, such as for toxic compounds, then destruction probably makes the most sense. Excluding value derived from VOC recovery, the oxidation option retains FBC system performance characteristics including reduction in energy usage.

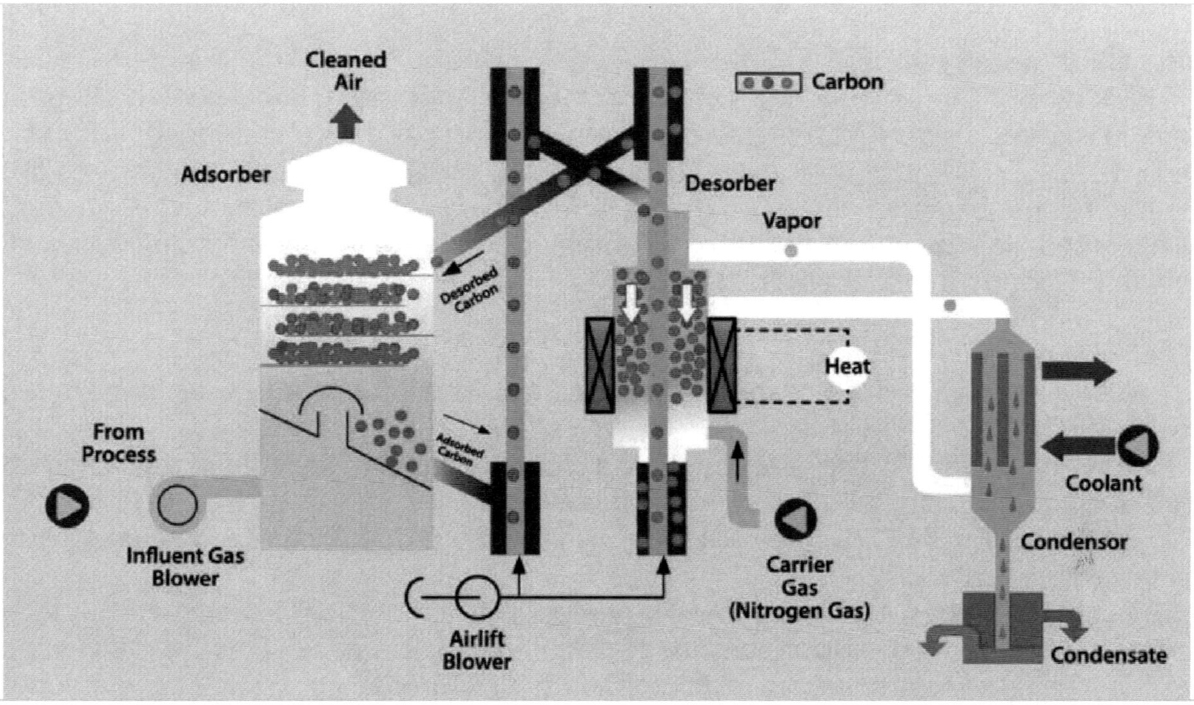

RECOVERY SYSTEM: Adsorbs, Desorbs, and Condenses VOCs

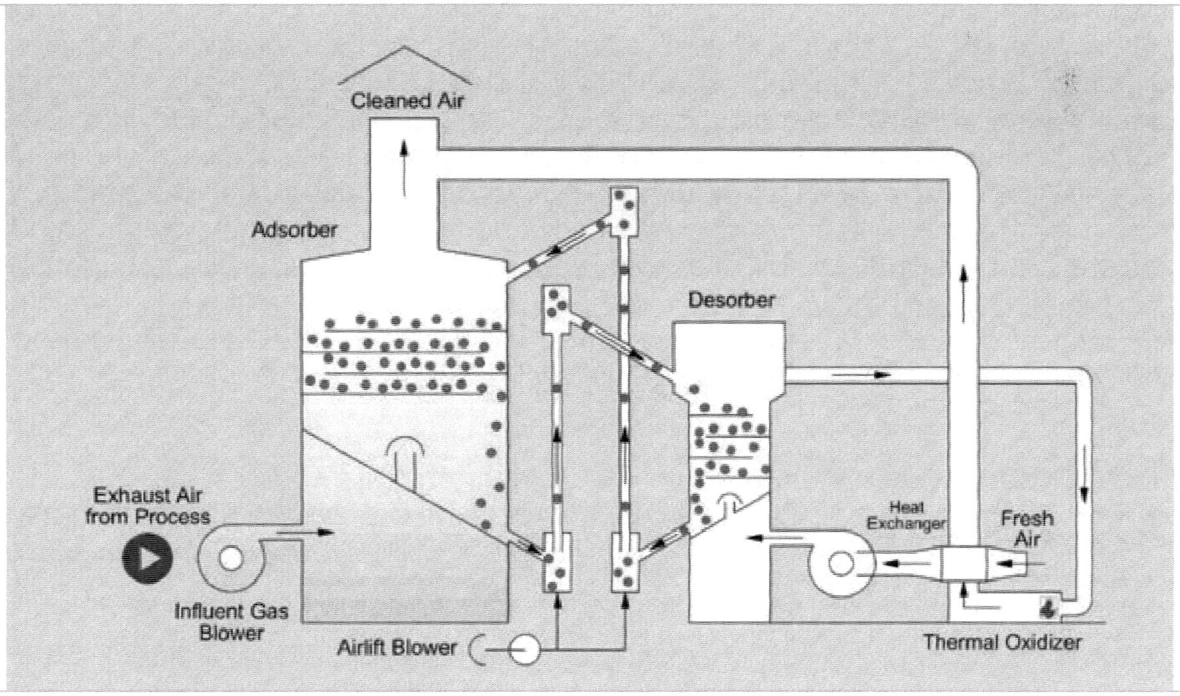

OXIDATION SYSTEM: Adsorbs, Desorbs, and Oxidizes VOCs

V. PROCESS PERFORMANCE CHARACTERISTICS:

1. The FBC reduces energy usage (i.e. natural gas) versus oxidation systems due to reduction or elimination of energy used to heat the air going through thermal oxidizers:

Baseline Oxidation Systems:

Thermal oxidizers burn VOCs by heating them in enclosed combustion chambers in the presence of excess oxygen. The precise temperature and residence time needed for effective reduction of VOC emissions varies with the nature of the VOC. However, temperatures of 1400-2000°F and residence times of 0.5-2 seconds are typical. Since auxiliary fuel is necessary to maintain combustion temperatures, most thermal oxidizers incorporate some form of heat recovery to lower auxiliary fuel costs.

FBC Technology:

The air going through the adsorber is not heated. Thus, the natural gas usage is reduced or eliminated.

Recovery – "All Electric" Option: For a recovery system, the control device can be "All Electric". It could eliminate the use of natural gas entirely.

Oxidation: The FBC oxidation system significantly reduces natural gas usage but does not eliminate it.

Energy Reduction Calculation: The amount of energy that is eliminated, that would have been used by the thermal oxidizer, can be calculated utilizing one of many credible sources. One of the best sources is the Institute of Clean Air Companies' Guidance Methodology[7]. Calculators are also available on-line including those provided by the EPA[8], Cycle Therm[9], and Captis Aire[10]. With these calculators, users can input their specific data and the calculators will calculate the estimated amount of energy required to run the thermal oxidizer. All these sources rely on a single primary factor to calculate energy required to heat the air. This factor is the specific heat of air which is the number of British Thermal Units (BTUs) required to heat one pound of air by one °F. This factor along with the system specific factors including the standard cubic feet per minute (sCFM) of air, the temperature increase, the density of air, and conversion factors provide the number of BTUs required to heat the air going through the device. These factors are used together with the energy efficiency and the heat generated by the oxidation of the emissions to calculate the BTUs of energy required to heat the air going through the RTO. Users can then multiply BTUs times the forecasted long-term cost of natural gas to estimate the long-term cost of natural gas required over the 20+ year life of the thermal oxidizer. This calculation provides the value of the natural gas elimination.

---

[7] "Institute of Clean Air Companies (ICAC) Guidance Method for Estimation of Gas Consumption in a RTO". Institute of Clean Air Companies website. https://cdn.ymaws.com/www.icac.com/resource/resmgr/RTO-F1.pdf (Accessed September 18, 2021)

[8] "Air Pollution Control Cost Estimation Spreadsheet". Environmental Protection Agency (EPA) website. https://www.epa.gov/sites/production/files/2018-01/oxidizers_calc_sheet_finalversion_1-16-2018.xlsm (Accessed Feb 3, 2021).

[9] "Operating Cost Calculator". Cycle Therm website. http://www.cycletherm.com/resources/operating-cost-calculator.aspx (Accessed Feb 3, 2021).

[10] "Calculate Engineering Estimate of Natural Gas Usage and Greenhouse Gas Generation". Captis Aire website. https://www.captisaire.com/Value-Calculator/ (Accessed Feb 3, 2021).

2. The FBC generates revenue from sales of recovered organics:

When using a recovery system, including the FBC, operators have the option to collect and sell or reuse the volatile organic compounds recovered. The liquid VOCs are collected from the FBC condenser for reuse or sale. The value of recovered VOCs will depend on the type, amount, and purity of VOCs collected.

3. The FBC reduces Greenhouse Gases (GHG) emissions versus oxidation systems due to the reduction or elimination of burning natural gas:

Due to the reduction or elimination of natural gas usage and reduction in electrical usage by the FBC, greenhouse gas emissions are also reduced. The greenhouse gas emissions reduction can be calculated based on the amount of carbon in natural gas eliminated and its stoichiometric conversion to carbon dioxide (CO_2). The Environmental Protection Agency (EPA) has an on-line calculator[11] that enables the user to calculate greenhouse gas emissions that are eliminated by eliminating natural gas used to heat the air in the thermal oxidizer.

4. The FBC enables creation of carbon offsets versus oxidation systems:

In conversations with American Carbon Registry[12] and ClimeCo[13], it may be possible to measure, register, and verify carbon offsets for a site converting to the FBC technology from an oxidation system due to the reduction in energy usage.[14]

The quantification methodology and value of carbon offsets is evolving quickly. But these carbon offsets can be sold or traded externally. Or they may be utilized for internal credit accounting purposes. Potential values for these carbon offsets vary widely.[15] However, as of early 2021, American Carbon Registry and ClimeCo estimated the value of carbon offsets for an FBC project at $5/ton.[16]

5. The FBC reduces electricity usage versus oxidation systems:

The horsepower required to run the fans to move the process exhaust air through the BAC media is reduced for the FBC due to the fact that it's BAC is fluidized (moving) rather than packed (non-moving). The pressure drop across the system is reduced from a typical 10-20" to 3.5". Thus, the horsepower required to run the fans for the FBC may be ~ ¼ of that required for the RTO.

To estimate the electrical savings, the sum of electrical demands from the FBC is subtracted from the sum of electrical demands of the thermal oxidizer. Net kW·h along with site-specific electricity costs and operating hours per year are used in combination to estimate electrical savings.

---

[11] "Greenhouse Gases Equivalencies Calculator - Calculations and References". EPA website. https://www.epa.gov/energy/greenhouse-gases-equivalencies-calculator-calculations-and-references (Accessed December 28, 2020).

[12] "Harnessing the Power of the Markets to Improve the Environment". American Carbon Registry website. https://americancarbonregistry.org/ (Accessed February 3, 2021).

[13] "Scalable greenhouse gas projects to earn real financial returns for your business." ClimeCo website. https://climeco.com/ (Accessed February 3, 2021).

[14] "AM0044: Energy efficiency improvement projects - boiler rehabilitation or replacement in industrial and district heating sectors --- Version 2.0.0". Clean Development Mechanism (CDM) website. https://cdm.unfccc.int/methodologies/DB/3HZ4USHZ2W449HMAXZN420E5PJB1QF (Accessed Sept 18, 2021)

[15] "Allowance Price Explorer". International Carbon Action Partnership website. https://icapcarbonaction.com/en/ets-prices (Accessed Jan 20, 2021).

[16] Eric Ripley (Director Industrial Programs, American Carbon Registry), in a phone conversation with Kim Tutin November 17, 2020.

6. <u>The FBC enables superior mass transfer of VOCs through counter current flow fluidized BAC media that is continuously regenerated:</u>

Baseline Adsorption Systems:

Adsorption: In sorbent treatment systems, process exhaust air moves across the sorbent. VOCs are removed via attraction (adsorption) of the VOCs on carbon or zeolite sorbent media in packed (non-moving) beds. This may be accomplished by concentrators that have wheels with sectors containing activated carbon or zeolite media. In this arrangement, most of the space on the wheel functions in adsorption mode, with one small sector being regenerated at any given time. Or the process exhaust air may be forced through large vessels containing sorbent media. In this arrangement, the VOCs are treated via a once-through process.

Regeneration / Reactivation: Regeneration is accomplished by heating the sorbent media with heaters, heated gases, steam, etc. If using steam, water may be a byproduct. This can be accomplished either via off-site regeneration or by alternating the flow between multiple vessels where one vessel functions in adsorption mode while the other is being regenerated at any given time. In both these arrangements, the sorbent media is being regenerated in batch or semi-continuous fashion. Reactivation may be accomplished by sending the sorbent media off site for reactivation. Or the sorbent media may be discarded when it becomes saturated.

Effect of water vapor: Especially for process exhaust air heavily laden with water vapor, i.e. 10% B_{ws} moisture by volume[17] or more, the sorbent media may become partially or fully saturated with water. At high moisture content, water molecules begin to compete with the hydrocarbon molecules for active adsorption sites. This reduces the capacity and the efficiency of the adsorption system.

FBC Technology:

Adsorption: The BAC media in the fluidized bed concentrator adsorber is fluidized (moving) by the process air moving opposite (counter current) to the BAC as it moves downward through the trays in the adsorber. This provides excellent mixing of the BAC and contact with the air. As this occurs, VOCs or odors are transferred from the process air to the BAC. The cleaned air exits through the top of the adsorber.

Regeneration / Reactivation: The BAC media is regenerated continuously via continuous recirculation to and from the desorber. It is reactivated continuously if system includes the Side Stream Reactivator (SSR). This ensures the BAC retains the capacity to maintain adsorption efficiency. This is done at temperatures well above the boiling point of water, often around 400°F or more. This minimizes or eliminates accumulation of water and/or organics on the BAC.

Effect of water vapor: Continuous regeneration and reactivation at temperatures above the boiling point of water, i.e. 400°F, volatilizes the water off the BAC media and prevents the BAC from becoming saturated with water. Thus, it maintains adsorption efficiency even when treating process exhaust streams with high moisture contents, as are typical for wood drying applications, i.e. ≥20% B_{ws} moisture by volume[18]. This system minimizes or eliminates need for dehumidification systems.

---

[17] "METHOD 4—DETERMINATION OF MOISTURE CONTENT IN STACK GASES". EPA website. https://www.epa.gov/sites/default/files/2016-06/documents/m-04.pdf. (Accessed October 17, 2021).
[18] "METHOD 4—DETERMINATION OF MOISTURE CONTENT IN STACK GASES". EPA website. https://www.epa.gov/sites/default/files/2016-06/documents/m-04.pdf. (Accessed October 17, 2021).

Effect of Fluidized (moving) vs Packed (non-moving) media: BAC media fluidization prevents pockets of high concentration VOCs, provides higher dry particulate tolerance (reduced plugging of interstitial spaces between media), and enables continuous on-site regeneration and reactivation.

TABLE I
PHYSICAL AND OPERATIONAL COMPARISON OF THE FBC VS THE RTO

Parameters:	RTO	FBC	Advantages of FBC
Treatment Temperature, °F	>1400°F	175°F	• Adsorption at process air temperature • No natural gas required for heating
Media Desorption Temperature, °F	>1400°F	400°F	• Much less energy required for desorption • Continuous media regeneration, above the boiling point of water, prevents the media from getting saturated with water. Thus, it maintains its ability to adsorb pollutants.
Media depth, feet (ft)	16 ft	1 ft	• Less Electricity (fan HP) required to move air through less media
Media Configuration	Packed (Not Moving)	Fluidized (Moving)	• Less Electricity (fan HP) required to move process air through fluidized media
Pressure Drop, Inches (")	10-20"	3.5"	• Less Electricity required to move process air against less pressure
Fan size, Horsepower, (HP)	2400 HP	600 HP	• Less HP due to media that is fluidized. • Less susceptible to plugging from particles
Emissions Treatment	Oxidation to CO^2	Adsorption (Capture)	• Emissions are NOT oxidized to CO_2, so reduces GHG emissions • Emissions collected can be sold or reused
Media Reuse	Discarded	Reused	• Media desorption with 'in unit' continuous regeneration. • Flowable bead shaped media can be removed and reused
Safety		Inherently Safer	• Lower operating temperatures • No large heavy moving parts • Easier cleaning and maintenance

VI. AWARDS:

The FBC has been recognized for excellence in environmental performance. Honda Motors of Alabama was recognized as *"Air Conservationist of the Year"* when they implemented the FBC to control emissions when they increased production.[19] Ford's "Fumes to Fuels" work was awarded the EPA's *"Clean Air Excellence"* award.[20] In this work, they utilized the FBC to capture solvent-laden air, strip the solvent from the air, concentrate the solvent up to 2,000 times, and send it to a fuel reformer. The fuel reformer used the concentrate to generate hydrogen-rich fuel, which was directed into a fuel cell to generate electricity.

VII. APPLICATIONS:

Concentration systems, including the FBC, are advantageously used in front of control devices when treating very high volumes of very dilute exhaust gas to reduce energy usage. Recovery systems are beneficial when the value of the recovered VOCs is high. The following are a few examples where the FBC technology has been utilized historically.

TABLE II
FBC APPLICATIONS

Application	Chemistry
Adhesive Tape Mfg	Toluene
Agricultural Products	Organic Acid & Ammonia Odor
Aluminum Casting	Phenol, Formaldehyde, Ammonia
Animal Lab	Odor Control
Atomic Power Plant	Styrene
Auto Parts Degreasing	Trichloroethylene
Bakery	Odor Control
Brewery	Odor Control
Car Wash	Kerosene
Cellophane Coating	Toluene, Ethyl Acetate, Butyl Acetate
Ceramic Condenser	Terpineol
Confectionery	Odor control
Chemical Production	Toluene, Higher Molecular Weight Alcohols, Carbon Tetrachloride

---

19 "HMA Honored as Air Conservationist of the Year". Honda website. https://www.hondaalabama.com/article/hma-honored-as-air-conservationist-of-the-year (Accessed September 21, 2021).

20 "Clean Air Excellence Award Recipients". Environmental Protection Agency (EPA) website. https://www.epa.gov/sites/default/files/2015-06/documents/clean_air_excellence_award_recipients_year_2003.pdf (Accessed September 21, 2021).

TABLE II (CONTINUED)
FBC APPLICATIONS

Application	Chemistry
Dye Production	Perchloroethylene
Electronics Mfg Plant	DMF, MEK, etc.
Electronics Parts Cleaning	Trichloroethylene, Toluene, 1,1,1-Trichloroethane
Fabric Washing	Perchloroethylene
Feed Processing	Odor Control
Film Coating	MEK, Methyl Cellosolve, Toluene, Tetrahydrofuran
Film Laminating	Toluene, Ethyl Acetate, n-Hexane
Gravure Printing	Toluene, Xylene, n-Propanol, n-Propyl Acetate, mixed
Iron Casting	Phenol, Formaldehyde, Ammonia
Lacquering of Film	Toluene, Ethyl Acetate, MEK
Landfill	Odor Control
LSI Mfg Plant	Phenol, Acetone, Methanol, Mixed Solvents
Lubrication Mfg	Mixed Solvent Odor Control Magnetic Tape, Chlorinated Hydrocarbon
Medicine Mfg Plant	Butanol, Methylene Chloride
Paint Booth	Thinner
Printing Ink Mfg	Mixed Solvents
PVC Resin Mfg Plant	Vinyl Chloride
Resin Mfg	N-Hexane, Methylene Chloride
Rubber Coating	Toluene
Rubber Vulcanizing	Odor Control
Sandpaper Mfg	Toluene, Xylene, Ethyl Acetate
Semiconductor Mfg	IPA, HMDS, NMP, Ethanol, Methanol, Ethyl lactate
Silk Screen Printing	Xylene, Mixed Solvents
Teflon Tubing Mfg	Perchloroethylene
Wastewater Aeration	Odor Control

TABLE III
EXAMPLES OF FBC SYSTEMS INSTALLED

FLUIDIZED BED CONCENTRATOR SYSTEMS INSTALLED: [21]	
Industry **Location** **Inlet Stream** **VOCs** **Performance Achieved** **Notes**	Semiconductor Chip Manufacturing Israel 20,000 CFM Semiconductor solvents, 50 ppmv 95% Reduction Efficiency System located in populated area of a major city has eliminated nuisance odor
Industry **Location** **Inlet Stream** **VOCs** **Performance Achieved** **Notes**	Semiconductor Chip Manufacturing Arizona 36,000 CFM Semiconductor solvents, 25-100 ppmv 90%+ Reduction Efficiency System met strict local compliance requirements while achieving Zero Unscheduled Downtime (ZUD)
Industry **Location** **Inlet Stream** **VOCs** **Performance Achieved** **Notes**	Teflon Tubing Manufacturing Massachusetts 1,800 CFM Perchloroethylene, 1500 ppmv 98% Reduction Efficiency Recovered solvent re-used in process. System payback in 1 year.
Industry **Location** **Inlet Stream** **VOCs** **Performance Achieved** **Notes**	Furniture Painting California 50,000 CFM Paint Solvents, 25-50 ppmv 90%+ Reduction Efficiency Recovered solvent and reformulated with paint
Industry **Location** **Inlet Stream** **VOCs** **Performance Achieved** **Notes**	Military Vehicles Michigan 150,000 CFM Paint Solvents 95% Reduction Efficiency Controlled multiple different interchangeable VOC sources.

[21] "Environmental C&C Inc". Environmental C&C website. https://www.environmentalcc.com/ (Accessed September 18, 2021).

Automotive Site, Alabama, 100,000 CFM
Paint Solvents, >90% Reduction Efficiency
Two adsorbers and one desorber

Image shows three adsorbers one desorber

Image shows ductwork that enabled control of any of the 3 sources to any of the 3 adsorbers.

Each adsorber has its own process fan.

System provided control of multiple different interchangeable VOC sources

Military Vehicles, Michigan, 150,000 CFM
Paint Solvents, 95% Reduction Efficiency

TABLE IV: VALUE CREATION CALCULATION EXAMPLE

Fluidized Bed Concentrator (FBC) vs Regenerative Thermal Oxidizer (RTO)	Value per Year
1) Reduce Energy Usage (Natural Gas) due to Elimination of Energy Used to Heat the Air in the RTO: [22] [23] a. $\frac{mmBTU}{Year}\ to\ heat\ air\ in\ RTO = (\Delta\ \mathrm{T})\ (sCFM)(1.08)(\%Heat\ Loss)(\frac{Operating\ Hours}{Year})/1000000$ b. $(1500^\circ F - 175^\circ F) * (300{,}000sCFM) * \left(\frac{1.08\ BTUs}{sCFM \cdot Hour \cdot Degree^\circ F}\right) * (7.5\%) * (8000\frac{Hrs}{Yr})/1000000$ c. $= \left(257{,}580\frac{mmBTU}{Yr}\right) - [\left(250\frac{lbs\ VOC}{Hour}\right) * \left(15{,}000\frac{BTUs}{lb\ VOCs}\right) * \left(\frac{8000\ Hrs}{Year}\right)] = \mathbf{227,580}\frac{\mathbf{mmBTUs}}{\mathbf{Year}}$ d. $\left(\frac{227{,}580\ mmBTUs}{Year}\right)(\$3.5\frac{Long\ Term\ Cost\ of\ Natural\ Gas}{mmBTU}) = \mathbf{\$796,530\ Savings/Year}$	Save by Eliminating Natural Gas **\$796,530**
2) Reduce Greenhouse Gases (GHG) Generated from Burning Natural Gas: [24] a. $Tons\ GHG = (mmBTU) * \left(\frac{117\ lbs\ CO2}{mmBTU}\right) * (\frac{1\ Metric\ Ton}{2205\ lbs})$ b. $\left(227{,}580\frac{mmBTUs}{Yr}\right) * \left(117\ lbs\frac{CO2}{mmBTU}\right) * \left(1\frac{Metric\ Ton}{2205\ lbs}\right) = \mathbf{12,076}\frac{\mathbf{Tons\ GHG\ Reduction}}{\mathbf{Year}}$	Sell Carbon Offsets **\$60,378**
3) Sell Carbon Offsets (Carbon Credits) Utilizing Carbon Accounting Service Providers, i.e. ClimeCo: [25] a. $\left(\frac{12{,}076\ Tons\ GHG}{Year}\right) * \left(\frac{\$5}{Ton\ GHG}\right) = \mathbf{\$60,378/Year\ Selling\ Carbon\ Offsets}$	
4) Reduce Electricity Usage by Reducing Fan Power to Move Process Air through Fluidized Media: $(2400RTO - 600FBC\ HP\ fans - 800\ HP) * \left(\frac{0.75\ kW \cdot h}{HP}\right)\left(\frac{\$0.07}{kW \cdot h}\right)\left(\frac{8000\ Hrs}{Year}\right) = \$420,000\frac{\mathbf{Savings}}{\mathbf{Year}}$	Save by Reducing Electrical Usage **\$420,000**
5) Sell Terpenes [26] **Yielded from Green Pine Wood** [27]**:** a. $\left(\frac{1\ Gallon\ Terpenes}{Oven\ Dry\ Ton\ Wood}\right) * \left(\frac{300{,}0000\ Dry\ Tons\ Wood}{Year}\right) * (75\%Yield) * \left(\frac{\$3}{Gal}\right) = \mathbf{\$675,000/Yr}$ **Selling Terpenes**	Sell Terpenes **\$675,000**
6) Reduce Maintenance Costs:[28] Media changeouts, large heavy moving part replacement, corrosion repairs, etc.	Save up to **~\$1 Million**
Total Value Creation per Year per Site:	**Up to \$3 Million**

[22] ICAC Guidance Method for Estimation of Gas Consumption in a RTO. Institute of Clean Air Companies website. https://cdn.ymaws.com/www.icac.com/resource/resmgr/RTO-F1.pdf (Accessed September 18, 2021)
[23] "Operating Cost Calculator". Cycle Therm website. http://www.cycletherm.com/resources/operating-cost-calculator.aspx (Accessed November 21, 2021).
[24] "Greenhouse Gases Equivalencies Calculator - Calculations and References". Environmental Protection Agency (EPA) website. https://www.epa.gov/energy/greenhouse-gases-equivalencies-calculator-calculations-and-references (Accessed December 28, 2020).
[25] "Allowance Price Explorer". International Carbon Action Partnership website. https://icapcarbonaction.com/en/ets-prices (Accessed Jan 20, 2021).
[26] "Price Chart for Gum Turpentine". PineChem website. http://www.pinechem.net/images/gall/ChartGNE.GIF Accessed Jan 3, 2021).
[27] "Sulfate Turpentine Recovery". Pulp Chemicals Association. Pages 9-12, 127. https://www.pinechemicals.org/store/viewproduct.aspx?id=1579188 (Accessed Jan 2, 2021).
[28] "Air Pollution Control Technology Fact Sheet". EPA website. https://www3.epa.gov/ttncatc1/dir1/fregen.pdf (Accessed October 17, 2021).

VIII. CASE STUDY: FBC TECHNOLOGY SPECIFIC TO WOOD DRYING

DEMONSTRATION DESCRIPTION:

Captis Aire LLC founder and Environmental C&C Inc. worked on a project together from 2015-2021 with a commercial oriented strand board (OSB) manufacturer to demonstrate the ability to utilize the FBC to control emissions from wood drying.

In wood products manufacturing, process gas exhaust streams released from drying wood contain VOCs, including turpentine (terpenes). These VOCs are considered pollutants/emissions and must be removed from the air, prior to release, to avoid air pollution emissions into our communities. Wood drying operations utilized in the manufacture of oriented strand board (OSB), lumber, wood to biofuels, i.e. wood pellets, plywood, particleboard, etc. are few examples of these processes that have process exhaust streams containing VOCs.

Work on this patent[29] pending[30] technology was executed in three phases. In the first phase, laboratory screening work was done to explore whether emissions from wood drying, primarily terpenes, could be adsorbed by Bead Activated Carbon (BAC) media at the expected operating temperatures. In the second phase, a small-scale pilot adsorber was installed at a commercial Oriented Strand Board (OSB) manufacturing site to demonstrate ability to adsorb emissions from wood drying under commercial manufacturing conditions. In the third phase, the pilot unit capabilities were expanded to include the adsorber, desorber, and condenser. The SSR performance was simulated by semi-continuous additions of virgin BAC during air emissions compliance testing.

In the third testing phase, the pilot unit ran for 6 months with over 1700 hours on-line. Ability to meet air emissions control compliance, per the Plywood and Composite Wood Products Maximum Achievable Control Technology (PCWP MACT) standard, was determined by two independent certified air emissions[31] stack testing companies[32] during two separate testing events.

---

[29] "METHODS AND SYSTEMS FOR RECOVERING TERPENE COMPOSITIONS FROM WOOD DRYING EXHAUST". World International Patent Office (WIPO) Website. https://patentscope.wipo.int/search/en/detail.jsf?docId=WO2019148208 (Accessed September 18, 2021).

[30] "METHODS AND SYSTEMS FOR CONTROLLING EMISSIONS FROM WOOD DRYING PROCESSES". WIPO Website. https://patentscope.wipo.int/search/en/detail.jsf?docId=WO2019148207 (Accessed September 18, 2021).

[31] "AIR Advanced Industrial Resources". AIR website. https://airtest1.com/ (Accessed April 16, 2020).

[32] "Testing and Modeling". John Zink Hamworthy Combustion website. https://www.johnzinkhamworthy.com/testing-modeling/ (Accessed Feb 3, 2021).

TABLE V
RESULTS OF AIR POLLUTION CONTROL COMPLIANCE TESTING (FBC UNIT):

Emissions Measured:	**FBC %Reduction Efficiency[33]**
Total Hydrocarbons (THC), as carbon including Methane	92%
Volatile Organic Compounds (VOCs), as carbon. (Total non-methane hydrocarbons – THC minus Methane)	92%
Formaldehyde	83%
Methanol	75%
GHC - Nitrogen Oxides (NOx)	55%
Particulate Matter (PM), total filterable	50%
Condensable Particulate Matter	59%
Total Particulate Matter, Filterable + Condensable	51%
Total PM2.5, Particle Size Analyses (PSA) Filterable + Condensable	60%
Total PM10, Particle Size Analyses (PSA) Filterable + Condensable	64%
Carbon Monoxide (CO)	7%
Sulfur Dioxide (SO2)	99%

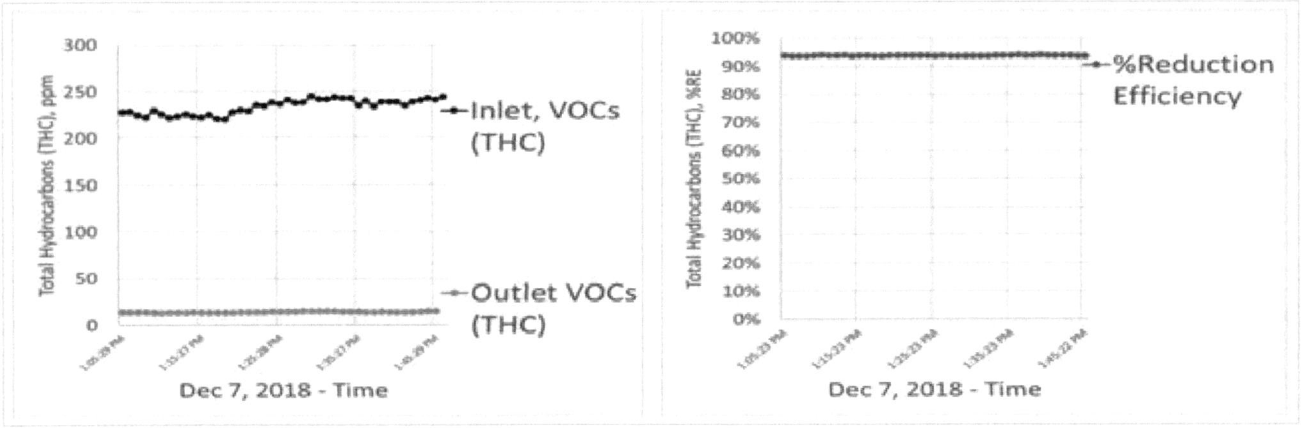

RESULTS OF VOC REDUCTION EFFICIENCY PERFORMANCE MEASURED BY FLAME IONIZATION DETECTOR (FBC DEMONSTRATION UNIT):
Achieved >90% Reduction Efficiency meeting PCWP MACT Standard[34]

[33] Emissions results FBC 2018 at commercial wood drying site done by Advanced Industrial Resources, Inc. Results were demonstrated during three 1-hour testing intervals. https://airtest1.com/ . (Accessed September 18, 2021)

[34] "Electronic Code of Federal Regulations (eCFR)". eCFR website. See https://www.ecfr.gov/cgi-bin/text-idx?node=sp40.13.63.dddd#ap40.14.63_12292.1 (Accessed April 16, 2020).

RESULTS OF VOLATILE ORGANIC COMPOUNDS CAPTURE:

Terpenes were collected from the FBC demonstration unit. These terpenes are a mixture of terpene hydrocarbons, primarily Alpha and Beta pinenes. These terpenes are responsibly-recovered, biobased materials. Potential applications for these terpenes include use in fuels, fragrances, solvents, adjuvants, cleaners, and polyterpene resins.

TERPENES COLLECTED FROM FBC DEMONSTRATION UNIT

VALUE OF TERPENES:

The Pine Chemicals Association has documented the gallons of terpenes per oven dry ton of wood in a variety of tree species. Typically, there is ~1 gallon of turpentine (terpenes)[35] per oven dry ton of wood product produced from green pine wood from the SE part of the United States, i.e. Loblolly Pine, Jack Pine, Slash Pine, Shortleaf Pine, Ponderosa Pine, Pitch Pine, Pond Pine, Red Pine, and Spruce Pine. If 75% of this is evaporated and captured from the drying process, the value of the terpenes per year can be calculated by multiplying this yield by the oven dry tons (ODT) of wood produced and a typical price per gallon[36] for terpenes. For example:

$$\left(\frac{1\ Gallon\ Terpenes}{ODT\ Wood}\right) * \left(\frac{ODT\ Wood}{Year}\right) * (\%Yield) * \left(\frac{\$\ Dollars}{Gallon\ Terpenes}\right) = \$\ /\ Year$$

TABLE VI
TERPENES TYPICAL PROPERTIES:

Property	Result
Appearance	Clear Liquid
Moisture	<0.1%
Weight per Gallon	7.2 pounds per gallon
Specific Gravity	0.86
Sulfur Content	< 10 ppm

---

[35] "Sulfate Turpentine Recovery". Pulp Chemicals Association. Pages 9-12, 127. https://www.pinechemicals.org/store/viewproduct.aspx?id=1579188 (Accessed Jan 2, 2021).

[36] "Price Chart for Gum Turpentine". PineChem website. http://www.pinechem.net/images/gall/ChartGNE.GIF Accessed Jan 3, 2021).

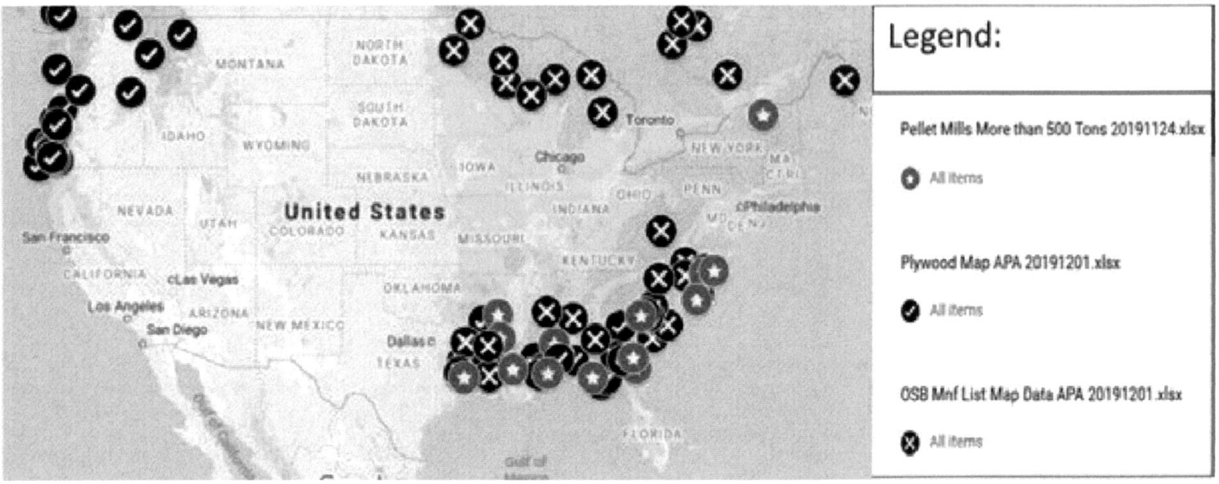

MAP OF POTENTIAL WOOD DRYING SITES FOR FBC USE:

This map shows potential sites where various wood products are produced where the FBC may be used to control emissions. Potential sites include but are not limited to Oriented Strand Board (OSB), lumber (sawmills)[37], plywood[38], wood to biofuels, i.e. wood pellets[39], particleboard, Medium Density Fiberboard (MDF), etc.

RESULTS OF DISCUSSIONS WITH ENVIRONMENTAL AGENCIES:

Official stack testing results were discussed with two state environmental agencies and two federal environmental protection agency (EPA) personnel. In all discussions, environmental agency personnel were supportive of providing a permit for full size commercial FBC unit.

FINAL RESULTS FROM WOOD DRYING DEMONSTRATION. THE UNIT:

- Provided salable terpenes, composed primarily of Alpha and Beta Pinenes.
- Reduced greenhouse gas (GHG) emissions.
- Reduced energy usage. It eliminated natural gas usage and reduced electrical usage.
- Met the PCWP MACT requirement of >90% Reduction Efficiency[40] in both tests. In the second test, the system met the compliance requirements three times in a row, over each of the three one-hour testing intervals.
- Demonstrated good operational reliability.
- Enabled State and Federal environmental personnel that reviewed these results to support permitting for a full commercial FBC system.

---

[37] "The mill map allows you to visually locate and drill down to find the mill you're interested in." Primary Forest Products Network website. https://primary.forestproductslocator.org/mill-map (Accessed September 18, 2021).

[38] "Manufacturer Directory". Engineered Wood Association website. https://www.apawood.org/manufacturer-directory (Accessed September 18, 2021).

[39] "Biomass Magazine". Biomass Magazine website. See http://biomassmagazine.com/plants/map/pellet/ (Accessed September 18, 2021).

[40] "Electronic Code of Federal Regulations (eCFR)". eCFR website. See https://www.ecfr.gov/cgi-bin/text-idx?node=sp40.13.63.dddd#ap40.14.63_12292.1 (Accessed April 16, 2020).

IX. CONCLUSIONS:

The FBC provides an affordable, reliable, and accessible clean technology for controlling VOCS, HAPS, and even particulate in some applications. This technology is well established in many applications including automotive and electronics manufacturing where it has been used successfully for many years. It is an emerging technology for use in biomass (i.e. wood drying) applications.

Especially for process exhaust streams with low VOC concentrations, the FBC technology is a patent[41] pending[42], cost competitive, industrial air pollution control technology that enables operators to advantageously:

1. Generate revenue via sales of recovered organic products (including biobased products such as terpenes in biomass applications) and sales of carbon offsets
2. Improve sustainability by reducing energy usage and thus greenhouse gas (GHG) emissions
3. Reduce operational costs by reducing energy usage (natural gas and electricity) and maintenance costs.
4. Enhance VOC mass transfer to the BAC media
5. Provide continuous in-situ BAC media regeneration and reactivation. This enables the media to continue to be reused.
6. Minimize water adsorbed on the BAC media thus maintaining its ability to adsorb VOCs, even in high moisture content process exhaust streams

Like any air pollution control technology installation, careful consideration of many design parameters, the unit's specific operating profile, and the operator's goals need to be taken into account during the design, installation, and startup/optimization processes. There are excellent resources, including Environmental C&C Inc.[43], Captis Aire LLC[44], and the Institute of Clean Air Companies (ICAC)[45] personnel, to assist with evaluating, implementing, and optimizing FBC applications.

---

[41] "METHODS AND SYSTEMS FOR RECOVERING TERPENE COMPOSITIONS FROM WOOD DRYING EXHAUST". World International Patent Office (WIPO) Website. https://patentscope.wipo.int/search/en/detail.jsf?docId=WO2019148208 (Accessed September 18, 2021).

[42] "METHODS AND SYSTEMS FOR CONTROLLING EMISSIONS FROM WOOD DRYING PROCESSES". WIPO Website. https://patentscope.wipo.int/search/en/detail.jsf?docId=WO2019148207 (Accessed September 18, 2021).

[43] "Environmental C&C Inc". Environmental C&C website. https://www.environmentalcc.com/ (Accessed September 18, 2021).

[44] "Emissions Control Technology Captures Emissions to Clean the Air". Captis Aire website. https://www.captisaire.com/ (Accessed November 21, 2021).

[45] "The Trusted Voice of the Clean Air Technology Industry". Institute of Clean Air Companies (ICAC) website. https://www.icac.com/ (Accessed September 18, 2021)

X. ACKNOWLEDGEMENTS:

The author recognizes inputs from a broad spectrum of personnel at the commercial OSB manufacturer and at Environmental C&C[46] including Harold Cowles, John Berger P.E., and Jim Starek.

CAPTIS AIRE CAPTURE-CONCENTRATE-CLEAN	Environmental C&C, Inc. Solvent Recovery & VOC Abatement Systems
Kim TUTIN Captis Aire LLC [47] 1612 W. Cleveland Ave, Suite 001 East Point, GA 30344-3213 Tel: +01 404 5802795, E-mail: kim@captisaire.com Website: www.captisaire.com	Harold COWLES Environmental C&C Inc. 16705 E. Calle de las Estrellas Road Scottsdale, AZ 85262-5982 Tel: +01 518 2250315, E-mail: halc@ecnc.com Website: https://www.environmentalcc.com/

[46] "Environmental C&C Inc". Environmental C&C website. https://www.environmentalcc.com/ (Accessed September 18, 2021).
[47] "Emissions Control Technology Captures Emissions to Clean the Air". Captis Aire website. https://www.captisaire.com/ (Accessed November 21, 2021).

631638 Conversion of Plastic Waste to Fuel: Assessment of Technologies and Economics

Abstract:

Plastics are extremely valuable to our society and are widely used across industries such as food, beverage, textile, healthcare, automotive, aerospace, construction, agriculture, and oil. The global production of plastics, accumulative from 1950 to 2017, is 8.3 billion metric tons, with 6.3 billion tons waste generated, of which only 9% has been recycled, 12% incinerated, with the remaining 79% landfilled or released into the environment. Mechanical recycling is used to recycle about 95% of plastic waste today, however, this method works well only for homogeneous and non-contaminated plastic waste, and for easily identifiable objects such as bottles and containers made of PET and PE. Chemical recycling can offer a solution in recycling many of the plastics which mechanical recycling fails to process. Chemical recycling involves the deconstruction of polymers to monomers for conversion back into polymers, or to intermediates that can be used as fuel or feedstock by the chemical industry. A number of chemical recycling technologies are being developed, and the processes are being tailored to handle specific feedstocks and/or produce desirable products. In this project we assess companies that are involved in the chemical recycling of plastic waste: their technology and/or key innovations, the feedstocks and product(s), and the current scale of operation. Further, we evaluate in this project the economics of chemical recycling with a specific interest on plastic-to-fuel conversion.

Copyright © American Institute of Chemical Engineers. All rights reserved.

623136 Continuous Back to Monomer Recycling of PET from Composite Materials Using a Twin-Screw Extruder

Abstract:

The current Packaging Act requires a 63% quota for the mechanical recycling of plastics from 2022 in Germany [1]. In order to meet this requirement, new efficient recycling technologies are necessary. Many plastics, such as polyethylene terephthalate (PET), are used in combination with other polymers. Apart from energy recovery, no recycling process for such multilayer material is available and it is therefore largely lost to the resource cycle.

The research project revolPET® specifically addresses this issue through the development of a continuous process for recovering the monomers terephthalic acid (TA) and ethylene glycol (EG) in primary raw material quality from mixed PET-containing waste [2]. While literature mainly reports batchwise depolymerization of PET in stirred tank reactors [3, 4], the revolPET® process uses an extruder as a continuous chemical reactor.

This contribution presents experimental results on the successful depolymerization of PET in composite materials into a TA salt and EG. The high flexibility of the extruder offers a wide range of options for process design in terms of thermal, mechanical and chemical conditions, so that suitable process windows are identified depending on the feed material. Key to this technology, the twin-screw extruder allows the two reagents to be mixed as solids. This innovative process intensification enables high degrees of depolymerization at average residence times below 2 minutes. In the experiments, a PET depolymerization degree of ≥ 97 % was achieved. Systematic parameter studies reveal the influence of temperature, filling degree of the screw chamber and residence time on the successful PET depolymerization. Figure 1 presents typical results for degree of depolymerization and residence time at varied screw revolutions.

In addition to the technical aspects, the contribution provides insight into the ecological assessment of this new recycling process. In particular, the challenge of the exothermic reaction to identify an optimal process window is shown. This topic is examined in more detail against the background of scaling. The presented experimental results will be discussed in detail and placed in the context of the current political debate.

Literature

[1] § 16 Absatz 2 VerpackG

[2] www.revolpet.eu

[3] A.M. Al-Sabagh, F.Z. Yehia, Gh. Eshaq, A.M. Rabie, A.E. ElMetwally, *Egyptian Journal of Petroleum* (2016) 25, 53-64, doi: 10.1016/j.ejpe.2015.03.001

[4] A. Oku, L.-C. Hu, E. Yamada, *J. Appl. Polym. Sci.* (1997) *63*, 595-601.

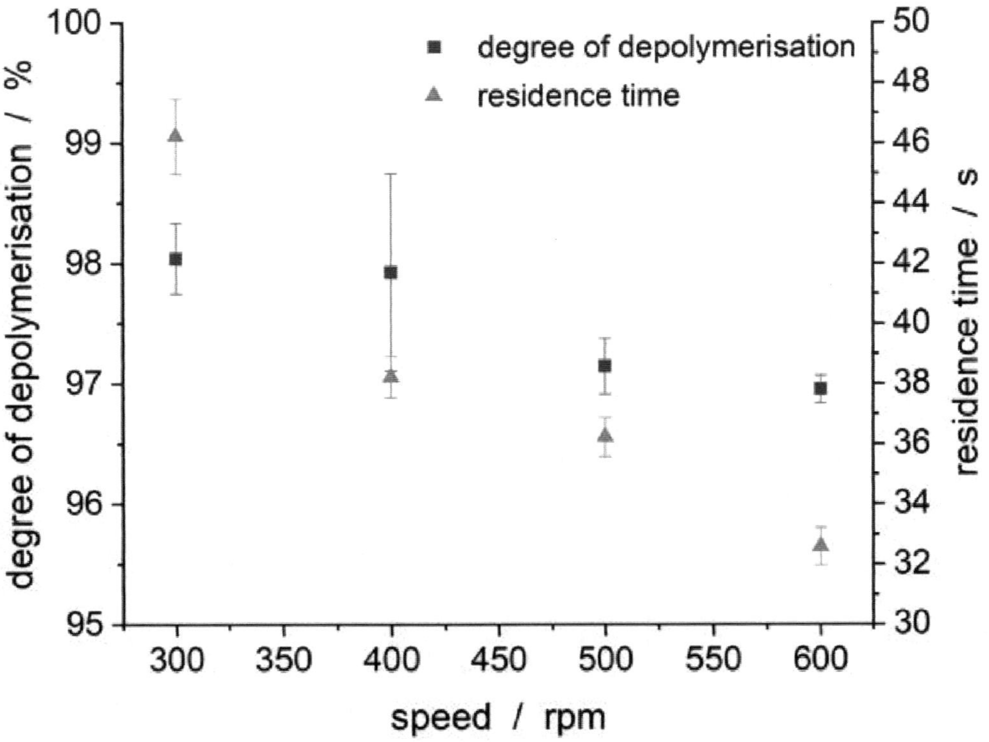

Figure 1: Degree of PET depolymerization in relation to the residence time at different speed

Copyright © American Institute of Chemical Engineers. All rights reserved.

629642 The Effect of Residence Time and Heterogeneous Catalysis on Continuous Hydrothermal Liquefaction of Food Waste

Abstract:

The United States releases a significant amount of methane through the decomposition of waste in landfills; methane is a greenhouse gas with a global warming potential 25 times greater than carbon dioxide [1]. According to the EPA, landfilling accounts for approximately 17% of the total US methane emissions [1], and food waste is the largest non-recyclable component of municipal waste [2]. Although food waste currently harms the environment, it also presents an opportunity to solve another looming threat: the US dependency on non-renewable fossil fuels. Mainstream Engineering, in collaboration with Worcester Polytechnic Institute, is developing technology to convert food waste to energy through hydrothermal liquefaction (HTL) that simultaneously addresses our reliance on fossil fuels and the environmental impact of food waste.

HTL converts food waste, and other wet wastes, into an energy dense biocrude by reacting the food mixture of carbohydrates, lipids, and proteins in near supercritical water. Traditionally, HTL reactions have been uncatalyzed or used a homogeneous catalyst (i.e., Na_2CO_3). We have shown previously that using a solid oxide, heterogeneous catalyst, $CeZrO_x$, in a batch reactor improves biocrude yield and energy recovery [3]. We have subsequently studied the effects HTL residence time and catalyst addition in a continuous stirred tank reactor (CSTR). A model food waste slurry was fed to the CSTR via a dual high pressure syringe pump system. The pressure of the system was maintained using a back pressure regulator rated to 300 °C and 5,000 PSI. Commercially available $CeZrO_x$ powder was compressed into tablets and immobilized in the CSTR using a catalyst cage. The catalyst tablets were able to withstand the hydrothermal environment without physically disintegrating. The stability of the catalyst tablets was characterized using XRD and BET before and after HTL. The energy recovery, product yields, and biocrude properties (HHV, TAN, and CHNO) were determined to characterize the effectiveness of $CeZrO_x$ in a continuous HTL reactor and evaluate the impact of residence time on continuous HTL of food waste.

References

[1] EPA. *Greenhouse Gas Emissions*. [cited 2021; Available from: https://www.epa.gov/ghgemissions/overview-greenhouse-gases.

[2] USDA. *Why should we care about food waste?* [cited 2021; Available from: https://www.usda.gov/foodlossandwaste/why.

[3] Maag, A.R., et al., *Catalytic hydrothermal liquefaction of food waste using CeZrOx*. Energies, 2018. **11**(3): p. 564.

Copyright © American Institute of Chemical Engineers. All rights reserved.

629415 Continuous Hydrotreatment of Sewage Sludge and *Spirulina* Algae Biocrudes from Hydrothermal Liquefaction: Different Catalysts for Different Organic Contaminants

Abstract:

Unprecedented global warming and raising public awareness are the driving factors to decarbonize long-haul transportation sector. In this regard, biomass could be a potential renewable source for the production of sustainable, carbon-neutral liquid fuels. Hydrothermal liquefaction (HTL) followed by catalytic hydrotreatment is one of the most promising pathways to covert varying feedstocks into drop-in fuels. The energetically dense black viscous product from HTL ("biocrude"), contains a diverse organic pool with considerable amount of oxygen (~5-18%), nitrogen (~1-8%) and metallic content (~0.1-0.6%). The presence of these organic contaminant results in poor miscibility with fossil fuels and poses new and exciting challenges e.g. thermal instability of biocrude at high temperatures which are indispensable for complete hydrodenitrogenation. This is why special attention is needed not only to realize this complex biocrude mixture but also to select different catalysts which performs best to saturate and remove certain organic contaminants under optimized conditions.

In this work, we utilized advanced analytical tools (i.e. FT-ICR MS etc.) and observed different classes of organometallics and nitrogen compounds from sewage sludge and *Spirulina* algae biocrude. From these results, we utilized three different types hydrotreating catalysts based on their porosity and nickel loading on γ-Al_2O_3 support and successfully demonstrate smooth continuous hydroprocessing operations (Figure 1) for hundreds of hours without coke formation, catalyst deactivation, and reactor plugging. Finally, 100% deoxygenation and ~96% denitrogenation were achieved. Furthermore, hydrotreated products underwent true-boiling point fractional distillation and showed on-spec metal content, density, pour point, cloud point and HHV for diesel and jet fuel.

Figure 1: Hydrotreated samples from the continuous hydrotreatment of sewage sludge (a) and *Spirulina* algae (b) biocrude.

Acknowledgement: This research has received funding from the European Union's Horizon 2020 research and innovation programme under grant agreement No 764734.

Copyright © American Institute of Chemical Engineers. All rights reserved.

627420 Review and Modeling of Organic Removal in Anaerobic Sludge Blanket Bioreactors for Methane Production and Energy Recovery from Domestic Wastewater

Abstract:

The current domestic wastewater treatment paradigm centers on aerobic wastewater treatment technologies such as activated sludge. While effective in meeting U.S. EPA regulated discharge standards for organics and suspended solids, providing oxygen to aerobic microorganisms is energy-intensive, typically accounting for ~50% of wastewater treatment facility costs. In total, wastewater conveyance and treatment accounts for ~3% of our nation's electrical energy usage. Direct anaerobic treatment of dilute wastewater is a viable methane-generating alternative to aerobic wastewater treatment. Anaerobic microorganisms convert wastewater organics (fats, carbohydrates, proteins, etc.) to methane-rich biogas, which can be harnessed to create energy via technologies such as combined heat & power. If more energy is generated from the methane produced than is used to run the process itself, then the process can be considered "energy-positive" (or a net energy producer). Anaerobic sludge blanket bioreactors, most notably upflow anaerobic sludge blanket (UASB) bioreactors and anaerobic baffled reactors (ABRs), have been studied at bench- and pilot-scale for several decades. UASBs are also currently operated at full-scale in several locations worldwide, mostly in warmer climates. Despite the number of completed studies, no current review critically examines performance data from different sized sludge blanket bioreactors collected under varying conditions (e.g., varying wastewater temperatures and hydraulic residence times). Further, no current review study conducts predictive modeling (i.e., Monte Carlo Simulation) on aggregated sludge blanket bioreactor data to better understand the operation of such bioreactors under varying conditions. This study aggregates and examines data from 145 UASBs studies and 56 ABRs studies to understand bioreactor performance (e.g., chemical oxygen demand and total suspended solids removal) and methane generation. This study also employs Monte Carlo Simulation using @Risk to better predict bioreactor performance under varying conditions. Additionally, this study identifies critical gaps in literature (e.g., a lack of consistently reported methane generation) that may impact full-scale implementation in certain climates (e.g., cold weather climates). Last, results from this study can help inform decision makers to better understand of anaerobic sludge blanket bioreactors can meet regulatory standards or generate methane for renewable energy at their wastewater treatment facilities.

Copyright © American Institute of Chemical Engineers. All rights reserved.

629993 Utilization of Marble Dust As a Concrete Additive to Enhance Its Properties

Abstract:

With the advancement in the building technology, each and every step to alter the cement properties for making it environmentally beneficial and equally strong is important. In this advancement, one always opts for an alternative which provides the strength to the building by reducing its cost and weight.

Marble often known as calcium carbonate ($CaCO_3$) has been a well known material used as a building material since the ancient times. Marble is often available in combination with Silicon Oxide (SiO_2), Aluminum Oxide (Al_2O_3), Iron Oxide (Fe_2O_3), Magnesium Oxide (MgO) and Calcium Oxide (CaO).

Marble is utilized at a number of operations and the waste obtained requires an extra effort for the adequate environmental disposal. The waste marble left in the environment can be harmful therefore must be utilized for an alternative process to reduce its harmful effect. The addition of marble dust to concrete is proving to be an effective alternative for the process. The process of marble addition is participating in the hydraulic reactions contributing significantly to the composition and microstructure of hydrated products. Another way of utilizing the marble dust is to utilize it in the geo-polymer concrete which will be strengthening the concrete properties.

The entire paper is evaluating the utilization of the marble dust in concrete so that the unsafe waste of marble is used and fewer efforts is to be taken for its disposal and also to accomplish viability, economy and environmental conservation.

Copyright © American Institute of Chemical Engineers. All rights reserved.

632259 Process Sustainability Metrics Based on the Safe and Just Operating Space for Humanity

Abstract:

Sustainability assessment methods aim to relate the environmental, economic, and social dimensions of sustainability to help with decision making in process designs. Sustainability assessment methods use metrics and indicators to quantify impact and performance of specific parameters to provide comprehensive information that can be communicated to decision makers. These metrics are critical to identify the shortcomings and allocate resources to improve process performance and design.

Integrating sustainability assessment in chemical process design has gained interest due to the major role of chemical industry in economic, societal, and environmental sectors. The chemical industry is considered one of the primary consumers of natural resources and generators of waste and emissions, which impose significant stress of ecosystem goods and services.

The triple bottom line concept (TBL) recognizes the environmental and social dimensions in sustainability assessment have been popular in the assessments of engineering decisions. Popular sustainability assessment frameworks in the chemical engineering literature such as the indexes developed by AIChE[1] and IChemE[2], and the GREENSCOPE[3] tool. These methods and tools evaluate different indicators based on the TBL dimensions separately, and then compare the alternatives in an aggregated manner.

One of the major shortcomings of these methods is ignoring the interactions and links between the different sustainability dimensions. That includes the true hidden costs of extracting resources and their impact on the natural ecosystems and social aspects such as justice and equity. Most methods only consider the demand of natural resources and ignore the capacity and limits of ecosystems to provide these goods and services, which can lead to the degradation of the ecosystems. Existing methods also do not consider social issues related to equitable access to resources and economic disparities.

The important role of ecosystems goods and services in sustainable process design have been discussed in the literature, and several efforts to quantify and include the supply and demand of ecosystems in process design have been developed[4,5]. Other frame works such as the planetary boundaries[6,7], which defined the "safe space" and upper limits of earth systems processes, have been implemented in assessment studies on a global scale. An extension of this framework which introduced the social dimension in terms of the "just space", propose lower limits that is necessary to meet the minimum social requirements for human wellbeing from natural resources[8].

Since most of the design decisions are made on a process level, assessment methods need to be implemented on this scale. This ensures correct evaluation with comprehensive information and identifies the direct stakeholders. The planetary boundary framework fails to identify the safe space on finer scales and downscaling of the framework does not provide proper assessment results. There is also a challenge in identifying the connection between the "just space" and the "safe space" due to the knowledge gap between disciplines.

Ecosystems provide goods and services that represent the foundation of all economic activities that is utilized by societies. This connection is key to protect natural resources and improve the quality of human life. This work proposes boundaries to identify the "safe space" in terms of the capacity of ecosystems to provide the related goods and services. This ensures no degradation of ecosystems and provides the flexibility to be applied on a local and service shed scales.

To achieve sustainability, basic human needs from ecosystems must be met, this includes water, food, and energy. Similar to the safe space, this work proposes boundaries to identify the "just space" in a way that insures human basic rights from ecosystem services are fulfilled without causing any stress on ecological systems. These needs can be quantified by assessing the goods and services stakeholders need from the ecosystem. For example, basic demand of water for drinking, cooking, and sanitation can be calculated based on the population utilizing a watershed. This demand is then subtracted from the supply of the ecosystem represented in the "safe space" to form a more comprehensive "safe and just operating space".

By considering the capacity of ecosystems, this work proposes sustainability metrics that combine the ecological and social boundaries to identify a sustainable design space that respects nature's limits and ensure essential human demands are met. This work does not attempt to quantify the minimum social demand for the objectives proposed by the just space framework, like gender equality and voice. However, this work does quantify the social objectives that are related to ecosystems goods and services, such as food, water, and energy required to meet basic human needs.

This work broadens the scope of current assessment methods to include the social dimension of sustainability. The developed metrics can be utilized by assessment methods to help with process design on different scales. These metrics will be applied on case studies of chemical manufacturing processes to demonstrate the applicability of the assessment methods on multiple scales, and to identify improvement or alternatives of the process design. The results from this work can assess the performance of chemical process and provide a communication tool to stakeholders on different scales and help in the decision-making process. This can be applied to evaluate existing designs, or to assess new technologies and more sustainable processes. This effort provides a big step towards holistic sustainability assessment by accounting for the minimum social requirements while respecting nature's limits, which contributes to human wellbeing and the prosperity of ecological systems.

References:

1- Calvin Cobb. The AIChE Sustainability Index: The Factors in Detail | AIChE. Technical report, 2009.

2- Inst. Chem. Eng. Sustainability Metrics: Sustainable Development Progress Metrics Recommended for Use in Process Industries. Technical report, Warwickshire, UK, 2014.

3- Gerardo J. Ruiz-Mercado, Raymond L. Smith, and Michael A. Gonzalez. Sustainability indicators for chemical processes: II. Data needs. Ind. Eng. Chem. Res.,feb 2012.

4- Bhavik R. Bakshi, Guy Ziv, and Michael D. Lepech. Techno-ecological synergy: A framework for sustainable engineering. Environ. Sci. Technol., 49(3):1752-1760, feb 2015.

5- Anders Bjorn, Manuele Margni, Pierre Olivier Roy, Cecile Bulle, and Michael Zwicky Hauschild. A proposal to measure absolute environmental sustainability in life cycle assessment. Ecol. Indic., 63:1-13, apr 2016.

6- Rockström, J., Steffen, (2009). A safe operating space for humanity. nature, 461(7263), 472-475.

7- Steffen, W., Richardson, K., Rockström, J. (2015). Planetary boundaries: Guiding human development on a changing planet. Science, 347(6223).

8- Raworth, K. (2012). A safe and just space for humanity: can we live within the doughnut?. Oxfam.

Copyright © American Institute of Chemical Engineers. All rights reserved.

625514 Multiscale Sustainability and Metrics Correlation Characterization and Utilization

Abstract:

Multiscale sustainability is receiving increasing attention in sustainability science and engineering. This is mainly due to the recognition that industrial sustainability problems should not be dealt with only at, for example, the product and/or process scale; rather, the system boundary should be extended so that certain critical sustainability issues in the other scales can be addressed simultaneously. The solutions derived in this way could be more systematic and effective. Recent advances in multiscale sustainability research show that a number of methodologies are developed to investigate the problems at the scales from the product/process/plant level to industrial zones or even ecosystems, and the solutions are more systematic and general. It is anticipated that more methodological and applied studies on the scales of process/plant and up will be reported in the near future.

It should be noted that certain types of industrial sustainability problems, especially those related to product design and manufacturing, could be originated to the issues that should be addressed at a finer scale, such as a mesoscale or even a microscale of length-time. There is a serious lack of methodological study on this type of multiscale sustainability problems. In this paper, we will discuss research opportunities and challenges in this challenging area and then describe a general methodology for developing sustainability indicators at the macro, meso, and nanoscales, for characterizing those indicators and identifying correlations among them. Succeedingly, we will introduce a methodology for selecting a minimum set of indicators and conducting coordinated multiscale sustainability assessment. Further, we will compare the effectiveness of the multiscale sustainability assessment with a usual single-scale based assessment in solution identification and implementation. A case study on developing sustainable nanocoating material and manufacturing will be illustrated to demonstrate methodological efficacy.

Copyright © American Institute of Chemical Engineers. All rights reserved.

625618 Eco-Efficiency Analysis of Different Hydrolysis Processes for p-Xylene Production

Abstract:

The role of bio-based chemicals in supporting industrial sustainability has become increasingly important in view of the rigid environmental regulations being enforced at a global level[1, 2, 3]. More specifically, the renewable nature of biomass used as raw material, and the low life cycle carbon emissions render bio-based chemicals a promising alternative to conventional petroleum-based products[4]. However, the high production cost, mainly related to the feedstock price and employed biomass conversion technology, seems to limit the broader deployment of bio-based products in the energy and chemical market. Efforts are currently focused on the development of more energy-efficient, cost-competitive, and less carbon-intensive technologies for the conversion of selected biomass feedstocks to different bio-products[5, 6].

A novel synthesis route of p-Xylene has been proposed which employs molten salt hydrate (MSH) for the hydrolysis of biomass[7]. The overall process comprises four major steps: 1) hydrolysis of biomass which commercially utilizes diluted acid (DA) and concentrated acid (CA) 2) dehydration of sugars 3) hydrodeoxygenation of 5-hydroxymethylfurfural and 4) cycloaddition of 2,5-Dimethylfuran. To evaluate the sustainability performance of the specific bio-based chemical which is assumed to be synthesized in a modular plant, an eco-efficiency analysis is carried out considering different supply chain logistics and lignocellulosic biomass feedstocks with respect to composition and crop production process. The employed eco-efficiency concept is based on the BASF methodology[8], which aggregates the relevant environmental impact category values of each selected process/product scenario into a single score through normalization and weighting – reflecting social preferences-, and combines that with the respective life cycle costs (LCC). Expanding on the previous studies of techno-economic and life cycle assessment of p-Xylene using hardwood/softwood biomass feedstock[7], the eco-efficiency scores of both novel and conventional hydrolysis techniques are estimated and presented in a two-dimensional portfolio matrix, enabling the identification of the best performing process design from both ecological and economic perspectives. The selected biomass feedstock is assumed to be delivered to the biorefinery plant by means of trucks at a distance of 100 km from the biomass conversion facility, while the production capacity is set at 400,000 metric ton of treated biomass feedstock per year[7].

In terms of the climate change impact category, the estimated CO_2-eq emissions of the MSH process are 75% lower and 9% higher than those of the DA and CA processes, respectively. In the case of water depletion impact category, the MSH process exhibits a lower value by 44% and 7% as compared to the aforementioned hydrolysis techniques. In terms of process economics, the minimum selling price of the p-Xylene produced by means of the MSH, DA and CA processes is estimated at \$1,480, \$2,320 and \$1,900 per metric ton[7], correspondingly. Upon weighting and normalizing the life cycle impacts of each synthesis route, the eco-efficiency profile of the MSH process has proved to be better with respect to both overall environmental impact and LCC as compared to the other processes, rendering the novel technology the most eco-efficient process design option under the examined operating conditions and selected biomass feedstock. The outcomes of this eco-efficiency analysis, evaluating various supply chain designs, and biomass feedstocks besides hardwood/softwood, as well as different sets of weighting factors used in the eco-efficiency methodology, provide important insights into the major cost drivers and environmental hotspots determining the sustainability profile of p-Xylene production.

REFERENCES

1. **Jong, E. de, A. Higson, P. Walsh, and M. Wellisch**, "Bio-based Chemicals - Value Added Products from Biorefineries," (2011).
2. **Hatti-Kaul, R., U. Törnvall, L. Gustafsson, and P. Börjesson**, "Industrial biotechnology for the production of bio-based chemicals - a cradle-to-grave perspective," *Trends in Biotechnology*, **25** (3), pp. 119–124 (2007).
3. **Song, J.H., R.J. Murphy, R. Narayan, and G.B.H. Davies**, "Biodegradable and compostable alternatives to conventional plastics," *Philosophical Transactions of the Royal Society B: Biological Sciences*, **364** , pp. 2127–2139 (2009).
4. **Zheng, J., and S. Suh**, "Strategies to reduce the global carbon footprint of plastics," *Nature Climate Change*, **9** (5), pp. 374–378 (2019).
5. **Lee, S.Y., R. Sankaran, K.W. Chew, *et al.***, "Waste to bioenergy: a review on the recent conversion technologies," *BMC Energy*, **1,** 4, pp. 1–22 (2019).
6. **Sankaran, R., R.A. Parra Cruz, H. Pakalapati, *et al.***, "Recent advances in the pretreatment of microalgal and lignocellulosic biomass: A comprehensive review," *Bioresource Technology*, **298**, p. 122476 (2020).
7. **Athaley, A., P. Annam, B. Saha, and M. Ierapetritou**, "Techno-economic and life cycle analysis of different types of hydrolysis process for the production of p-Xylene," *Computers and Chemical Engineering*, **121**, pp. 685–695 (2019).
8. **Grosse-Sommer, A.P., T.H. Grünenwald, N.S. Paczkowski, R.N.M.R. Van Gelder, and P.R. Saling**, "Applied sustainability in industry : The BASF eco-eEfficiency toolbox," *Journal of Cleaner Production*, **258**, p. 120792 (2020).

Copyright © American Institute of Chemical Engineers. All rights reserved.

625466 Reinforced Sustainability Assessment and Decision Making in the Post-COVID-19 Manufacturing Era

Abstract:

The COVID-19 pandemic has ushered in various unprecedented challenges and severely disrupted economic and social systems worldwide. In the U.S., one of the hardest-hit segments by the pandemic is the manufacturing industry. As the negative effects of the COVID-19 pandemic start to recede, industries look beyond the fog of uncertainty towards long-term changes, and develop strategies for more sustainable and resilient development in the post pandemic era. However, their endeavors have encountered a number of challenges. Answers have been sought to the questions about the effectiveness and completeness of traditional sustainability metrics used in sustainability assessment, and impacts on the decision making process due to a possibly extended sustainability assessment scope as well as sustainability goal resetting.

In this paper, we will review the scope of traditionally practiced metrics in economic, environmental, and social sustainability categories as well as their sub-categories and discuss the insufficiency for the sustainability assessment of systems when experiencing severely uncertain, disruptive events. We will then propose a number of new types of sustainability metrics and a method for necessary consolidation of new and existing metrics to avoid overlapped coverage of measurement aspects. It is conceivable that adoption of a reinforced sustainability metrics system will impact the decision making process, especially due to the involvement of severe uncertainty. We will introduce an Analytical Hierarchical Process (AHP) based decision-making method to derive more systematic, transparent, and robust decisions. The method is designed for reproducibility. It does not over burden those using the framework, as it gives detailed instructions on the information that is needed. The efficacy of the framework will be demonstrated by the study on small and medium sized electroplating facilities for sustainable development in the post COVID-19 manufacturing era.

Copyright © American Institute of Chemical Engineers. All rights reserved.

628162 Data-Driven Life Cycle Assessment Accounting Uncertainties

Abstract:

Ideally, primary data collection is recommended for every life cycle assessment (LCA) study. However, due to limited availability or accessibility to first-hand data, related sources of secondary data can be a good alternative in practice. In this work, various sources of uncertainty in using secondary data from the ecoinvent Life Cycle Inventory (LCI) database [1] are illustrated with an LCA case study on global air travel.

Among the three linked system models ecoinvent provides [2], the cut-off model is used in this work with raw data retrieved from the database to build technology and intervention matrices [3]. Inside the database, the basic uncertainty associated with parameters' intrinsic variability can be modeled with one of the seven different statistical distributions provided by the database while the additional uncertainty associated with imperfect data quality is quantified with the five criteria in pedigree matrix. The two types of uncertainty can be combined via the pedigree approach [4] and then assessed with a Monte Carlo simulation. Given the recent effort of ecoinvent to derive uncertainty factors for the pedigree matrix based on empirical studies [5], the effect of updated pedigree matrix coefficients is also evaluated.

Furthermore, since unit processes in the database provide information on gate-to-gate LCIs, analysis on emission hotspots along the supply chain of air travel can be conducted by recursively expanding upstream processes until desired level of details is achieved. Such analysis also quantifies the sensitivity with respect to the choice of system boundary and helps to prune away insignificant processes during subsequent data collection and/or refinery efforts.

When using ecoinvent as a source of secondary data for LCA studies on real world processes, uncertainty arises not only from the data and parameters inside the database, it can also come from the imperfect mapping between processes available in the database and the real-world process of interest. Concretely, for the LCA case study on air travel, flight specific parameters, e.g., plant type and occupancy level, are not explicitly modeled in the database. Instead, global averages or fixed-value assumptions are used, which could introduce additional uncertainty when using ecoinvent to calculate CO_2 emissions for a real flight. To combat this issue, the International Civil Aviation Organizations (ICAO) carbon emissions calculator [6] can be adopted, but it also suffers from the following limitations. Firstly, the ICAO calculator only provides direct CO_2 emissions during air travel. Other types of environmental flows and indirect emissions are not considered. Secondly, only a single value is estimated by the calculator without any uncertainty quantification. Therefore, a two-step method is proposed to leverage on the advantages of both sources. In particular, the more accurate data from ICAO calculator is used for fuel consumption while the more comprehensive ecoinvent database is used for calculating LCIs (including but not limited to fossil CO_2) per unit fuel consumed with uncertainty evaluated from ratio estimators using the technique of Monte Carlo simulation [7]. The developed framework can be generally used for any process when conducting LCA, with specific results in this work informing global air travel related life cycle impact.

References

[1] Wernet G, Bauer C, Steubing B, Reinhard J, Moreno-Ruiz E, Weidema B. The ecoinvent database

version 3 (part I): overview and methodology. Int J Life Cycle Assess 2016;21:1218–30. https://doi.org/10.1007/s11367-016-1087-8.

[2] ecoinvent. System Models in ecoinvent 3 2013. https://www.ecoinvent.org/database/system-models-in-ecoinvent-3/system-models-in-ecoinvent-3.html (accessed April 10, 2020).

[3] Guillaume B. Building matrices from spold files. 2013.

[4] Muller S, Lesage P, Ciroth A, Mutel C, Weidema BP, Samson R. The application of the pedigree approach to the distributions foreseen in ecoinvent v3. Int J Life Cycle Assess 2016;21:1327–37. https://doi.org/10.1007/s11367-014-0759-5.

[5] Ciroth A, Muller S, Weidema B, Lesage P. Empirically based uncertainty factors for the pedigree matrix in ecoinvent. Int J Life Cycle Assess 2016;21:1338–48. https://doi.org/10.1007/s11367-013-0670-5.

[6] International Civil Aviation Organizations (ICAO). ICAO Carbon Emissions Calculator 2016. https://www.icao.int/environmental-protection/Carbonoffset/Pages/default.aspx (accessed November 30, 2020).

[7] Art B. Owen. Monte Carlo theory, methods and examples. 2013.

Copyright © American Institute of Chemical Engineers. All rights reserved.

633053 A Comparison of Environmental Impact of Various Silicas Using Dozn Green Chemistry Evaluator

Abstract:

Silica production constitutes a multi-billion USD market, which continues to grow and include a range of different silica grades/types (e.g. gel, colloids, precipitated, mesoporous, etc.).[1] The increasing demands of precipitated silica from low-value industries (e.g., rubber manufacturing) requires greener production routes which are also economical and efficient. While the high value silicas, such as mesoporous MCM-41 and SBA-15, are highly desired in medicine, pollution control and catalysis, their synthesis is resource and energy intensive.[2] Learning from biology, bioinspired and potentially sustainable synthesis of high value silica was reported recently with applications in air and water decontamination, biocatalysis, carbon capture, energy storage and drug delivery.[3] However, currently, there is a significant lack of data that can inform operational decisions based on sustainability of these different grades of current and emerging silicas.

In order to address this issue, in this study, we compare the environmental impact of a variety of low and high value silicas including fumed, precipitated, sol-gels, mesoporous and bioinspired products. By employing the DOZN™ tool, which is based on 12 Principles of Green Chemistry, to evaluate different forms of synthesis of chemically identical products, we aim to illustrate the impact of synthetic pathways on the sustainability of a product. Moreover, we quantify the impact of operational decisions, such as calcination vs solvent extraction for purification and drying temperatures. Thus, showing the sustainability costs of operational conditions when working at large scale.

We have found that, although all processes compared resulted in the same chemical product and even started off from largely the same precursors, their synthetic pathways and operational conditions still impacted all major categories of the 12 principles of green chemistry: resource use improvement, energy efficiency and hazard reduction. The results enabled comparison across each of the 12 principles for different silicas in order to identify key differences and reasons behind underpinning the greenness of these products. Therefore, we believe this comparative evaluation represents an ideal case study for the DOZN™ tool, showing its thoroughness against the simpler E-factors and specific yield analyses, as it includes important considerations like the renewability of feedstocks and the impact of catalysts on the atom economy of processes. Further, it also helps in identifying research areas to focus on in order to make a given synthesis greener.

References:

1. Manning, Brambilaa and Patwardhan, *Unified mechanistic interpretation of amine-assisted silica synthesis methods to enable design of more complex materials*. Molecular Systems Design & Engineering, 2021, Advance Article DOI: 10.1039/D0ME00131G.
2. Patwardhan, Manning and Chiacchia, *Bioinspired synthesis as a potential green method for the preparation of nanomaterials: Opportunities and challenges*. Current Opinion in Green and Sustainable Chemistry, **12**, 110-116, 2018.
3. Patwardhan and Staniland, *Green Nanomaterials*, IoP Publishing, Bristol, 2019.

Copyright © American Institute of Chemical Engineers. All rights reserved.

626304 Reduction of the Generated Sugarcane Vinasse in Distillation through HEAT Integration.

Abstract:

Despite the well-known benefits of ethanol as a renewable, environmentally friendly and fossil-free source of energy, the enormous and still increasing volume of sugarcane processed over the years has been generating impacts. Sugarcane vinasse is the most voluminous by-product generated in ethanol production and is characterized by high minerals and organic matter concentrations. Highly nutritional, it has the potential to replenish most of the minerals taken by sugarcane plants during the season, promoting natural development of the soil, reducing costs with chemical fertilizers. However, despite its benefits, it can also be highly pollutant if treated or disposed incorrectly, injecting excessive levels of compounds and organic matter in the soil, compromising crops efficiency by overfertilization, polluting rivers and groundwater, generating a high environmental impact. To mitigate such impacts in Brazil, environmental standards such as CONAMA 430/11 and CETESB P 4.231/15 defines criteria for the correct handling, storing and application of vinasse in the crop, leading the need to distribute it throughout wider areas, demanding a larger and more expensive infrastructure, thus shortening its economic radius for application. Vinasse is composed almost entirely by water (~ 95% in volume) being a fraction of it removed by evaporation in the end of the process, however vinasse reduction systems as applied today have poor efficiency, demanding extra steam from the plant, making it unfeasible in most cases. The solution proposed by this work uses pinch analysis and heat integration techniques between distillation columns and evaporators in order to efficiently reduce the volume of vinasse still during distillation step, using the heat generated by the system and demanding no extra steam. Preliminary process simulations showed promising results, dropping vinasse volume rate from 12:1 to 5:1 without the need of a dedicated evaporation system nor demanding extra steam from the plant.

Copyright © American Institute of Chemical Engineers. All rights reserved.

628178 European Energy Transition: Pathway to Net – Zero By 2050

Abstract:

With the European Green deal initiatives, Europe advances towards becoming the first carbon neutral continent by 2050. Using policy initiatives set by the European Commission, each member state is committed towards this 2050 goal. Along this commitment, the European Green deal also emphasises on the importance of interconnecting energy systems between member states. This is to ensure reaching the 2050 goal with a better energy infrastructure in the continent. To establish a net-zero carbon infrastructure in the European continent, many studies suggested relying on intermittent renewables. However, this raises the concern related to grid stability as energy production from intermittent renewables is not a stable process[1]. Hence, this brings the question of to what extent the EU can integrate these intermittent renewables and what better alternative technologies can be deployed instead. In terms of better alternatives, many studies have proved the effectiveness of deploying carbon capture and storage (CCS) and negative emissions technologies (NETs) for securing a stable grid[2], especially when an energy mix contains intermittent renewables. With this, a key question rises about how the energy system mix should evolve in the EU and what pathway is most economic, and how the pathway for each individual member state will vary. Thus, the objective of this work is to study the evolution of the electricity system in the EU to identify the economic pathways towards carbon-neutrality. To perform this case study, we chose the electricity sector as a representative and collected the necessary data. Such data consisted of installed capacity, demand profile, electricity prices, intermittent renewables availability, grid features, interconnections...etc. for each member state. These parameters were used as an input to the Electricity Systems Optimisation (ESO)[3] tool that optimises the system in terms of system cost. In this tool NETs and CCS are considered as technologies, and each member state is treated as a node connected to the other. The unique feature of this modelling tool to other EU electricity systems literature, is the consideration of grid features and grid expansion in terms of interconnections for each member state. The novelty of this work is shown through considering various factors in the model. First of which is the new carbon target; net-zero, where most studies in the literature considered the old carbon target (85% - 95% reduction) with a reference point of 2015. In addition to this, country level policies such as early decarbonisation and nuclear retirement were not included in most of the modelling work. Hence, this work addresses these common literature gaps by considering the most recent available data, country level policies and third countries that EU member states are connected to.

To conduct our assessment, we first assessed the current status of the EU in terms of previous carbon targets. This is to gain an image of how each member state's energy mix has evolved and whether previous carbon targets were met or not. From this, it was strongly observed that each member state is moving in a different direction as some were becoming more fossil fuel intense. This in return showed us how EU member states were not working collaboratively to meet their targets and hence, we next studied whether this was to their advantage or not. To quantify this in the ESO tool, we used grid constrains and emissions target to build our scenarios. In our first scenario, we implemented member states' grids connection in a way that they form one EU grid and each member state can contribute to the other's grid requirements. With this scenario, we also dedicated the net-zero target to the continent itself, (e.g. some countries can deploy NETs to offset emissions from other member states). In the second scenario, we separated the grids of each member state and connected them via interconnection lines. This is to ensure that expansion of the grid occurs within the member state itself and grid requirements are met solely by the member state. In terms of emissions, the member states in the second scenario were to target net-zero emissions separately. To asses which grid configuration is better, the results from these scenarios were compared based on total system cost and carbon intensity. The conclusion drawn from this analysis is the importance of EU member states to move forward by working together to reach the 2050 net-zero target. This is due to the benefits gained from sharing resources, as it was observed within the model how these resources vary based on geographical location.

References:

1. Heuberger CF, Mac Dowell N. (2018) ''Real-world challenges with a rapid transition to 100% renewable power systems,'' Joule, March, pp. 367 – 370.
2. Fajardy M, Mac Dowell N. (2017) ''Can BECCS deliver sustainable and resource efficient negative emissions?,'' Energy & Environmental Science, 1389 – 1426.
3. Heuberger CF, Bains PK, Mac Dowell N. (2020) ''The EV-olution of the power system: A spatio-temporal optimisation model to investigate the impact of electric vehicle deployment,'' Applied Energy, January, pp. 113715.

Copyright © American Institute of Chemical Engineers. All rights reserved.

623643 *The Potential Role and Value of Power-to-Gas Storage in the UK Energy System*

Abstract:

As the power system in the UK undergoes decarbonisation, the transformation of the capacity mix remains a topic of discussion. Not only novel generation technologies, such as power plants with carbon capture and storage (CCS), but also storage technologies are to be considered. The increasing share of intermittent renewables in the energy mix creates the need for higher storage capacity and/or more flexible plants. Furthermore, the additional requirements of other sectors of the energy system, specifically the added demand from electric vehicles (EVs) and from heat pumps (HP), may impact the power demand profile and not only raise overall demand but also increase peak demand. This raises the question which storage technologies and how much storage capacity is required to satisfy the power demand at all times.

In the UK context, pumped hydro storage (PH) is already a part of the system capacity. Interconnection with neighbouring countries is also used to balance demand peaks and renewables fluctuations. These technologies, however, are mainly used for short-term storage, as in, storage within a day or between days. It is of interest if long-term storage, such as power-to-gas storage technologies, could also have a place in the capacity mix. In this work, we therefore examine under which circumstances power-to-gas storage has value in the future UK energy system under a net zero carbon target by 2050. Specifically, we model a route of power to synthetic methane to salt cavern storage to CCGT (P2M) with recycling of CO_2, which has been determined to be a promising power-to-gas storage route based on techno-economic analysis by Yao *et al.*[1]. We use the electricity systems optimisation model with capacity expansion (ESO-X) developed by Heuberger *et al.*[2]. The model constraints include build rates, ancillary services requirements, plant flexibility constraints, etc. We use one year of full-hourly renewables availability (solar, onshore wind and offshore wind) and demand profiles. This allows the analysis of seasonal storage. The demand is considered to be comprised of three parts, power demand, added demand from electric vehicles, and added demand from heat pumps as shown in Figure 1. Owing to the high amount of uncertainty regarding the level of electrification for both transport and heat, we consider four scenarios for both with varying levels of electrification ranging from minimal to total. The same applies to the generating technologies allowed in the system. We analyse scenarios with and without gas-fired power plants with CCS (CCGT-CCS) as a form of stable, low-carbon generation, and bioenergy with CCS (BECCS), as negative emissions technology, since the deployment of these technologies is uncertain. Furthermore, we analyse the value of key technology parameters of P2M such as input-to-output power ratio and round-trip efficiency and perform sensitivity analysis regarding the CAPEX of the technologies.

Excerpts of the dispatch schedules for scenarios with and without CCS and P2M are shown in figure 2. Our results suggest that in a scenario where all technologies are available for deployment with historical build rate limits, the increasing seasonality of the demand curve and the fluctuation of the increased share of renewables would optimally be absorbed by a combination of all archetypes of technologies. Renewables contribute a large share of power on days with high availability. CCGT-CCS and CCGT provide load-following and flexibility on peak days, respectively. The continuous operation of BECCS offsets emissions from gas-fired power plants. Using our central numbers for technology features and costs, P2M would have value in a cost-optimal technology mix by absorbing excess renewable energy and releasing it on peak days and hours (around 6 GW with 3.6 TWh storage capacity in 2050).

When CCS is removed from the capacity mix, P2M becomes a key technology. In our scenario without CCS or P2M, meeting demand with renewables and short-term storage (PH and battery storage) requires large amounts of curtailment, incurs lost load, and hence doubles the total system cost. When P2M as seasonal storage is added (26 GW and 13 TWh installed in 2050), it is used in the system to absorb power from renewables instead of curtailing it and providing flexible generation on low-availability days. This reduces the necessary amount of renewables capacity substantially. Notably, in scenarios with P2M available, it is deployed in greater amounts than battery storage despite the higher CAPEX and lower round-trip efficiency, indicating the increased need for seasonal as opposed to short-term storage.

Our sensitivity analysis shows a link between the deployment of onshore wind and power-to-gas storage. Lower seasonal storage CAPEX enables the deployment of both in conjunction, raising the renewables capacity as shown in Figure 3. Furthermore, our model suggests that renewables in combination with seasonal storage could be cost-competitive with CCGT-CCS in case of continuously high renewables build rates. We further identified the ratio of charging power to discharging power of the power-to-gas technology as a parameter with impact on its deployment potential and found that cost reductions of the technology may provide more value to the system than technology improvements.

In summary, grid-level power-to-gas storage may have a place in the future energy system in the UK. Whether or not it has value depends heavily on the build rates of renewables and on the relative cost of power-to-gas storage and low-carbon dispatchable power such as CCGT-CCS. Our results underscore the value of a diverse energy system with a mix of multiple technologies for generation and storage.

References

1. Yao JG, Bui M, Mac Dowell N. Grid-scale energy storage with net-zero emissions: comparing the options. *Sustainable Energy & Fuels*11 (2019): 3147-3162.
2. Heuberger CF, Rubin ES, Staffell I, Shah N, Mac Dowell N. Power capacity expansion planning considering endogenous technology cost learning. *Appl Energy*. 2017;204:831-845.

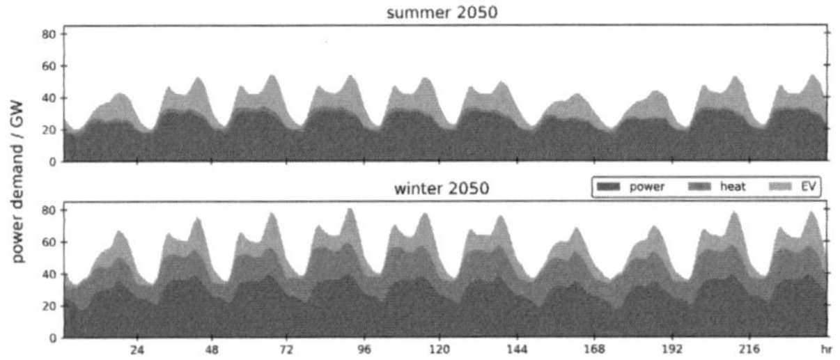

Figure 1: Example of the demand profile used in the model. Hourly power demand, added demand from heat pumps and electric vehicles, for ten days in summer and winter.

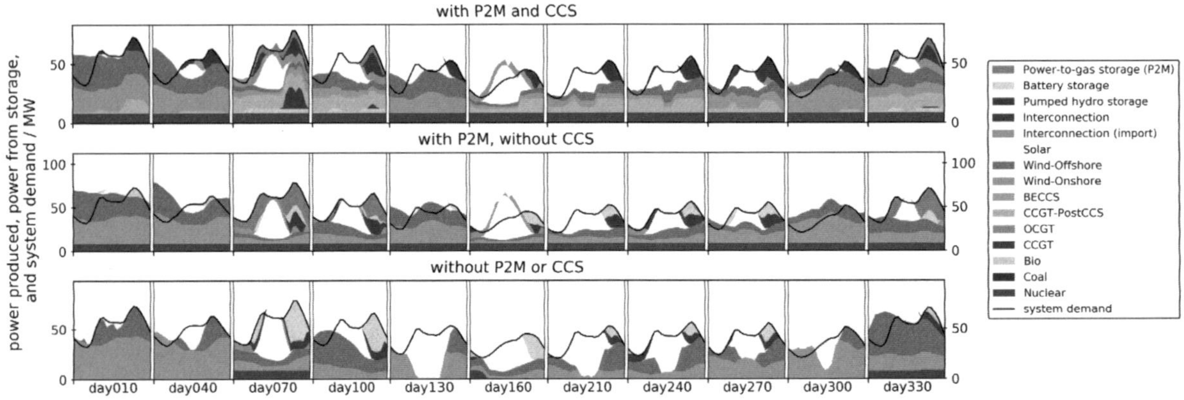

Figure 2: Excerpts of hourly dispatch schedules throughout the year (power produced and power from storage) for central electrification scenario. Stack higher than demand indicates power is fed to storage.

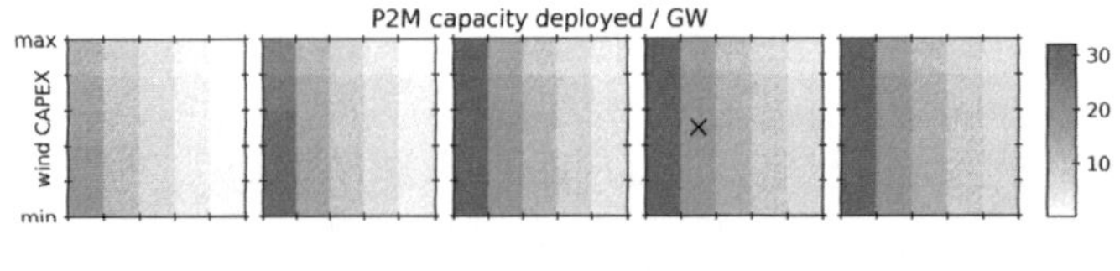

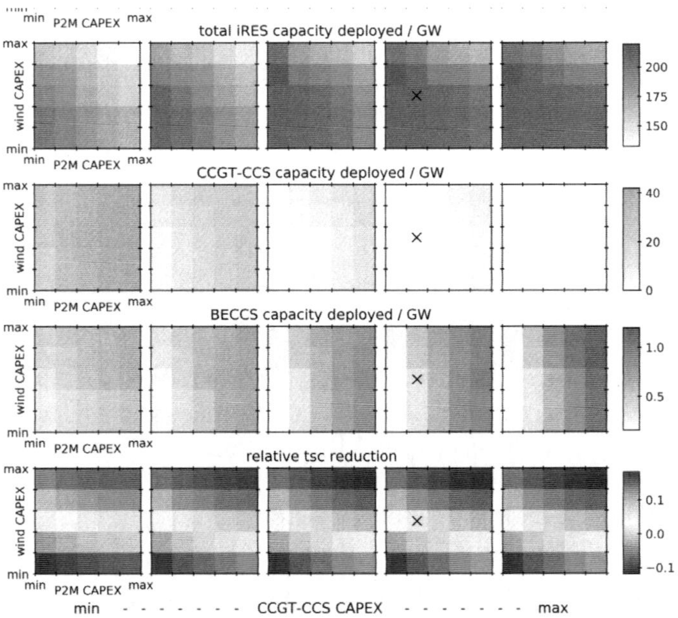

Figure 3: Deployment of power-to-gas storage (P2M), intermittent renewables (iRES), CCGT-CCS, and BECCS, and relative reduction of total system cost (tsc) depending on technology CAPEX. "min" values are estimated lower bounds, "max" values are 1.5 of the current CAPEX. Build rate limits set to 2x historical values. Cases marked with (x) indicate model did not converge within allocated time; values are interpolated based on neighboring cases.

Copyright © American Institute of Chemical Engineers. All rights reserved.

631387 Design of an MEA-Based Carbon Capture System for Cement Industry

Abstract:

This work presents the optimal design of an MEA-based carbon capture system applied to industrial source of CO_2. This study presents a sensitivity analysis of the cement plant flue gas generation as a function of raw materials and operating conditions. Using those results to obtain an optimal design for the MEA system using derivative-free optimization solvers. Finally, the MEA system design is evaluated using different operating conditions in the kiln model.

Cement production is one of the largest and most energy-intensive industries in the world. In 2018, the production of cement accounted for about 7% of global CO_2 emissions [IEA, 2018]. In 2019, the cement production was responsible for 40.9 MMT CO_2 equivalent in the U.S., accounting for 24.4% of U.S. total industrial CO_2 emissions [EPA, 2021]. In cement plants, CO_2 emissions originate from the calcination of limestone where $CaCO_3$ is converted to CaO and CO_2, and from the combustion process of the fossil fuels in the calciner and rotary kiln [Voldsund et al., 2019]. Thus, the operating conditions and the final products impact the CO_2 generated at the plant (both flow rate and CO_2 mol%), which depends on the fuel used, air/fuel mass ratio, composition and flow rate of raw materials for cement production, flame temperature for combustion and flame temperature location. The CO_2 concentration ranges from 14 to 33 mol% [Hasan et al., 2012], assuming that flue gas is an ideal gas, then the vol% is equal to mol%. The DOE/NETL-2002/1163 report shows a typical CO_2 composition of 18.9 mol%, which will be considered as the baseline for this study. CO_2 emissions associated with cement production are targeted for reduction, but will inevitably come at reduced efficiency and increased costs, which may affect product quality. To properly evaluate the required balance between GHG emissions, utility cost, and product quality, an integrated uncertainty quantification analysis and system optimization are required.

This work leverages the open-source Carbon Capture Simulation Initiative's (CCSI) toolset (https://github.com/CCSI-Toolset) and uses the baseline MEA system model [Morgan et al., 2018] developed for flue gas from a Supercritical Pulverized Coal-fired plant as a basis for this study. A kiln model to estimate the CO_2 emissions originated from the calcination process and an Aspen Plus model have been developed to estimate the CO_2 emissions from the combustion processes.

Finally, all the models are integrated and simulated using the Framework for Optimization, Uncertainty Quantification, and Surrogates (FOQUS) [Miller et al., 2016]. We leverage the FOQUS capabilities for performing uncertainty quantification and system optimization (aiming at minimizing regeneration reboiler duty considering the uncertainties from upstream cement production process) using derivative-free optimization tools [Larson et al.]. The uncertainty quantification study explored flue gas conditions as a function of kiln operating conditions, including raw material conditions, cement quality, flame temperature, and flame temperature location. The optimization study aims to obtain the optimal design of an MEA system and operating conditions that minimize the reboiler duty while achieving 90% CO_2 capture target and a given cement quality. The results show the optimal design of the absorber and regeneration columns (diameter, packing height, and flooding %), optimal heat integration, and number of parallel trains for a baseline cement plant operating condition.

References

DOE/NETL-2002/1163. Cement kiln flue gas recovery scrubber project: A DOE assessment. Nov. 2001. Document from Inner communications.

EPA. Inventory of U.S. Greenhouse Gas Emissions and Sinks: 1990-2019- Chapter 4. Industrial Processes and Product Use, 2021. https://www.epa.gov/ghgemissions/draft-inventory-us-greenhouse-gas-emissions-and-sinks-1990-2019. [Accessed Mar. 15, 2021].

Hasan, M.F., Baliban, R.C., Elia, J.A. and Floudas, C.A., 2012. Modeling, simulation, and optimization of postcombustion CO_2 capture for variable feed concentration and flow rate. 1. Chemical absorption and membrane processes. Industrial & engineering chemistry research, 51(48), pp.15642-15664.

IEA. Technology Roadmap – Low-Carbon Transition in the Cement Industry: Foldout, 2018, https://webstore.iea.org/technology-roadmap-low-carbon-transition-in-the-cement-industry-foldout. [Accessed Mar. 15, 2021].

Larson, J., Menickelly, M., & Wild, S.M. Derivative-free optimization methods. 2019. arXiv preprint arXiv:1904.11585, https://arxiv.org/pdf/1904.11585.pdf. [Accessed Mar. 26, 2021].

Morgan, J.C., Soares, C.A., Omell, B., Bhattacharyya, D., Tong, C., Miller, D.C. (2018). Development of a rigorous modeling framework for solvent-based CO2 capture. Part 2: Steady-state validation and uncertainty quantification with pilot plant data. Industrial & Engineering Chemistry Research 57: 10464-10481.

Miller, D.C., Agarwal D.A., Bhattacharyya D., Boverhof J., Cheah Y-W., Chen Y., (2016). Innovative computational tools and models for the design, optimization, and control of carbon capture processes. 26th European Symposium on Computer Aided Process Engineering – ESCAPE 26. 2391-2396.

Voldsund, M., Gardarsdottir, S.O., De Lena, E., Pérez-Calvo, J.F., Jamali, A., Berstad, D., Fu, C., Romano, M., Roussanaly, S., Anantharaman, R. and Hoppe, H., 2019. Comparison of technologies for CO2 capture from cement production—Part 1: Technical evaluation. Energies, 12(3), p. 559.

Disclaimer: This project was funded by the United States Department of Energy, National Energy Technology Laboratory, in part, through a site support contract. Neither the United States Government nor any agency thereof, nor any of their employees, nor the support contractor, nor any of their employees, makes any warranty, express or implied, or assumes any legal liability or responsibility for the accuracy, completeness, or usefulness of any information, apparatus, product, or process disclosed, or represents that its use would not infringe privately owned rights. Reference herein to any specific commercial product, process, or service by trade name, trademark, manufacturer, or otherwise does not necessarily constitute or imply its endorsement, recommendation, or favoring by the United States Government or any agency thereof. The views and opinions of authors expressed herein do not necessarily state or reflect those of the United States Government or any agency thereof.

Copyright © American Institute of Chemical Engineers. All rights reserved.

629225 Energy and Exergy Analysis of the Novel Supercritical CO_2 Brayton Cycle Using Different Designs of the Precooler

Abstract:

As part of the nationally determined contributions (NDCs) defined within the United Nations Framework Convention on Climate Change (UNFCCC), each country's ambitious approach is required to decrease the world's greenhouse gas emissions owing to their obligations towards the Paris Agreement. The major segment of these greenhouse gas emissions comes from the power generation and industrial sector, which alone comprises 69.8% of the total gas emissions worldwide. Given this, a novel supercritical power cycle, namely the carbon dioxide Brayton cycle (sCO_2-BC), could be of great value. The cycle with supercritical carbon dioxide (sCO_2) as a working fluid combines the best feature of both Rankine and Air Brayton cycles. This cycle operates near the critical point of sCO_2, resulting in significantly higher thermal efficiency than the cycles mentioned above [1] even at moderate values of turbine inlet temperature. Also, due to the high density of the working fluid in the supercritical region, heat exchangers and turbomachinery components are up to ten times smaller in comparison with the Rankine cycle, consequently reducing the overall system size up to four times and the environmental footprint of the planet by 25% at least [2]. Furthermore, as the critical temperate of sCO_2 is very close to ambient, this makes (sCO_2-BC) well suited for dry-air cooling, reducing the cycle's water consumption by 100% [1]. To meet the tight emission standards being imposed, an urgent need to enhance renewable energy sources has risen. In this regard, integration of the (sCO_2-BC) with concentrated solar power plants (CSP) is highly pertinent and cost effective for the next-generation CSP plants operating at temperature above 600° C [3]. Additionally, (sCO_2-BC) has been accounted for as an ideal power block for the next generation nuclear reactor [4] and industrial waste heat recovery system [1]. Visibly it is the future of the power cycle.

In the novel supercritical carbon dioxide Brayton (sCO_2-BC), the pre-cooler (Fig. a) role is critical [1,5]. It serves not only as a sink to the power cycle but it also regulates the conditions at the compressor's inlet. The compressor's inlet temperature is intended to be maintained close to the critical temperature of carbon dioxide (CO_2) to achieve greater cycle efficiencies[6]. However, exceptionally higher values of the specific heat capacity of near its critical point (up to 40 times higher than water) requires exceedingly high water flow rates [7] on the cold side to achieve the desired exit temperatures of CO_2. Consequently, the pre-cooler's pumping power requirements become high enough to deteriorate the cycle's performance. This problem can only be mitigated by exploring new channel geometries with enhanced thermohydraulic characteristics. Moreover (sCO_2-BC) currently lacks investigations on enhancing the thermohydraulic performance of pre-cooler using efficient channel geometries. Currently installed (sCO_2-BC), facilities and available numerical work on the topic focus on printed circuit heat exchangers (PCHEs) with straight channel geometries. However, it has been reported [8,9] frequently in the literature that thermal and hydraulic characterizes of the PCHEs with straight channels exhibit significantly low thermal and hydraulic characteristics compared with zigzag and airfoil channel-geometries. One of the reasons why pre-cooler's designs with the aforementioned channel-geometries were never explored was the unavailability of their thermal and hydraulic characteristics in the pre-cooler's operating regime due to limited lab facilities and intricacies linked with abrupt variations in the thermophysical properties near the critical point.

Regarding the discussion above, the current work deals with the numerical evaluation of pressure drop and heat transfer characteristics of the pre-cooler with zigzag channels and airfoil channels. Computation geometries are shown in Fig. a. Hexahedral mesh was generated using ICEM-CFD that was solved employing ANSYS-CFX. Real gas property tables through RGP files were utilized to implement the abrupt variations in the working fluid's thermophysical properties. Furthermore, a pitch-averaged data post-processing methodology is introduced and opted for the precise post-processing of the data[10].

Computed data was used to train the machine learning model based on the artificial neural network (ANN) to predict Nusselt number (Nu) and friction factor (f) as shown in Fig. d. Later, an in-house pre-cooler design and analysis code (PCDAC) was developed and coupled with cycle design point code (CDPC). The precooler code (PCDAC) employs (Fig. b and Fig. c) the developed Nusselt number (Nu) and friction factor (f) correlations based on the CFD data to achieve different precooler designs using various design values of heat exchanger's effectiveness (ϵ_D), and design value of the inlet Reynolds number (Re_D). Later an impact of the different designs of the heat exchanger was accessed on the cycle's performance using both energy and exergy analysis. Moreover, a multi-object optimization study was conducted to locate the best compromise between the cycle's performance and the pre-cooler's size.

Results suggest that the performance of (sCO_2-BC)can be enhanced substantially by employing zigzag and airfoil channel geometries for the pre-cooler design (Fig. e). Cycle's highest efficiency was found with airfoil channel geometry of the pre-cooler at $20k > Re_D > 25k$ and $0.92 > \epsilon_D > 0.95$. However, the pre-cooler's most compact designs were found with zigzag channel geometry with design values of Re_D and ϵ_D ranging from 35k to 40k and 0.98 to 0.99, respectively. Optimization results suggest that replacing straight channels with airfoil channels could reduce the pre-cooler's size up to 2.3 times and bring about a considerable enhancement in the power cycle's efficiency.

References

[1] K. Brun, P. Friedman, R. Dennis, Fundamentals and Applications of Supercritical Carbon Dioxide (sCO_2) Based Power Cycles, Woodhead Publishing an imprint of Elsevier, 2017.

[2] Y. Ahn, J.I. Lee, Study of various Brayton cycle designs for small modular sodium-cooled fast reactor, Nucl. Eng. Des. 276 (2014) 128–141. https://doi.org/10.1016/j.nucengdes.2014.05.032.

[3] M. Binotti, M. Astolfi, S. Campanari, G. Manzolini, P. Silva, Preliminary assessment of sCO_2 cycles for power generation in CSP solar tower plants, Appl. Energy. 204 (2017) 1007–1017.

[4] F. Crespi, G. Gavagnin, D. Sánchez, G.S. Martínez, Supercritical carbon dioxide cycles for power generation: A review, Appl. Energy. 195 (2017) 152–183.

[5] M. Kulhánek, V. Dostál, Supercritical carbon dioxide cycles thermodynamic analysis and comparison, in: SCO2 Power Cycle Symp., Boulder, Colorado, 2011: pp. 24–25.

[6] M. Saeed, S. Khatoon, M.-H. Kim, Design optimization and performance analysis of a supercritical carbon dioxide recompression Brayton cycle based on the detailed models of the cycle components, Energy Convers. Manag. 196 (2019) 242–260. https://doi.org/10.1016/j.enconman.2019.05.110.

[7] V. Dostal, D. MJ, H. P.A, A supercritical carbon dioxide cycle for next generation nuclear reactors, MIT-ANP-TR-100, advanced nuclear power technology program report. Cambridge (MA): Massachusetts Institute of Technology, Massachusetts Institute of Technology,Ph.D Thesis, 2004.

[8] M. Saeed, A.S. Berrouk, M. Salman Siddiqui, A. Ali Awais, A. S.Berrouk, M.Salman Siddique, M.A. Awais, Effect of Printed Circuit Heat Exchanger's Different Designs on the Performance of Supercritical Carbon Dioxide Brayton Cycle, Appl. Therm. Eng. 179 (2020) 115758. https://doi.org/10.1016/j.applthermaleng.2020.115758.

[9] C. Huang, W. Cai, Y. Wang, Y. Liu, Q. Li, B. Li, Review on the characteristics of flow and heat transfer in printed circuit heat exchangers, Appl. Therm. Eng. 153 (2019) 190–205. https://doi.org/10.1016/j.applthermaleng.2019.02.131.

[10] M. Saeed, A.S. Berrouk, M. Salman Siddiqui, A. Ali Awais, Numerical investigation of thermal and hydraulic characteristics of sCO2-water printed circuit heat exchangers with zigzag channels, Energy Convers. Manag. 224 (2020) 113375. https://doi.org/10.1016/j.enconman.2020.113375.

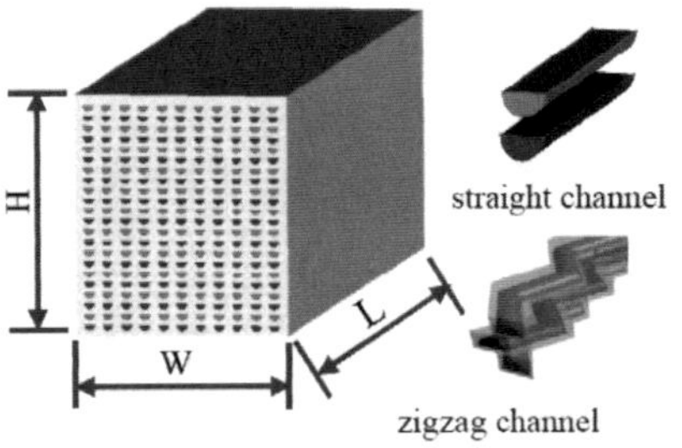

a) Printed circuit heat exchangers (PCHEs)

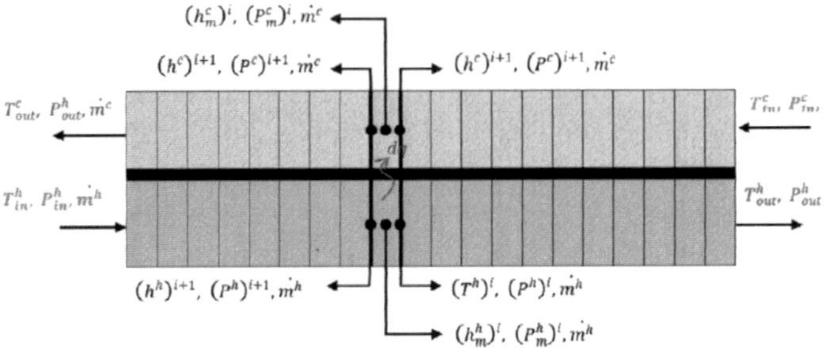

b) Discretized model

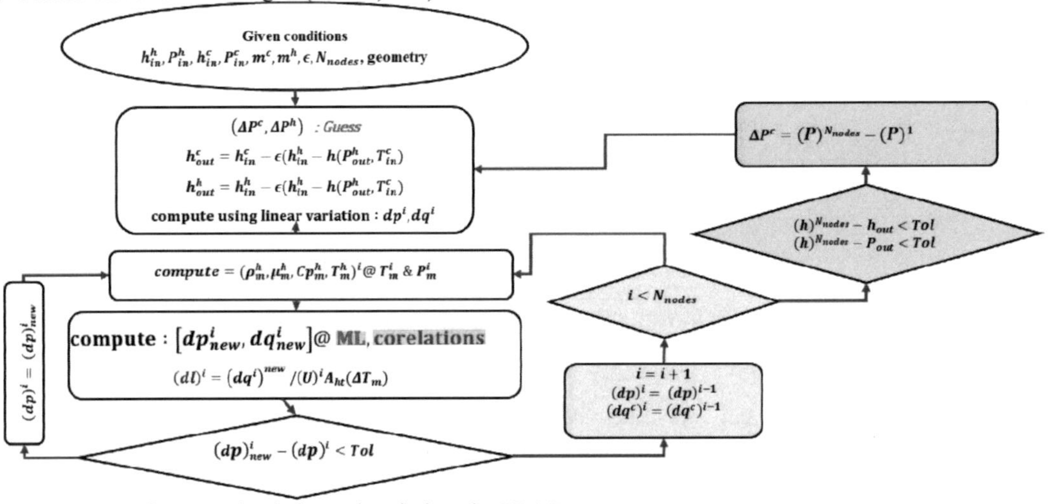

c) Flow diagram for precooler design and analysis code (PDAC)

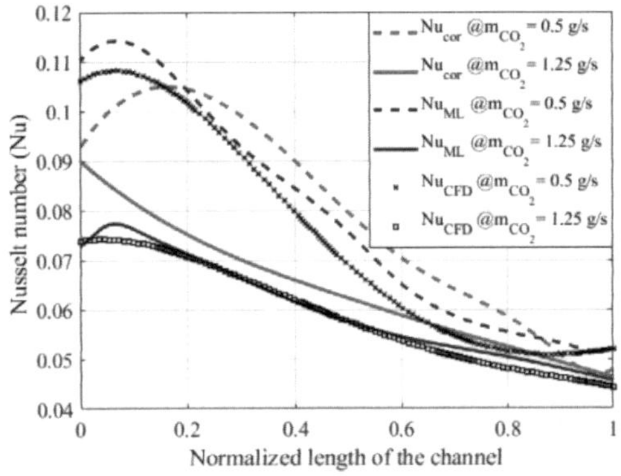

d) Comparison of CFD results and PDAC results based on the trained ML model

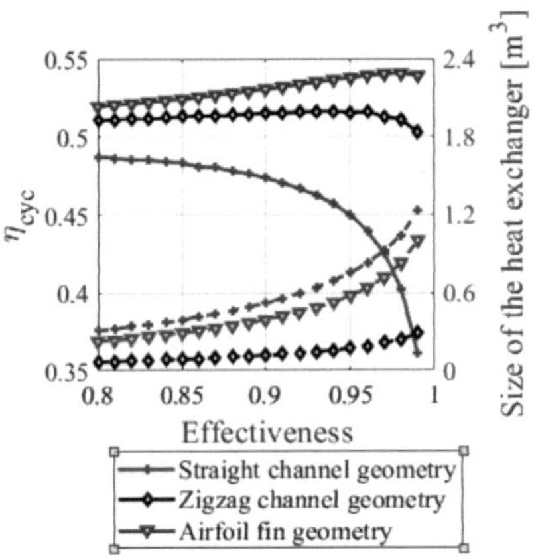

e) Cycle's performance utilizing precooler with different channel geometries

Copyright © American Institute of Chemical Engineers. All rights reserved.

628171 Achieving Carbon-Neutral Supply of Power, Methanol and Ammonia through the Energy-Chemical Nexus

Abstract:

In order to combat climate change, the Paris Agreement has aimed to keep the global temperature rise in this century well below 2 °C, and preferably 1.5 °C, above pre-industrial levels [1]. Despite the controversy in exact definition of pre-industrial periods, the annual average global surface temperature in 2019 has already increased by 0.98 °C relative to 1951-1980 average temperatures which is clearly after the (First) Industrial Revolution [2]. Furthermore, according to the Intergovernmental Panel on Climate Change (IPCC), the trend of increasing annual anthropogenic CO_2 emission persists in the recent years [3]. Among the various sources of global greenhouse gas (GHG) emission, energy sector has become the world's largest GHG emitter [4].

The decarbonization of energy sector at national scale is usually country-specific in order to fully realize the potential of inter-sectorial synergies [5]. In this work, China is selected as the country under investigation considering its large amount of CO_2 emission [6] and the government's determination in emission reduction [7,8]. It is believed that the concept of energy-chemical nexus could potentially help the country achieve a low-carbon future by integrating the development of energy and chemical sectors that lead in national CO_2 emission [9]. Therefore, this study aims to evaluate the preliminary feasibility of forming such nexus in China across a timeframe of over 40 years in the future from temporal, spatial and sectorial perspectives. For the concreteness of this analysis, electricity is selected as the representative product for energy sector as it currently constitutes more than 45% of China's total energy related CO_2 emission [10]. For chemical sector, both methanol and ammonia are selected considering China's current role as the world's largest producer and consumer of both chemicals [11,12]. Also, the recent development of active, stable and selective heterogeneous catalysts for methanol and ammonia synthesis further improve their competitiveness as future liquid energy carriers [13,14].

In this study, electricity generation from both fossil fuels and renewable resources, with or without carbon capture and storage (CCS) facilities, are considered. For methanol and ammonia production, apart from conventional fossil-based synthesis, hydrogen supplied by water electrolysis and carbon dioxide captured from power plants can also be used as feedstock to facilitate inter-sectorial integration. In order to quantitatively assess the impact of energy-chemical nexus on climate change, two relevant planetary boundaries, namely atmospheric CO_2 concentration (ppm) and energy imbalance at top of atmosphere (W/m^2), are involved. Unlike conventional comparative life cycle assessments whose results are most meaningful when a comparison between different options is concerned, planetary boundaries analysis provides a set of evaluation criteria for absolute sustainability based on the safe operating spaces of fundamental Earth system processes [15]. Finally, subject to the operation constraints of various technologies and emission constraints from the carbon neutrality target, a nexus optimization model is developed to study the optimal planning of energy-chemical nexus in China from 2018 to 2060 with forecasted product demands, resource capacities and technology capital and operating costs.

References

[1] United Nations Framework Convention on Climate Change (UNFCC). The Paris Agreement 2015.

https://unfccc.int/process-and-meetings/the-paris-agreement/the-paris-agreement (accessed April 5, 2020).

[2] NASA: Climate Change and Global Warming. Global Temperature 2019. https://climate.nasa.gov/vital-signs/global-temperature/ (accessed April 5, 2020).

[3] Intergovernmental Panel on Climate Change (IPCC). AR5 Synthesis Report: Climate Change 2014 2014. https://www.ipcc.ch/report/ar5/syr/ (accessed April 5, 2020).

[4] United States Environmental Protection Agency (US EPA). Global Greenhouse Gas Emissions Data 2010. https://www.epa.gov/ghgemissions/global-greenhouse-gas-emissions-data#Trends (accessed April 5, 2020).

[5] Wang X, Guo M, Koppelaar RHEM, Van Dam KH, Triantafyllidis CP, Shah N. A Nexus Approach for Sustainable Urban Energy-Water-Waste Systems Planning and Operation. Environ Sci Technol 2018;52:3257–66. https://doi.org/10.1021/acs.est.7b04659.

[6] Netherlands Environmental Assessment Agency (PBL). CO2 time series 1990-2015 per region/country 2017. https://edgar.jrc.ec.europa.eu/overview.php?v=CO2ts1990-2015 (accessed February 10, 2020).

[7] State Council. Work Plan of Controlling Greenhouse Gas Emissions During the 13th Five-Year Plan 2016. http://www.gov.cn/zhengce/content/2016-11/04/content_5128619.htm (accessed February 10, 2020).

[8] Matt M. Climate change: China aims for "carbon neutrality by 2060" . Online 2020. https://www.bbc.com/news/science-environment-54256826 (accessed April 11, 2021).

[9] Li Y, Lan S, Pérez-Ramírez J, Wang X. Achieving a low-carbon future through the energy-chemical nexus in China. Sustain Energy Fuels 2020;4:6141–55. https://doi.org/10.1039/d0se01337d.

[10] IEA. CO2 Emissions from Fuel Combustion 2016. OECD; 2016. https://doi.org/10.1787/co2_fuel-2016-en.

[11] Alvarado M. The changing face of the global methanol industry. IHS Chem Bull 3 2016:10–1.

[12] Zhang WF, Dou ZX, He P, Ju XT, Powlson D, Chadwick D, et al. New technologies reduce greenhouse gas emissions from nitrogenous fertilizer in China. Proc Natl Acad Sci U S A 2013;110:8375–80. https://doi.org/10.1073/pnas.1210447110.

[13] Shih CF, Zhang T, Li J, Bai C. Powering the Future with Liquid Sunshine. Joule 2018;2:1925–49. https://doi.org/10.1016/j.joule.2018.08.016.

[14] Guo J, Chen P. Catalyst: NH3 as an Energy Carrier. Chem 2017;3:709–12. https://doi.org/10.1016/j.chempr.2017.10.004.

[15] Steffen W, Richardson K, Rockstrom J, Cornell SE, Fetzer I, Bennett EM, et al. Planetary boundaries: Guiding human development on a changing planet. Science (80-) 2015;347:1259855–1259855. https://doi.org/10.1126/science.1259855.

Copyright © American Institute of Chemical Engineers. All rights reserved.

627121 Economically Viable CO_2 Electroreduction: Retrofit Technology for Chemical Manufacturing

Abstract:

Use of carbon dioxide (CO_2) as a chemical feedstock for bulk products would facilitate the transition to a carbon-neutral economy. While electrocatalytic methods for converting CO_2 into commodity chemicals have been discovered, their technical limitations, the high costs of CO_2 capture and required logistics make them uncompetitive with petrochemical-based manufacturing. Instead, we demonstrate that integrating CO2R as a means to recycle waste CO_2 and regenerate feedstocks is both technically feasible and profitable in the current market. As a specific use case, integration of CO2R with copper catalysts into ethylene oxide production will simultaneously reduce operational costs and up to 80% of direct CO_2 emissions. Based on our analysis of chemical manufacturing processes suitable for this type of CO2R retrofit, the approach has the potential to become a widely adapted and transformative decarbonization method across different industries, such as e.g. ammonia, synthetic fuels, cement and steel production.

Copyright © American Institute of Chemical Engineers. All rights reserved.

625328 Mechanisms and Kinetics of Biochar Mediated Sorption of Linuron, Diuron and Atrazine from Aqueous Media

Abstract:

The continued intensive use of pesticides for agricultural and non-agricultural purposes has led to their presence detection in various environmental compartments- surface water, groundwater, soil, air. Whereas the maximum permissible concentration of an herbicide such as diuron in the drinking water for instance, is 100 ng/L, research has shown that 70% of water samples collected from rivers around Europe contains concentration as high as 864ng/L. Consequently, it is aimed to facilitate effective removal of pesticides in water. Adsorption - as a surface phenomenon, has been found to hold a great potential for this task. Due to the amount and type of functional groups found on biochars, and their relatively low cost, these materials have shown great potentials for sorption of several organic micropollutants from aqueous solution. To this end, two biochars, B1 and B2 derived from potato peels and livestock feeds residues respectively, were tested as adsorbents for the removal of two phenylurea herbicides (diuron and linuron) and one triazine (atrazine) from aqueous media. The physicochemical properties of the biochars were studied using proximate analysis, CEC, %C, %N, C:N, specific surface area, methylene blue number, Fourier Transform Infrared Spectroscopy, and X-ray diffractometry. Batch adsorption was performed by varying adsorbent dosage, initial solute concentration, and solution pH. The equilibrium data were analysed with Langmuir and Freundlich equations. The physicochemical studies revealed specific surface areas of 1.35 (B1) and 132.60 (B2) m^2/g, pH values of 8.95 (B1) and 10.02 (B2), CEC of 3.15 (B1) and 20.64 (B2) mmolc/100g, %C 91.4 (B1) and 79.9 (B2), %N 0.51 (B1) and 0.54 (B2), with C:N ratio 181 (B1) and 148 (B2). The FTIR spectra of the biochars before and after adsorption of the herbicides revealed some similarities in their prominent spectra bands. The following surface functional groups were identified: aromatic and aliphatic C-H stretching, polysaccharide C-O stretching, aromatic C=C vibration, C=O stretching (indicative of esters, aldehyde, ketone, and carboxylic acids), C≡C vibration, O-alkyl groups, and O-H stretching of hydroxyl groups from alcohol, phenols and carboxylic acids. the adsorption of the herbicides onto the biochars, caused some vibration peaks to increase in intensity, some low intensity peaks to appear, and some peaks to become less intense (a decrease in the C=O peak intensity was more pronounced in B2 after adsorption). Also, it can be seen that the polar groups (-COOH, O-alkyl, C=O, and/or phenolic -OH) on the biochars favorably influenced pesticides adsorption. This may be ascribed to strong H-bonding interactions occurring between the herbicides and these polar groups on the biochars. In addition, the primary adsorption mechanisms could involve n-π interactions, π-π stacking, pore fillings and some van der Wal's interactions. For both biochars, isotherm data fitted better to Langmuir equation (R^2 = 0.94-0.99) than the Fruendlich (0.92-9.97), to deliver the maximum monolayer adsorption capacities, Q_{max} 0.48, 3.22, and 10.5 mg/g respectively for atrazine, diuron, and linuron on B1; and Q_{max} 18.1, 71.43, and 333.37 mg/g, respectively for atrazine, diuron and linuron on B2. The study showed that B2 has far much higher adsorption capacities for the three compounds than B1, thus pointing out the great influence of the physicochemical properties (especially the specific surface area and CEC) of the biochars on adsorption. The corresponding adsorption coefficients was in the order linuron > atrazine > diuron, which is the same as the order of their octanol-water partition coefficients (log K_{ow}) also points to the influence the molecular property of the compounds exerts on their adsorptions. In conclusion, B2, more especially, has shown to be a very promising low-cost adsorbent for these three moderately polar herbicides from two molecular groups, and by extension, for other pesticides from the groups. Available literature data have also shown that B2 competes more favourably than many other adsorbents for the removal of these compounds. However, its effectiveness for compounds outside these groups needs to be investigated.

Copyright © American Institute of Chemical Engineers. All rights reserved.

HIGHLY ACTIVE HETEROGENEOUS CATALYST M-CFA (M = Co, Fe) DERIVED FROM POWER PLANT FLY-ASH FOR THE DEGRADATION OF ACRYLONITRILE AN EMERGING POLLUTANT

ARVIND KUMAR[1, *], BASHESWAR PRASAD[1]

[1]*Department of Chemical Engineering, Indian Institute of Technology Roorkee,*

Roorkee-247667 (Uttarakhand) India

***Corresponding Author:** Arvind Kumar (akumar6@ch.iitr.ac.in)*

Keywords: Catalytic oxidation, Acrylonitrile, M-Al_2O_3/SiO_2 catalyst

ABSTRACT

In the present study, M-CFA (M = Fe and Co) catalyst was employed as a heterogeneous catalyst and activated with peroxymonosulfate (PMS) for the degradation of acrylonitrile. The catalyst was synthesized by co-precipitation method and further characterized by various techniques viz., XRD, FTIR, and FE-SEM. The acrylonitrile degradation study was carried out for various parameters viz., catalyst dose (200-1000 mg/L); PMS dosage (0.5-2.5); pH (2.5-10); temperature (30-70 ºC) and initial acrylonitrile concentration (100 mg/L). The kinetic study for the degradation of acrylonitrile at various temperatures were performed and well-represented by 1st order and nth order kinetic models.

Introduction

Acrylonitrile is an important raw material for the production of various chemical products viz., acrylonitrile butadiene styrene (ABS) resin, acrylic fibers, nitrile rubber and other synthetic resin (Kumar and Prasad, 2020; Popuri et al., 2011) Acrylonitrile is very highly toxic compound and also known as carcinogenic and refractory pollutant. USEPA has cited the acrylonitrile 3rd rank most toxic pollutant in the list of 129 priority pollutant list (USEPA, 2001). ABS resin is the major source of acrylonitrile release from aqueous solution during the production. In the ABS resin production huge amount of toxic wastewater generated which contain the acrylonitrile and acrylonitrile family compounds such as 3, 3-oxydipropiononitrile, 3, 3-imminodipropiononitrile and acetonitrile (Kumar et al., 2021a; Lai et al., 2013). All the aforementioned pollutants are known to be carcinogenic, mutagenic, or teratogenic nature (Krichevskaya et al., 2009).The chronicle exposure of acrylonitrile leads to adverse impact on human being such as mild jaundice, nausea, severely irritate the respiratory system, headache (Kumar and Prasad, 2020; Léonard et al., 1999). Environmental and Water Bureau, Michigan USA, has set the allowable discharge permissible limit for acrylonitrile 0.0026 mg/L in ground water (Wastewater Disposal Permits, Michigan USA, 1999). Considering all the aforementioned fact it is urgently need of sophisticate economical and environ friendly technique to removal of acrylonitrile from aqueous solution so that adverse impact on flora fauna could be prevented. Therefore, a novel heterogeneous catalyst was synthesized from metal incorporated coal fly ash (CFA) and activated with peroximonosulfate (PMS). CFA is the byproduct of coal-fired power plant, which has rich composition of Al_2O_3, SiO_2, Fe_2O_3 and other constituent MgO and CaO present. Every year, huge amount of coal fly ash produces from burning of coal about 130 million tons in India and it is projected to reach 130 million ton in year 2030 (Haque, 2013). Therefore, it would be great concern of safe disposal of coal fly ash and would be required large area of land for dispose it. Various authors have been used coal fly ash as a supporting material for synthesis of catalyst for various environmental applications and reported the remarkable results. In this investigation, authors have synthesized the Cu and Co supported CFA catalyst and successfully activated with PMS. Peroxymonosulfate (PMS) has emerged as a high potential oxidant due to its excellent stability, low cost, and easily activated with transition metals (Lu et al., 2018; Zheng et al., 2019), heat (Yang et al., 2010), ultrasound (Soumia and Petrier, 2016), and ultraviolet irradiation (Sharma et al., 2016). However, no study has been reported for treatment of acrylonitrile aqueous solution by Cu and Co supported CFA catalyst and activated by PMS so far. In this investigation, M-CFA (M = Fe and Co) were successfully synthesized and their PMS activation performance were evaluated by acrylonitrile degradation. The synthesized catalyst was characterized by various analytical techniques viz., X-ray diffraction (XRD), Fourier transform infrared spectra (FTIR), Field emission scanning electron microscope (FE-SEM) coupled with energy dispersed X-ray (EDX).

Materials and method

Coal fly ash collection and simulated wastewater sample preparation

Coal fly ash sample was collected from Renusagar power plant, Utter Pradesh (India). The chemical composition of fly ash was quantified by x-ray fluorescence spectrometer (XRF) as shown in Table 1. Acrylonitrile stock solution (100 mg/L) was prepared in double distilled water and preserve at 4 °C to prevent unwanted microbial growth. In this study, all the chemical reagent used were of analytical grade. Acrylonitrile and sulphuric acid were purchased from SRL Chemies, Mumbai (India). Copper (II) nitrate hexahydrate, cobalt nitrate hexahydrate, polyvinyl pyrrolidone, sodium hydroxide, peroximonosulfate (PMS) were procured from Himedia, Mumbai, India. Double distilled water was used throughout the whole experiments.

Table 1. Chemical composition of fly ash

Chemical Constituents	SiO_2	Al_2O_3	Fe_2O_3	SO_3	Na_2O	K_2O	TiO_2	CaO
Percentage (wt.%)	56.8	25.9	5.8	3.4	2.9	2.1	1.9	1.2

Catalyst synthesis

Prior to synthesis of catalyst, fly ash was thermally activated at 750 °C for 3 h to remove the unwanted impurity such as carbon and sulphur. Subsequently, resulting sample was chemically activated with NaOH and further treat with sulphuric acid. Thereafter, resulting sample was washed several times with double distilled water until the pH of the sample become nearly neutral and subsequently dried at 110 °C for 12 h. The resulting sample was known as activated fly ash and denoted as A-CFA (Gao et al., 2020). Fe-CFA and Co-CFA catalyst were synthesized by wet impregnation method. Initially, 10 wt.% of metal precursors $Cu(NO_3)_2.6H_2O$ and $Co(NO_3)_2.6H_2O$ and known amount of prepared activated fly ash were dissolve separately in double distilled water. Subsequently, metal precursors separately added dropwise in continuous stirring of support (fly ash) solution at 250 °C until the solution become viscous gel. The resulting gel was dried at 110 °C for 20 h, subsequently calcined at 750 °C for 7 h. The obtained samples were denoted as Co-CFA and Fe-CFA.

Experimental procedure

All the batch experiments were conducted in lab scale 250 ml three-neck glass reactor. Reactor was placed inside an oil bath for maintaining the desire temperature. In each batch experiments, desired amount of catalyst dose was introducing in 100 ml of acrylonitrile aqueous solution and subsequently add the known amount of oxidizing agent dosage (PMS) and homogeneously mixed through continuous stirring on magnetic stirrer. Reaction temperature was maintained by PID controller in oil bath. pH of the reaction was regulated by 1 N NaOH and 1 N H_2SO_4 as required accordingly. In a regular interval, 1 ml of sample was withdrawn and filtered through 0.22 μm syringe membrane filters prior to analysis. Subsequently, filtered sample was analysed by various analytical techniques.

Instrumentation and analytical techniques

Synthesized catalyst was characterized by XRD Bruker D8 AXS diffractometer Germany, in the range of 2θ values 10-90° with scanning rate 1° min^{-1} and step size of 0.02°. The morphology of catalyst was analysed by FE-SEM (FEI Quanta 200 FEG, USA) with magnification range 12X-1000k and EDX and EELS detector. FTIR spectra of catalyst was recorded by FTIR, Nicolet Avatar 370 CSI spectrometer in the range of 4000-400 cm^{-1}. The concentration of acrylonitrile was analysed by HPLC (HPLC, Waters India, Ltd.) equipped with C18 column (3.9 mm × 150 mm) at 196 nm wavelength. Pure Milli-Q water was used as a mobile phase with 0.5 mL/min.

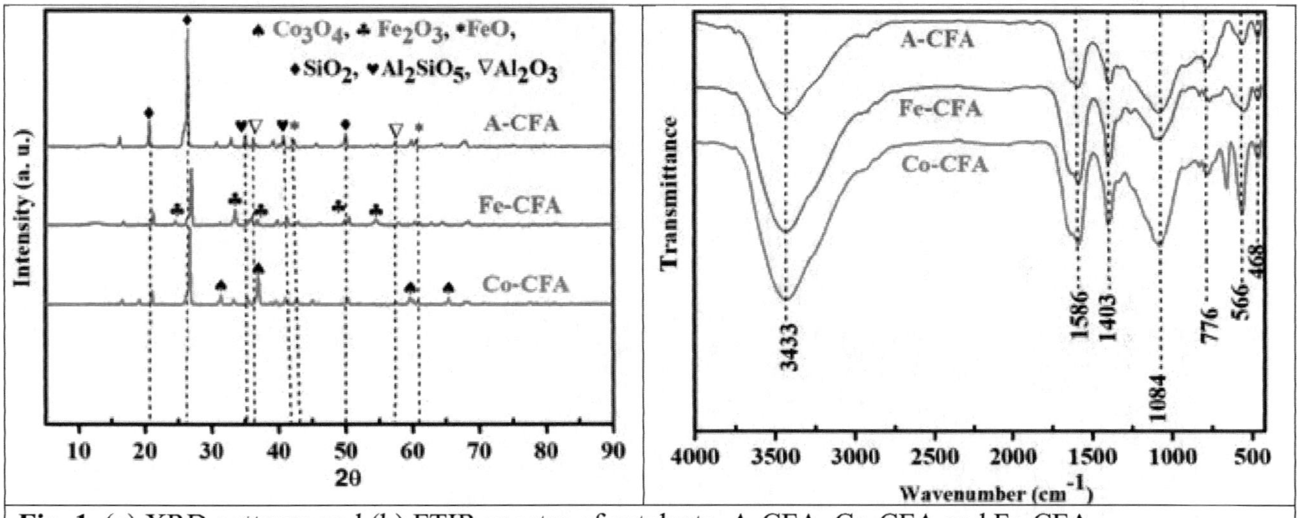

Fig. 1. (a) XRD patterns and (b) FTIR spectra of catalysts, A-CFA, Co-CFA and Fe-CFA.

Results and discussions

Catalyst characterization

X-ray diffraction (XRD) patterns of synthesized catalyst M-CFA (M = Fe and Co) were recorded to identify the crystalline phases in the range of 2θ values 10-90°. XRD results as shown in Fig 1(a)., exhibited the various crystalline phases viz., CoO_3O_4, Fe_2O_3, FeO, Al_2O_3, SiO_2 and Al_2SiO_5. Fourier transform infrared (FTIR) spectra of catalyst M-CFA (M = Fe and Co) were recorded in the range from 4000 to 500 cm^{-1} to identify the functional group present in the sample. FTIR spectral peak around 3433 cm^{-1} attributed to stretching mode of free or bounded hydroxyl group as shown in Fig 1(b). Next spectral peak around 1586 cm^{-1} and 1403 cm^{-1} attributed to bending vibration mode of C=C, and –CH stretching vibration, respectively. Remaining strong spectral peaks around 776 cm^{-1}, 566 cm^{-1} and 468 cm^{-1} ascribed to bonding between metal and oxygen M-O, M-OH, respectively. In order to analysed the surface morphology of the synthesized catalysts and their corresponding composition, FE-SEM was performed with EDAX analysis. As shown in Fig 2(a), the morphology of chemically treated A-CFA catalyst showed non-uniform agglomerated particle similar to reported by (Kumar et al., 2021b). While loading of cobalt on A-CFA catalyst, morphology of Co-CFA catalyst was not effectively changed and surface of the catalyst become rough as shown in Fig 2(b). Catalyst Fe-CFA showed rough web like morphology as shown in Fig 3(c). Moreover, chemical composition of elements presents in the catalyst was analysed by EDX, which shows that stoichiometric composition is close to weight composition.

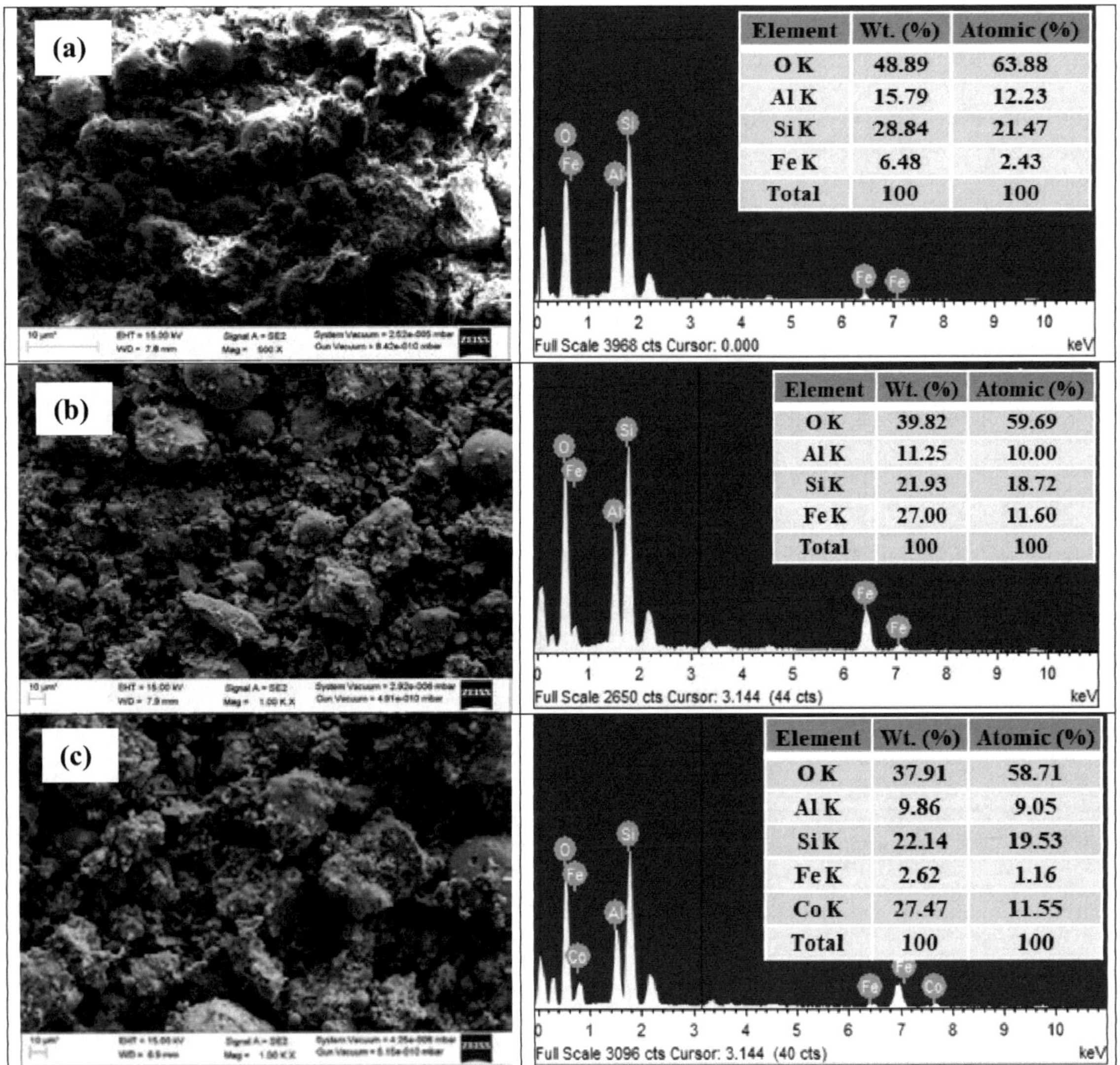

Fig. 2. FE-SEM images and EDAX analysis of catalysts, (a) A-CFA, (b) Co-CFA and (c) Fe-CFA.

Effect of parameters on acrylonitrile removal:

Effect of catalyst dosage

In order to enhance the degradation of acrylonitrile, optimization of catalyst dosage is very necessary and it play a crucial role. Over dose of catalyst can enhance the cost of treatment process. Accordingly, catalyst dosage for all the three type of was varied from 200 mg/L to 1000 mg/L. Removal efficiencies of acrylonitrile increased with increase in catalyst dosage up to 500 mg/L, subsequently removal efficiency decreased while introducing over dosage. The maximum removal of acrylonitrile was observed i.e., 29.84%, 43.54% and 34.42% with catalyst A-CFA, Co-CFA and Fe-CFA reaction condition (PMS dosage 1 mM, pH 6.5, reaction temperature 30 °C) as shown in Fig 3(a). Removal efficiency decreased while using the over dosage might be causes of agglomeration of catalyst particle which leads to reduce the active sites on the catalyst surface (Kumar and Prasad, 2020).

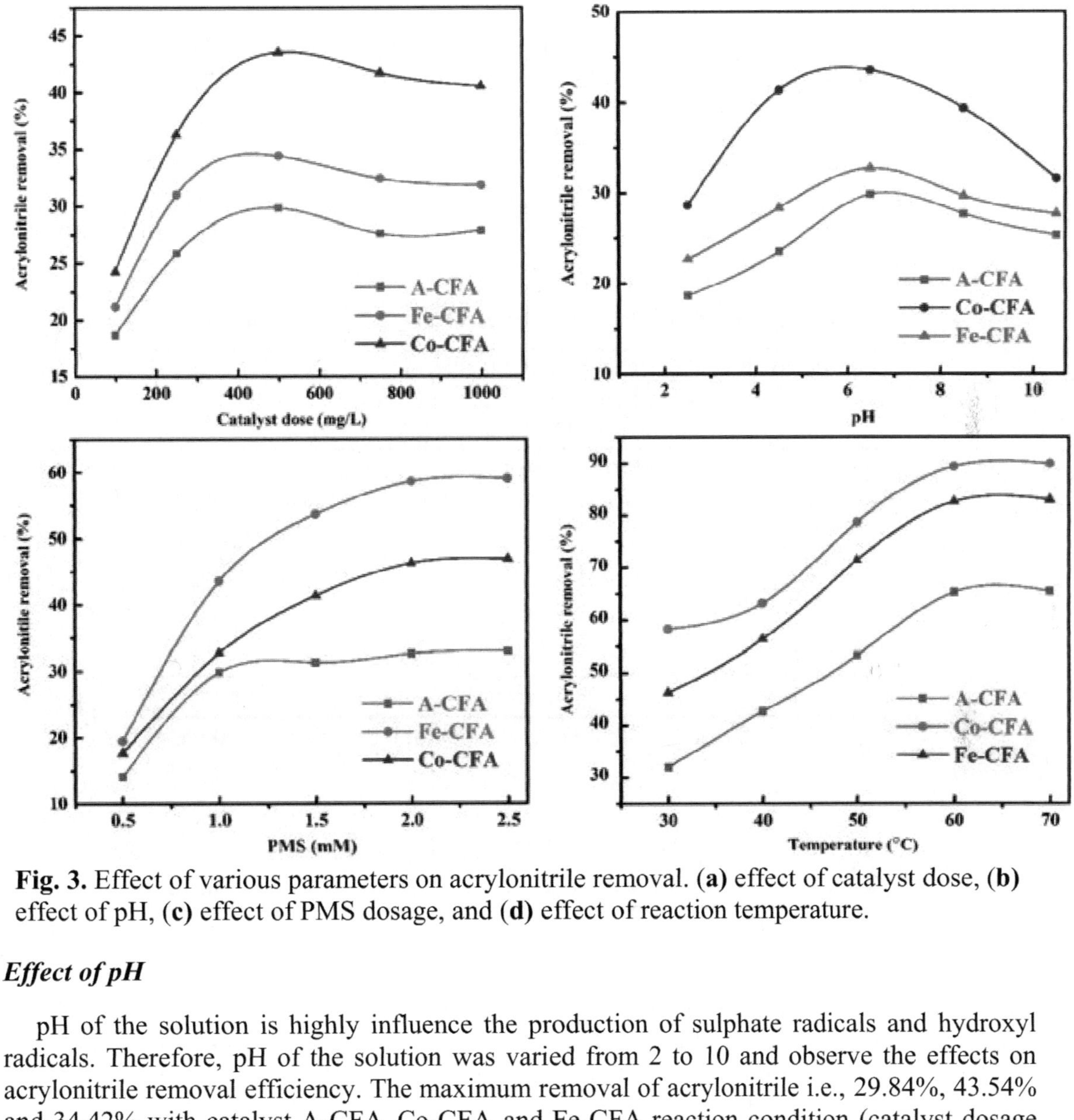

Fig. 3. Effect of various parameters on acrylonitrile removal. (**a**) effect of catalyst dose, (**b**) effect of pH, (**c**) effect of PMS dosage, and (**d**) effect of reaction temperature.

Effect of pH

pH of the solution is highly influence the production of sulphate radicals and hydroxyl radicals. Therefore, pH of the solution was varied from 2 to 10 and observe the effects on acrylonitrile removal efficiency. The maximum removal of acrylonitrile i.e., 29.84%, 43.54% and 34.42% with catalyst A-CFA, Co-CFA and Fe-CFA reaction condition (catalyst dosage 500 mg/L, PMS dosage 1 mM, reaction temperature 30 °C) as shown in Fig 3(b). Removal efficiency decreased in both medium acidic and alkaline medium.

Effect of PMS dosage

Effects of PMS dosage on acrylonitrile removal was analysed in the range of 0.5 Mm to 2.5 Mm. The removal of acrylonitrile gradually enhanced with raising the PMS dosage up to 2 mM subsequently, no significant changes were observed on removal efficiency while introducing over dosage of PMS. The maximum removal of acrylonitrile i.e., 32.56%, 58.64% and 46.25% were observed with catalyst A-CFA, Co-CFA and Fe-CFA at reaction condition (catalyst dosage 500 mg/L, pH 6.5, reaction temperature 30 °C) as shown in Fig 3(c). The over dosage of PMS scavenges the generated sulphate radicals ($SO_4^{\bullet-}$) and hydroxyl radical ($^{\bullet}OH$) which leads to reduce the removal efficiency.

Effect of reaction temperature

Effect of reaction temperature on acrylonitrile removal was varied from 30 °C to 70 °C. The removal of acrylonitrile gradually enhanced with raising the reaction temperature up to 60 °C subsequently, no significant changes were observed on removal efficiency while increasing the over temperature. Increase in reaction temperature accelerate the PMS decomposition and enhance the production of sulphate radicals ($SO_4^{\bullet-}$) and hydroxyl radical ($^{\bullet}OH$). The maximum removal of acrylonitrile i.e., 65.26%, 89.36% and 82.62% were observed with catalyst A-CFA, Co-CFA and Fe-CFA at reaction condition (catalyst dosage 500 mg/L, PMS dosage 2 mM, and pH 6.5) as shown in Fig 3(d).

Table 2: Values of kinetics parameters at various temperatures.

Temperature (°C)	Rate constant (K_n) [(mg/L) $^{(1-n)}$ min^{-1}]	Order of reaction (n)	ARE	R^2
	1st order			
30	4.3×10^{-03}	1	3.69	0.986
40	5.6×10^{-03}	1	2.48	0.991
50	6.7×10^{-03}	1	2.80	0.995
60	9.5×10^{-03}	1	2.07	0.998
70	10.6×10^{-03}	1	4.36	0.975
	nth order			
30	2.36×10^{-03}	1.23	3.67	0.981
40	2.56×10^{-03}	1.28	3.64	0.985
50	1.87×10^{-03}	1.43	5.12	0.986
60	1.26×10^{-03}	1.55	3.83	0.994
70	1.66×10^{-03}	1.66	4.27	0.972

Kinetic study

The kinetic study of acrylonitrile degradation was performed at optimum operating conditions and kinetics data reported on Table 2. The acrylonitrile degradation was considered as a nth order reaction, rate of acrylonitrile degradation can be written as in Equation (1) as following.

$$-\frac{dC}{dt} = k_n C^n \qquad (1)$$

Where, n is reaction order, k_n is nth order rate constant and C is acrylamide concentration. Integrating the above equation within limit, t=0, $C=C_0$ and t=t, $C=C_t$, n^{th} order equation was obtained.

$$C_t^{1-n} - C_0^{1-n} = (n-1)k_n t \qquad (2)$$

Above Equation (2) was solved by nonlinear regression techniques. The average relative error (ARE) function was used to minimize the error as given in Equation (3).

$$ARE = \frac{100}{N}\sum_{i=1}^{n}\left|\frac{C_{t,exp}-C_{t,cal}}{C_{t,exp}}\right|_i \qquad (3)$$

Where, subscript "cal" and "exp" denoted the calculated and experimental values, respectively.

Conclusions

This study demonstrates that M-CFA (M = Fe and Co) catalyst synthesized by co-precipitation method exhibit very high catalytic activity towards the degradation of acrylonitrile. The maximum removal of acrylonitrile 89.36% were observed with catalyst Co-CFA at optimum operating condition (catalyst dosage 500 mg/L, PMS dosage 2 mM, and pH 6.5, reaction temperature 60 °C). The acrylonitrile degradation kinetics was performed and well fitted by first order and n^{th} order kinetic model.

References

1. Gao, S., Liu, N., Liu, J., Chen, W., Liang, X., Yuan, Y., 2020. Synthesis of higher alcohols by CO hydrogenation over catalysts derived from $LaCo_{1-x}Mn_xO_3$ perovskites: Effect of the partial substitution of Co by Mn. Fuel 261, 116415.
2. Haque, M.E., 2013. Indian Fly-Ash: Production and consumption scenario. Int. J. Waste Resour. 3, 22–25.
3. Krichevskaya, M., Jõks, S., Kachina, A., Preis, S., 2009. Gas-phase photocatalytic oxidation of acrylonitrile. Photochem. Photobiol. Sci. 8, 600–603.
4. Kumar, A., Prasad, B., 2020. Catalytic peroxidation of acrylonitrile aqueous solution by Ni-doped CeO_2 catalysts: characterization, kinetics and thermodynamics. Int. J. Environ. Sci. Technol. 17, 1809–1824.
5. Kumar, A., Prasad, B., Garg, K.K., 2021a. Enhanced catalytic activity of series $LaCu_xFe_{1-x}O_3$ (x = 0.2, 0.4, 0.6, 0.8) perovskite-like catalyst for the treatment of highly toxic ABS resin wastewater: Phytotoxicity study, parameter optimization and reaction pathways. Process Saf. Environ. Prot. 147, 162–180.
6. Kumar, A., Prasad, B., Sandhwar, V.K., Garg, K.K., 2021b. Mechanistic insight into heterogeneous Fenton-like catalysis with M-Al_2O_3/SiO_2 (M = Fe, Co and Ni) for acrylonitrile mineralization from real ABS resin wastewater: Optimization and toxicity assessment. J. Environ. Chem. Eng. 9, 105177.
7. Lai, B., Zhou, Y., Yang, P., 2013. Treatment of wastewater from acrylonitrile-butadiene-styrene (ABS) resin manufacturing by biological activated carbon (BAC). J. Chem. Technol. Biotechnol. 88, 474–482.
8. Léonard, A., Gerber, G.., Stecca, C., Rueff, J., Borba, H., Farmer, P.., Sram, R.., Czeizel, A.., Kalina, I., 1999. Mutagenicity, carcinogenicity, and teratogenicity of acrylonitrile. Mutat. Res. Mutat. Res. 436, 263–283.
9. Lu, S., Wang, G., Chen, S., Yu, H., Ye, F., Quan, X., 2018. Heterogeneous activation of peroxymonosulfate by LaCo1-xCuxO3 perovskites for degradation of organic pollutants. J. Hazard. Mater. 353, 401–409.
10. Popuri, S.R., Chang, C.Y., Xu, J., 2011. A study on different addition approach of Fenton's reagent for DCOD removal from ABS wastewater. Desalination 277, 141–146.
11. Sharma, J., Mishra, I.M., Kumar, V., 2016. Mechanistic study of photo-oxidation of Bisphenol-A (BPA) with hydrogen peroxide (H_2O_2) and sodium persulfate (SPS). J. Environ. Manage. 166, 12–22.

12. Soumia, F., Petrier, C., 2016. Effect of potassium monopersulfate (oxone) and operating parameters on sonochemical degradation of cationic dye in an aqueous solution. Ultrason. Sonochem. 32, 343–347.
13. Yang, S., Besson, M., Descorme, C., 2010. Catalytic wet air oxidation of formic acid over Pt/$Ce_xZr_{1-x}O_2$ catalysts at low temperature and atmospheric pressure. Appl. Catal. B Environ. 100, 282–288.
14. Zheng, H., Bao, J., Huang, Y., Xiang, L., Faheem, Ren, B., Du, J., Nadagouda, M.N., Dionysiou, D.D., 2019. Efficient degradation of atrazine with porous sulfurized Fe_2O_3 as catalyst for peroxymonosulfate activation. Appl. Catal. B Environ. 259, 118056.

622842 Two-Step Process on Persistent Organic Pollutants Removal – Insight into the Adsorption Loss during the Electro-Fenton Regeneration Process

Abstract:

Electro Fenton (EF) processes are proven effective for the mineralization of organic pollutants yet suffering from the practical challenge in applications. In this study, a cyclic process combining adsorption and electro Fenton process was developed to improve the practicality of the decontamination of organic pollutants. During the adsorption-electrochemical regeneration cycles, a sharp adsorption loss at the beginning of the cycles was observed, and the adsorption capacity further yielded to a plateau (52.4% of the initial capacity of ACF) without continuous loss. The analysis via Fourier-transform infrared spectroscopy (FTIR), X-ray photoelectron spectroscopy (XPS), and point of zero charge (pH_{PZC}) confirmed the stability of the ACF in the process. The further pore structure analysis showed a complete recovery of the pores with sizes above 12 Å, with poor recovery (only 31.52%) of pores with a diameter below 12 Å. We deduced from the results that the loss in adsorption is due to the formation of iron complexes ferrioxalate, which makes micropores inaccessible for the ferric catalyst. Furthermore, the release of organic substances from the spent ACF electrodes during the alkaline elution confirmed the incomplete removal of degradation intermediates accumulation. Overall, our study demonstrated the high feasibility of the cyclic process towards organic pollutants removal and provided a comprehensive image of the EF-based activated carbon regeneration process.

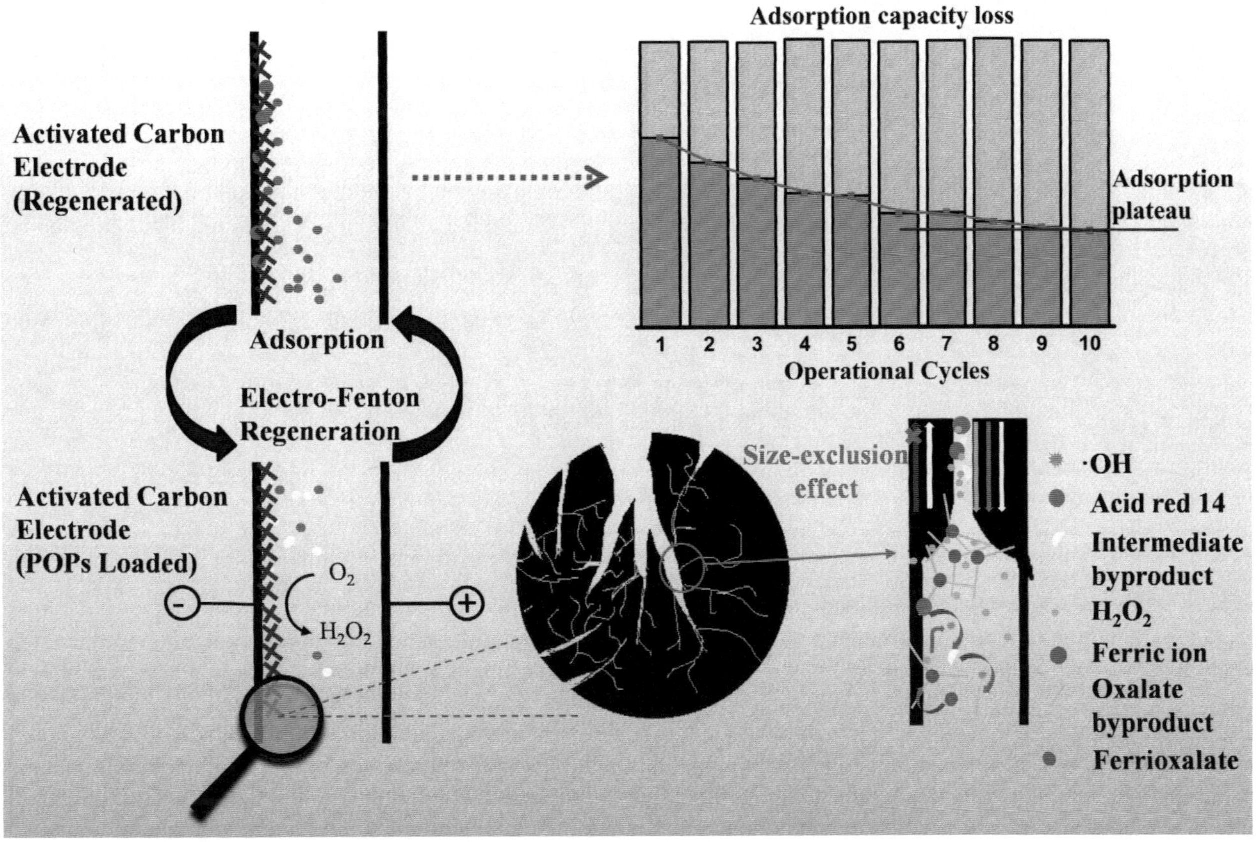

Copyright © American Institute of Chemical Engineers. All rights reserved.

627241 Removal of Aluminium Ions in Drinking Water Using Bentonite in Egypt

Abstract:

Aluminum ions traces in drinking water have been detected in El Khanka Water Treatment plant in Egypt because of the lean sludge generated from the use of Alum in water treatment. Experimental investigation of using bentonite to remove aluminum salts from lean sludge was performed at different pH values. 98.8% removal of Al ions was achieved at pH 11, and 0.6 ppm concentration of bentonite in the sludge solution. Kinetic model of pseudo-second-order described the reaction accurately, and the rate constant was calculated at 11.922 $mol.^{-1}$ s^{-1}. An industrial process was designed to remove Aluminium ions from the drinking water in Egypt. Additionally, a brief economic estimate of applying the water treatment process was prepared.

Copyright © American Institute of Chemical Engineers. All rights reserved.

Ammonia Removal from Wastewater Using Both Activated Carbon and Carbon Nanotube

Basim Ahmed Abussaud, KFUPM, Dhahran, Saudi Arabia

Abstract

The removal of ammonia from wastewater has been studied using both Activated Carbon (AC) and Carbon Nanotube (CNT) at different conditions i.e., pH, amount of dosage, contact time, shaking speed, and temperature. To enhance the removal process, the adsorbent materials have been impregnated with Fe, Al, and Cu. The maximum removal that has been achieved was 40% with using 2 g of activated carbon impregnated with aluminum oxide at a pH of 10.5 and a shaking speed of 200 RPM for 2 hours. It was shown that the temperature has a noticeable effect on the removal. While only around 21% removal was achieved at 25°C, 40% removal was achieved when the temperature was increased to 70°C with aluminum oxide impregnated AC. These results shows that this method of removal can be used as a pre-treatment since the removal percentage is lower than other available techniques.

Introduction

Due to the fast growth in the chemical industry, many toxic pollutants can be found in the industrial wastewater. Since the regulations are becoming stricter, the concentration of these contaminants should not exceed certain predetermined limits before it can be either discharged to the sea or reused. Among these contaminants is ammonia which is one of the toxic materials that can be found in water [1-4]. The most common methods for the treatment of ammonia are chlorination, biological oxidation, ion exchange, membranes, and adsorption. The choice from these methods will depend mainly on the cost of the treatment method. Adsorption is one of the cheapest methods for the removal of the contaminants from water. In this method, an adsorbent is used to adsorb the containment from water. One of the most important factors when choosing the adsorbent is the surface area. Usually, as the surface are increases, the efficiency of the adsorbent increases. Activated Carbon (AC) is considered to be one of the adsorbents that is proven to be excellent in removing many of the contaminants from water like phenol [5]. It has a surface area ranging from 500-1500 m^2/g [6]. The Carbon Nano Tube (CNT) is another excellent adsorbent that was discovered in 1991. It has a very high surface area. It was used successfully for the removal of many contaminates like benzene, toluene, para-xylene, and phenol [7-10].

Literature Review

Ammonia can be found in the waste water effluent from different industries like refining, food production, and fertilizers. There are many methods that have been reported for the removal of ammonia. The Ion Exchange (IE) method is one of those methods. Cooney et.al shows that the IE method has a high removal efficiency with a low cost. Other methods that have been reported for the removal of ammonia include wet air oxidation (WAO) [12], ozonation [13], nitrification and denitrification [14].

Adsorption is considered an effective and low-cost method that is commonly used for the removal of contaminant from wastewater. In this method, an adsorbent is used to adsorb the containment from the wastewater. Activated carbon (AC) and the Carbon Nano Tube (CNT) are an example of these adsorbents. The main difference between them is in their surface areas. The CNT has much more surface area than AC. The CNT can exist as either single walled carbon nano tube or multi walled carbon nano tube [15]. ACs can be classified into different groups based on their physical characteristics: powdered activated carbon, granular activated carbon, extruded activated carbon, impregnated carbon, and polymer coated carbon [16].

Modification of both AC and CNT surface with different groups and oxides has shown improvement in the adsorption capacity due to an increase in active site availability and the enhancement in the bonding interactions [17].

Methodology

3200 ppm of ammonia is prepared and a 100ml flask is used. Since the ammonia is a volatile compound, it is covered to prevent vaporization. All experiments were conducted at room temperature. The effect of pH, agitation speed, dosage, contact time, and temperature on the percent removal of the ammonia has been studied. The concentration of the ammonia is measured at the beginning of the experiment (C_i)and at the end (C_f). The percent removal is then calculated as:

$$\% \, Removal = \frac{C_i - C_f}{C_i} \times 100$$

Where:

C_i = *Initial concentration (ppm)*

C_f = *Final concentration (ppm)*

In addition, the effect of the impregnation of different ions on the adsorbent capacity is studied using four different oxides, namely copper oxide, silver oxide, aluminum oxide, and iron oxide.

Results

1- Effect of pH

In the aqueous solution, the total ammonia exists in two forms, namely ionized form (NH_4^+) and unionized form (NH_3) according to the pH of the solution. Since the unionized ammonia (NH_3) exists at the high pH, this study will consider the pH range from 9-11. The experiment was performed at 25°C with 200 RPM for two hours. Figures 1 and 2 shows the effect of both AC and CNT on the removal of the ammonia. It can be seen clearly that there was no significant effect of pH on the removal percentage.

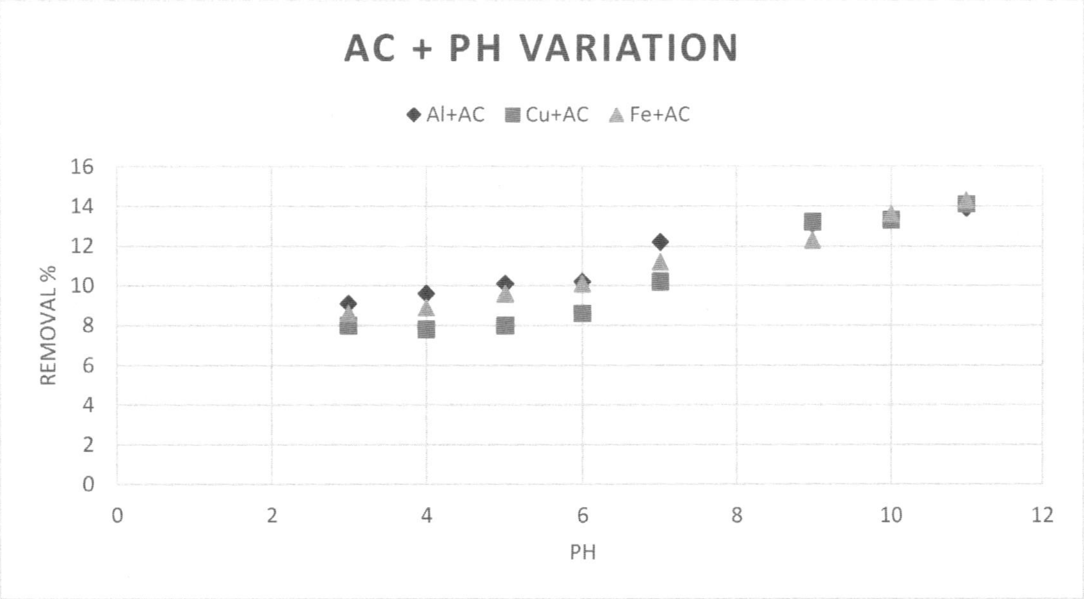

Figure 1: Effect of pH on ammonia removal using AC

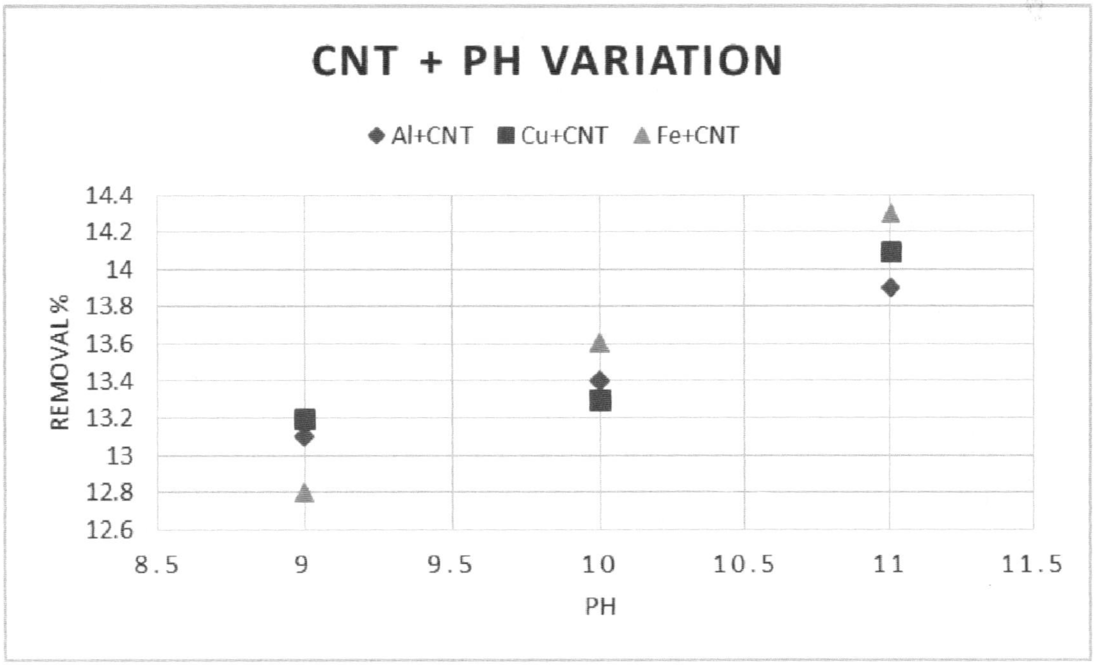

Figure 2: Effect of pH on ammonia removal using CNT

2- Effect of Agitation Speed

The agitation speed was varied from 100 to 250 RPM while keeping all other values of pH, dosage, temperature and contact time constant at 10.5, 1 g, 25°C and 2 h. It can be seen from Figures 3 and 4 that as the agitation speed increases from 100 to 200, the percentage removal increases. After that, it is almost constant. By increasing the agitation speed, the resistance for ammonia to be in contact with the adsorbent surface will be reduced and hence the removal will be increased.

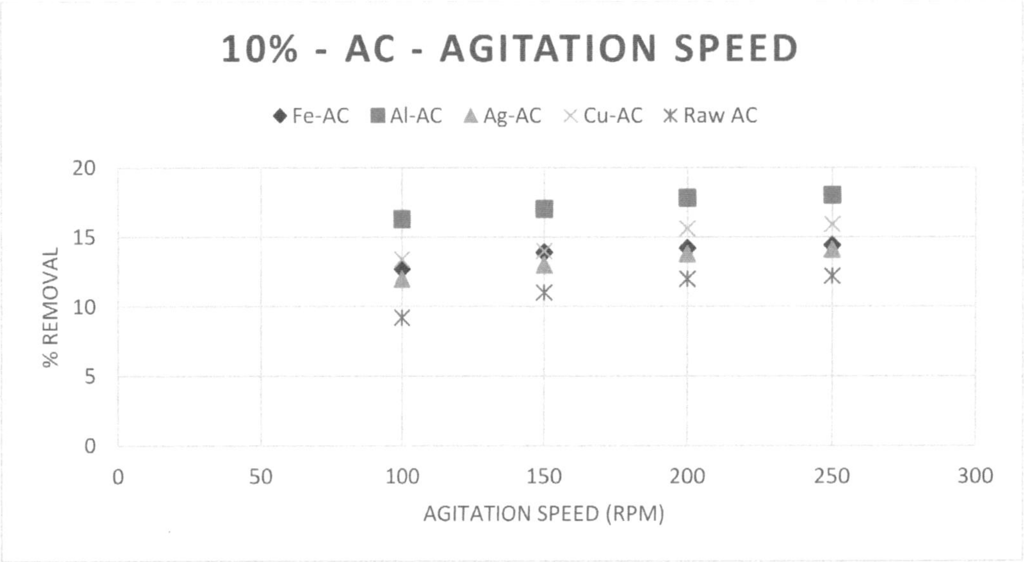

Figure 3: Effect of agitation speed on ammonia removal using 10% AC

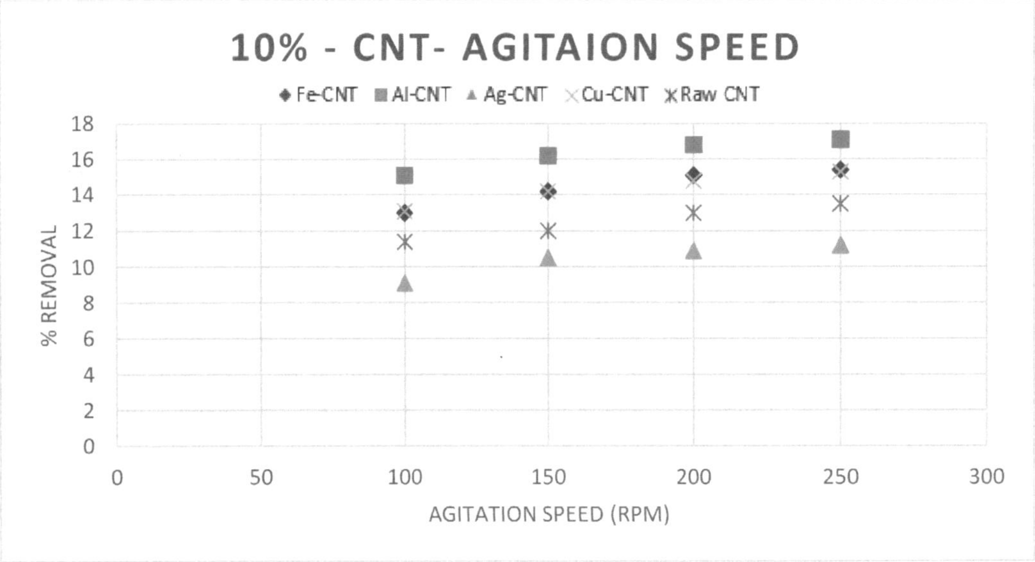

Figure 4: Effect of agitation speed on ammonia removal using 10% CNT

3- Effect of Adsorption Dosage

The amount of dosage used was varied from 0.05-2 g in 100 ml for both AC and CNT. The pH was 10.5, the contact time was 2 h, the agitation speed was 200 RPM, and the temperature was 25°C. Both Raw and impregnated AC and CNT were investigated.

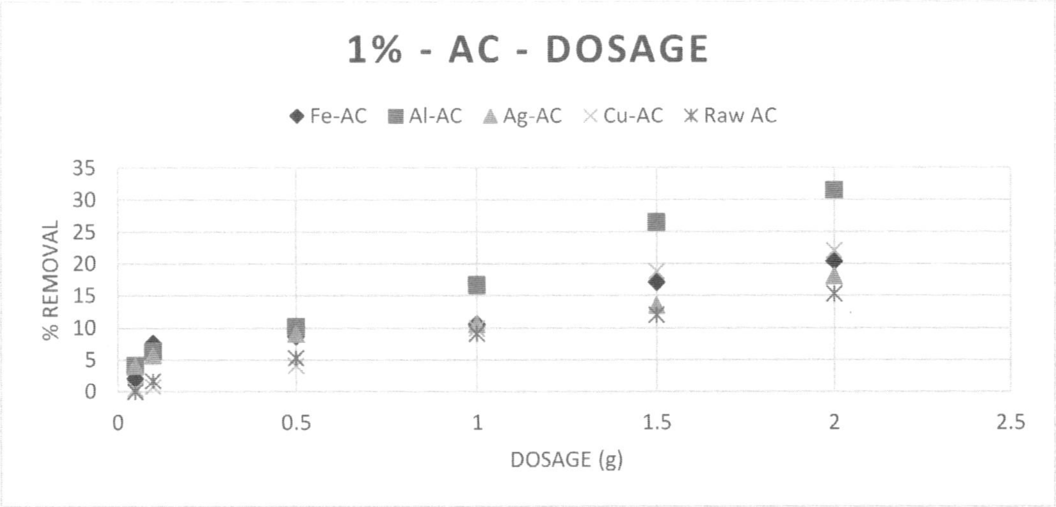

Figure 5: Effect of Dosage on ammonia removal using 1% AC

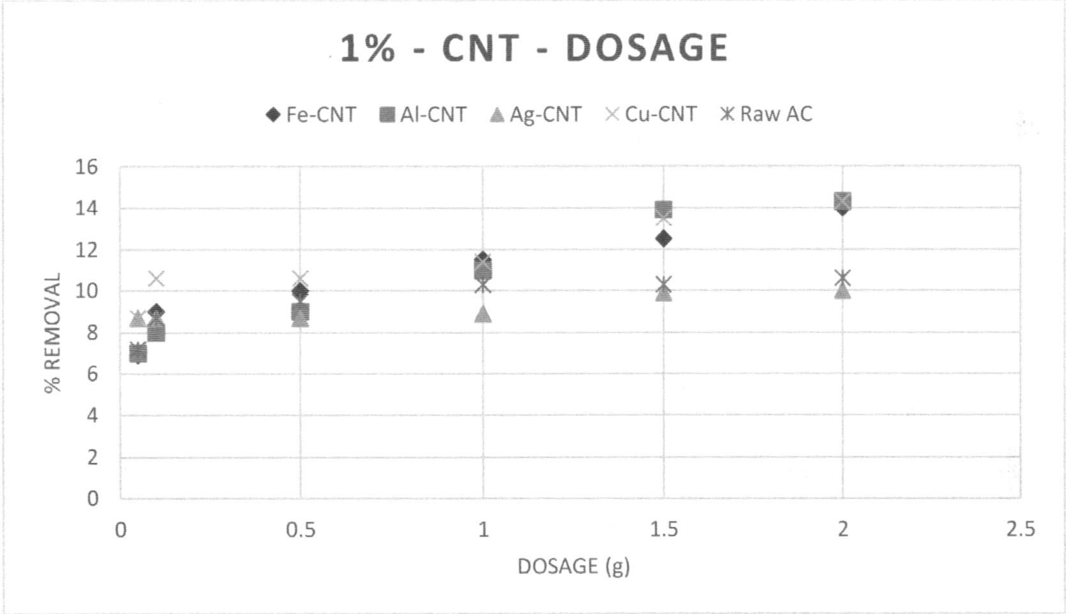

Figure 6: Effect of Dosage on ammonia removal using 1% CNT

Both Figures 5 and 6 shows the effect of the dosage on the percentage removal of ammonia for 1% metal loading. It was found that as the amount of dosage increases the percentage removal also increased, due to the availability of more active sites. The AC impregnated with aluminum gave the best removal. For CNT, all types of the modified CNT gave the same results except for the AG impregnated CNT which gave the worse results. Figures 7 and 8 shows that when the loading of element is increased to 10% the removal was enhanced with both AC and CNT. The best removal that was achieved was 40.3% with the AC impregnated with Al. These results shows that the AC gave better results than the CNT. It is clear from the figures that as the dosage increases, the percentage removal keeps increasing and did not reach equilibrium which means that if we keep increasing the dosage there is still a possibility to have better removal. However, from an economical point of view, we cannot just continuously increase the dosage.

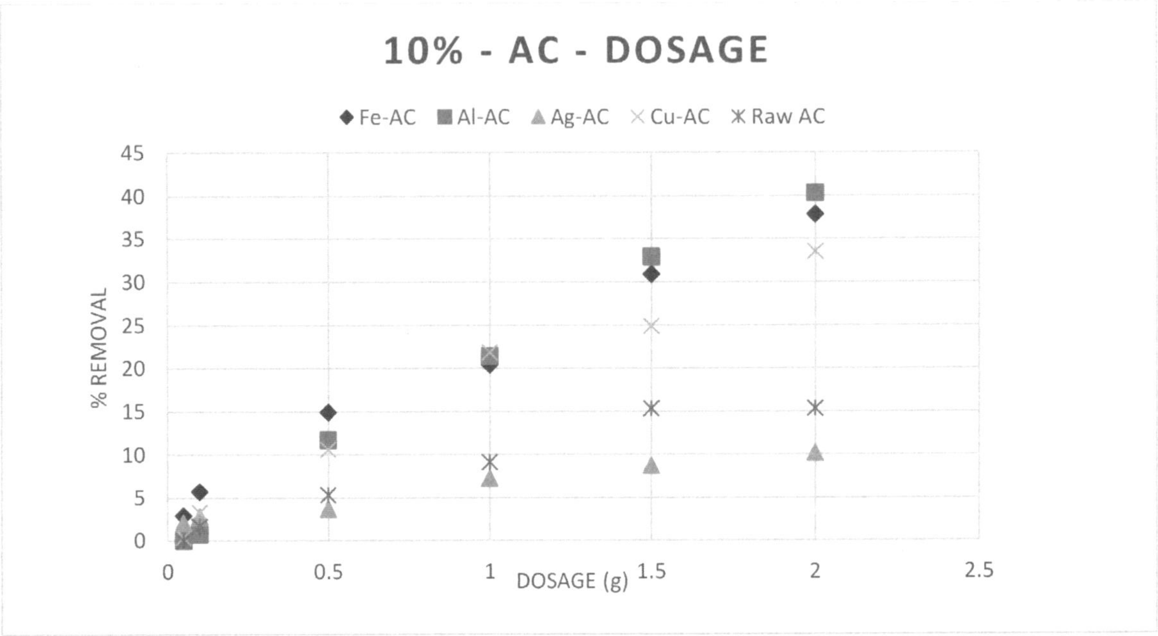

Figure 7: Effect of Dosage on ammonia removal using 10% AC

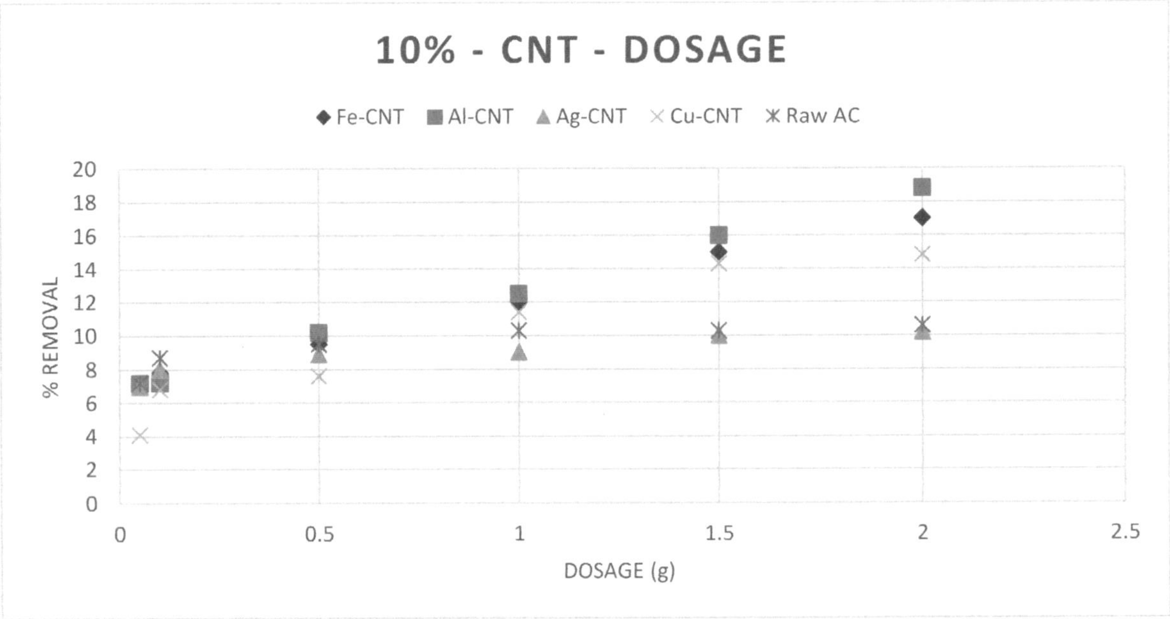

Figure 8: Effect of Dosage on ammonia removal using 10% CNT

4- Effect of contact time

To study the effect of time, pH, agitation time, temperature, and dosage were fixed at 10.5, 200 RPM, 25°C and 1 g respectively. The time interval was varied from 30 – 1440 minutes.

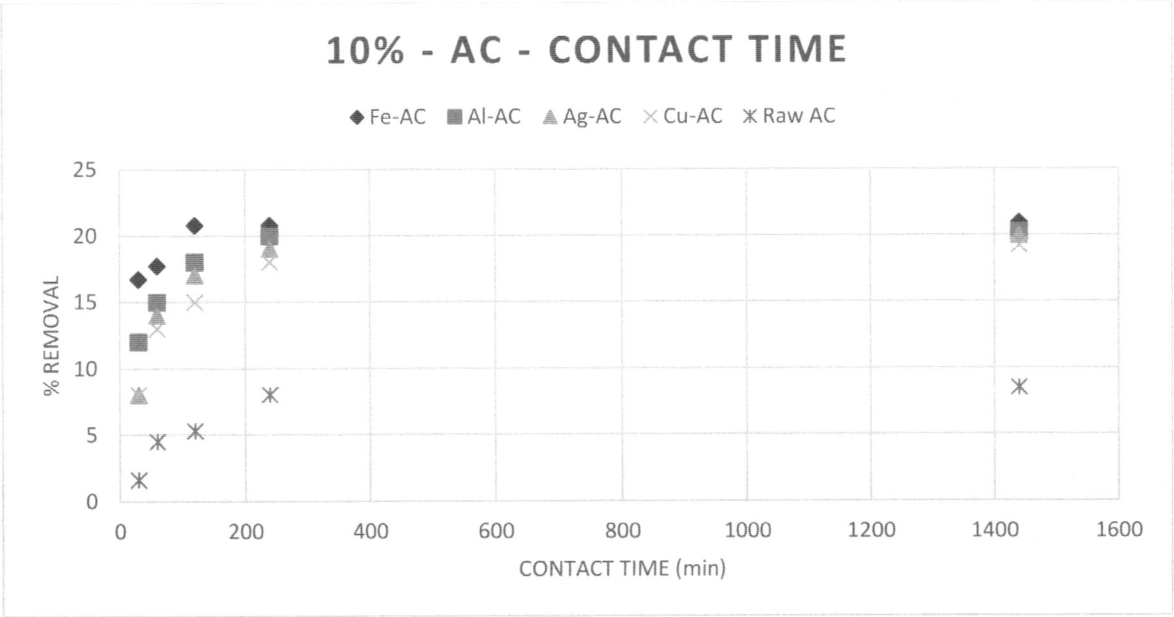

Figure 9: Effect of Contact time on ammonia removal using 10% AC

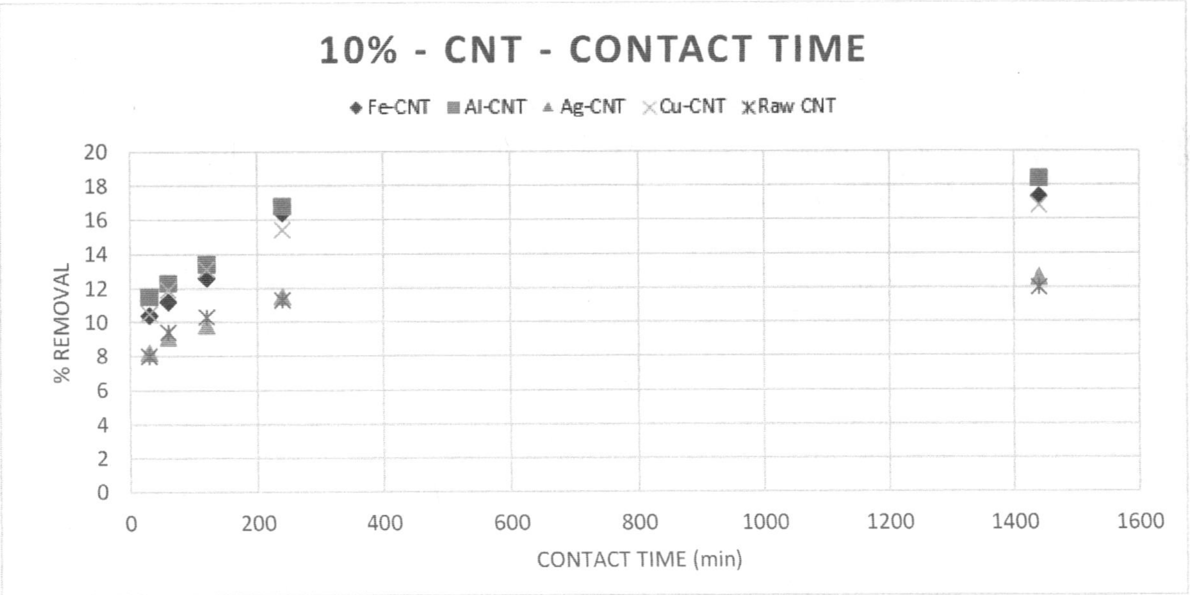

Figure 10: Effect of Contact time on ammonia removal using 10% CNT

It can be noticed from both Figures 9 and 10 that the removal increased rapidly in the first 120 minutes after which it starts to slow down between 120-240 minutes. After that it reached equilibrium since no change was noticed in the remaining time and this can be attributed to the saturation of the adsorbent. It was also clear that the impregnation has enhanced the removal for both AC and CNT.

5- Effect of Temperature

To study the effect of temperature, the temperature was increased within the range of 25°C to 70°C, keeping pH, dosage, contact time, and agitation speed constant at 10.5, 1 g, 2h and 200 RPM respectively.

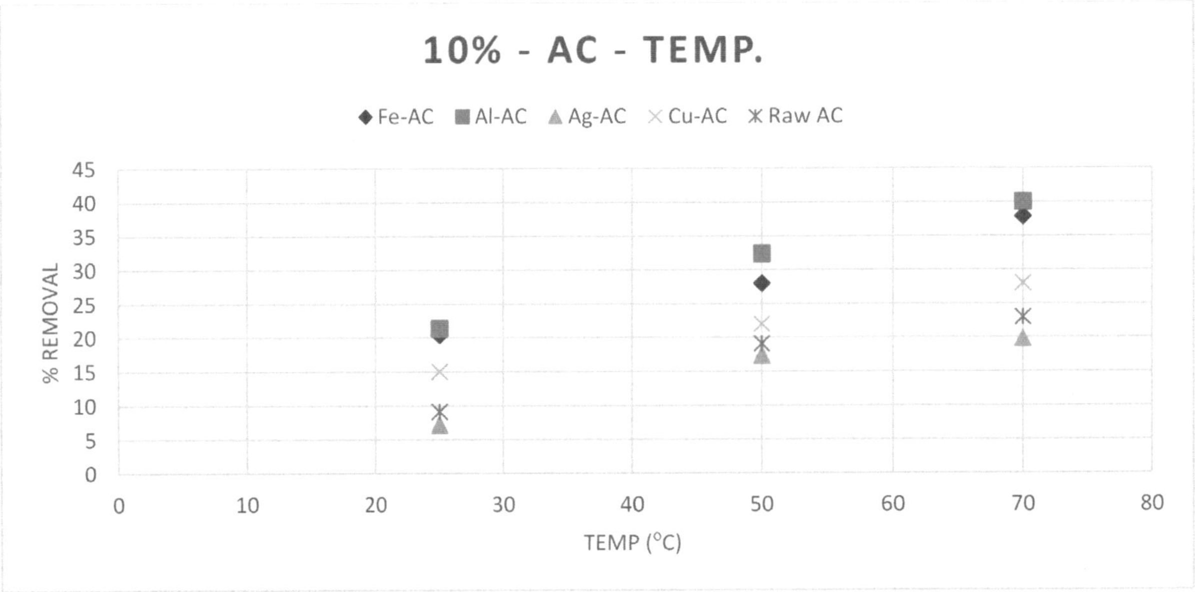

Figure 11: Effect of Temperature on ammonia removal using 10% AC

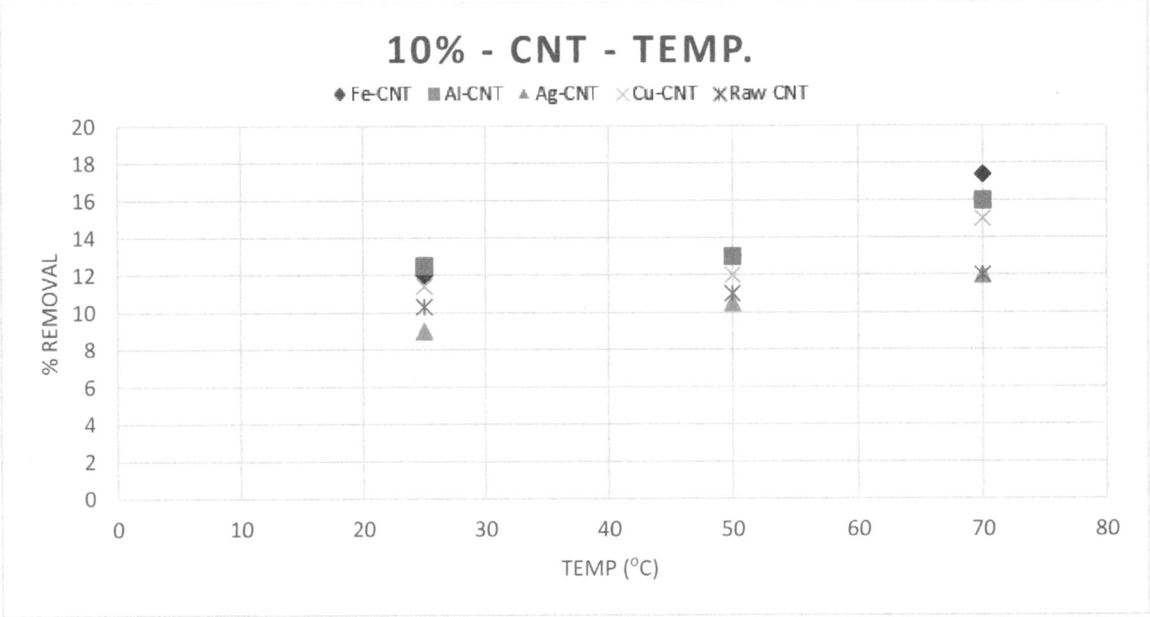

Figure 12: Effect of Temperature on ammonia removal using 10% CNT

Both Figures 11 and 12 Shows that as the temperature increases the percent removal was increased. The reason behind this is that as the temperature increases, the molecular movement will increase which will give a better chance for ammonia molecules to interact with adsorption surface. Also, due to the increase in the movement of the molecules, more active sites will be accessible which in turn will enhance the removal of ammonia. In addition, the resistance of both H+ and OH- ions of water are less with the increase of the temperature which will increase the external mass transfer.

Conclusion

This study shows the effect of pH, dosage, contact time, agitation speed, and temperature on the removal of ammonia from wastewater. Both activated carbon and carbon nano tube, raw and impregnated were used. It was shown that the pH has almost no effect of the removal of ammonia. On the other hand, increasing the dosage of adsorbents, contact time, agitation speed, and temperature will each enhance the removal of ammonia. Activated carbon shows better removal than CNT. When both AC and CNT were impregnated with metal ions, the removal capacity was enhanced. Among Al, Ag, Fe and Cu, Al was the best. From the results, it can be noticed that the removal percentage achieved using this technique was less than many of the available conventional methods. Accordingly, this method can be used as a pre-treatment technique due to its economic practicality.

Acknowledgments

The author is thankful to the Department of Chemical Engineering, King Fahd University of Petroleum and Minerals for providing the support to conduct this research. In addition, the author acknowledges the deanship of scientific research at KFUPM as they are supporting the conference attendance.

References

1. EPA, Process design manual for nitrogen control U.S.EPA, Technology Transfer, 1975.
2. Luebs, R.E. and Laag, A.E. Ammonia and related gases emanating from a large dairy area. Calif. Agric., 1973,27(2).
3. Hazardous Substances Data Bank: Ammonium Chloride, Bethesda, M. D., National brary of Medicine 1990.
4. Environment Canada. National inventory of natural and anthropogenic sources and emissions of Ammonia (1980) Rep. No. EPS 5/IC/1, Environmental Protection Programs Directorate (1985)
5. Sorption of phenol from waters on activated carbon impregnated with iron oxide, aluminum oxide and titanium oxide. B Abussaud, HA Asmaly, TA Saleh, VK Gupta, MA Atieh, Journal of Molecular Liquids 213, 351-359
6. Yin, Y.C., Aroua, M.K. Wan Daud, W.M. (2007). Review of modifications of activated carbon for enhancing contaminat uptakes from aqueous solutions. Sepration and Purification technology, Vol.52, 403-415.
7. Ferric oxide nanoparticles decorated carbon nanotubes and carbon nanofibers: from synthesis to enhanced removal of phenol, HA Asmaly, B Abussaud, TA Saleh, VK Gupta, MA Atieh, Journal of Saudi Chemical Society 19 (5), 511-520
8. Evaluation of micro-and nano-carbon-based adsorbents for the removal of phenol from aqueous solutions, Hamza A Asmaly, Basim Abussaud, Ihsanullah, Tawfik A Saleh, Alaadin A Bukhari, Tahar Laoui, Ahsan M Shemsi, Vinod K Gupta, Muataz Ali Atieh, Toxicological & Environmental Chemistry 97 (9), 1164-1179

9. Adsorption of toluene and paraxylene from aqueous solution using pure and iron oxide impregnated carbon nanotubes: kinetics and isotherms study, A Abbas, BA Abussaud, NAH Al-Baghli, HH Redhwi, Bioinorganic chemistry and applications 2017
10. Benzene removal by iron oxide nanoparticles decorated carbon nanotubes, A Abbas, BA Abussaud, NAH Al-Baghli, M Khraisheh, MA Atieh, Journal of Nanomaterials 2016
11. Cooney, E.L.; Booker, N.A.; Shallcross, D.C.; Stevens, G.W., (1999).Ammonia removal from wastewaters using natural Australiannzeolite. II. Pilot-scale study using continuous packed column process, Sep. Sci. Technol., 34(14): 2741-2760
12. Bernardi, M.; Le Du, M.; Dodouche, I.; Descorme, C.; Deleris, S.; Blanchet, E.; Besson, M., (2012). Selective removal of the ammonium-nitrogen in ammonium acetate aqueous solutions by catalytic wet air oxidation over supported Pt catalysts, Appl. Catal., B(128): 64-71
13. Schroeder, J.P.; Croot, P.L.; Von Dewitz, B.; Waller, U.; Hanel, R., (2011). Potential and limitations of ozone for the removal of ammonia, nitrite, and yellow substances in marine recirculating aquaculture systems, Aquacult. Eng., 45(1): 35-41
14. Feng, S.; Xie, S.; Zhang, X.; Yang, Z.; Ding, W.; Liao, X.; Chen, C., (2012). Ammonium removal pathways and microbial community in GAC-sand dual media filter in drinking water treatment, J. Environ. Sci., 24(9): 1587-1593
15. Carbon Nanotubes and Related Structures: New Materials for the Twenty-First Century, by Harris, Peter, Cambridge University Press, 2001
16. Oubagaranadin, J.U.K., Murthy,Z.V.P. Activated carbons: Classifications, properties and applications, pp.239-266, 2011.
17. S. Zhang, T. Shao, H.S. Kose, T. Karanfil, Adsorption kinetics of aromatic compounds on carbon nanotubes and activated carbons., Environ. Toxicol. Chem. 31 (2012) 79–85. doi:10.1002/etc.724.

629159 Removal of Mixture of Dyes Using Water Stable Metal Organic Framework

Abstract:

Metal Organic Frameworks (MOFs) are advanced materials finding diverse applications in several areas including catalysis[1], wastewater treatment[2], sensing[3], etc. Owing to their high capacity, these materials may be effectively used in the removal of dyes from aqueous streams. MOFs are three dimensional networked compounds consisting of metal centres and linkers. The metal centre, along with their coordinated groups, constitutes the secondary building unit (SBU). These SBUs hold the 3 dimensional network and the linkers form the bridge between different SBUs in a unit cell. The MOF under investigation viz. MIL-101(Cr) is chromium centred and consists of the linker terephthalic acid. This MOF synthesis was intensified using an optimized solvothermal procedure. The XRD, BET and FTIR studies revealed that the MOF produced was similar to the one synthesized solvothermally by S. Karmakar et al [4]. The results will be presented.

The MOF synthesized was used for adsorption of sunset yellow and coomasie brilliant blue g250 from aqueous solutions. These dyes are the pollutants obtained from household as well as textile industry. They are found to be anionic in nature as they have SO_3^- functional groups and are found to have low pK_a values and, hence, will exist in aqueous solution as anionic and thus the operation of pH was maintained from 3 to 7 because beyond pH of 8 the MOF was found to be unstable as the carboxylate groups present in the linker got dissociated.

Preliminary adsorption studies were performed and it was found that the MOF was quite significantly removing the dyes from the aqueous stream but was found to take 3 days to equilibrate, thus, it was planned to alter the linker to improve the adsorption kinetics by adding functional groups.

[1] J. Liu, L. Chen, H. Cui, J. Zhang, L. Zhang, and C. Y. Su, "Applications of metal-organic frameworks in heterogeneous supramolecular catalysis," *Chemical Society Reviews*, vol. 43, no. 16. Royal Society of Chemistry, pp. 6011–6061, Aug. 21, 2014, doi: 10.1039/c4cs00094c.

[2] J. Canivet, A. Fateeva, Y. Guo, B. Coasne, and D. Farrusseng, "Water adsorption in MOFs: Fundamentals and applications," *Chemical Society Reviews*, vol. 43, no. 16. Royal Society of Chemistry, pp. 5594–5617, Aug. 21, 2014, doi: 10.1039/c4cs00078a.

[3] L. Huang, H. Wang, J. Chen, Z. Wang, and J. Sun, "Synthesis, morphology control, and properties of porous metal–organic coordination polymers," *Microporous Mesoporous Mater.*, 2003, Accessed: Apr. 30, 2018. [Online]. Available: https://www.sciencedirect.com/science/article/pii/S1387181102006091.

[4] S. Karmakar, D. Roy, C. Janiak, and S. De, "Insights into multi-component adsorption of reactive dyes on MIL-101-Cr metal organic framework: Experimental and modeling approach," *Sep. Purif. Technol.*, vol. 215, no. October 2018, pp. 259–275, 2019, doi: 10.1016/j.seppur.2019.01.013.

Copyright © American Institute of Chemical Engineers. All rights reserved.

625367 Removal of Congo Red Dyes from Aqueous Solutions By Porous γ-Al_2O_3 Nanoshells

Abstract:

Alumina coating on carbon black (alumina/CB) was prepared by a condensed layer deposition (CLD) coating process to enable nanoscale structure with high surface areas. Al_2O_3 nanoshells were derived after burning the alumina/CB under oxygen gas at 500 °C, with a high specific surface area. Both Al_2O_3 nanoshells and alumina/CB were used as adsorbents to remove Congo Red dye from aqueous solutions in batch adsorption processes. The adsorbents were characterized using various techniques, such as N_2 adsorption-desorption isotherms (BET), X-ray diffraction (XRD), thermo-gravimetric analysis and differential thermal analysis (TGA - DTA), transmission electron microscopy (TEM), Fourier-transform infrared spectroscopy (FTIR), and zeta potential. Characterization results indicated that the alumina/CB and Al_2O_3 nanoshells yield high surface area of 290 and 217 m^2/g, respectively. While wide angle XRD and FTIR studies confirmed the phase transformation of γ- Al_2O_3 after the removal of CB at 500 °C. A series of parameters were investigated in adsorption experiments, including contact time, initial dye concentrations, ionic strength and pH of the solutions. It was found that equilibrium for Congo red adsorption can be reached within 30 min, much faster than reported by other studies in the literature on similar adsorbents. The adsorption mechanism for both samples followed the Langmuir isotherm with a pseudo second order reaction. The adsorption isotherm studies showed that Al_2O_3 nanoshells achieved a 44.8 % higher adsorption capacity than that of alumina/CB. The Al_2O_3 nanoshells have a maximum adsorption capacity of 370.4 mg/g (25 °C; pH 7; no salt added), better than or comparable to those previously reported adsorbents. The adsorption capacity of Al_2O_3 nanoshells was more favorable at lower pH, and the optimal adsorption ability was achieved at pH 4.0 with a removal efficiency at 98.6%. The adsorption process was associated with the H-bonding and electrostatic attraction between the Congo red dye and Al_2O_3 nanoshells surface. Overall, these results suggest that the low cost Al_2O_3 nanoshells is a very efficient adsorbent for anionic Congo red dye removal from contaminated water due to their fast kinetics along with high adsorption capacity.

Copyright © American Institute of Chemical Engineers. All rights reserved.

631724 Downscaling Camx Using a Reduced Complexity Model: A Multiscale Modeling Approach for Urban Air Pollution

Abstract:

Exposure to air pollution is associated with an increased risk of respiratory and cardiovascular disease. Urban populations are often exposed to high air pollution levels and are at greater risk of developing these health outcomes. Unfortunately, air pollution remains a complex problem in urban areas. Emissions from sources present in urban areas result in an irregular distribution of pollutants. Many chemical and dynamic processes that affect urban air pollution also occur on a wide range of spatial and temporal scales. The nature of urban air pollution requires models capable of high spatial and temporal resolution to capture local and regional interactions. Eulerian grid chemical transport models are employed and can provide estimates of pollutant concentrations at the regional scale. These models may also estimate concentrations at some urban scales. However, the highest resolution cannot adequately capture the spatiotemporal variability of air pollutants needed to understand urban air pollution.

Therefore, in this work we propose a multiscale modeling system to overcome this limitation of regional chemical transport models. We use the Comprehensive Air Quality Model with Extensions (CAMx), a common regional scale model used in the United States, to predict concentrations at a regional scale (12 x 12 km resolution) for the Contiguous United States. We further resolved these estimates using a reduced complexity model, InMAP, that uses grid adaptations to produce a variable horizontal resolution grid. The reduced complexity model provides higher spatial resolution, down to 500-meters, in densely populated areas and lower resolution in more sparsely populated areas. This provides additional spatial resolution in the urban areas that experience more spatial heterogeneity in air pollutant concentrations. We use source-specific spatial surrogates to calculate the emissions in each grid cell. We compare the simulated concentrations to monitoring data provided by the EPA. The proposed approach is the first step toward understanding the spatial and temporal variability of air pollution in urban areas.

Copyright © American Institute of Chemical Engineers. All rights reserved.

624154 The Single Parameter Hygroscopicity of Biodegradeable Polymers

Abstract:

The water-uptake ability of different compounds determines the final size and the fate of the aerosols in a supersaturated environment. Hygroscopicity is an important physicochemical property that compares the water-uptake ability of different substances. The traditional single parameter hygroscopicity indicates that the water-uptake ability depends on the molecular volume of the solute. This is true for substances with molecular weight less than 200g/mol. The Flory-Huggins Kohler (FHK) theory can be applied to predict the growth of a larger molecular weight polymeric aerosol. FHK shows that the water uptake ability of the polymer no longer depends on the molar volume of the solute. In this study, we develop the expression for single parameter hygroscopicity from FHK, and show that the hygroscopicity depends on both the dry particle size and the water-polymer interaction parameter in the FHK model. We also measure the hygroscopicity of four different hydrophilic biodegradable compounds, mannitol, lactose, gelatin and polyethylene glycol. The molecular weight of the four compounds are all significantly different and range in the order of 50 to 10,000 g/mol. The hygroscopicity of the four compounds derived from both traditional Köhler and FHK is compared; for very large molecules traditional methods can underpredict hygroscopicity up to 200%. A new size-dependent parameterization for hygroscopicity of polymers is provided. In summary, the water affinity plays an important role of the water uptake of the organic nano-particles.

Copyright © American Institute of Chemical Engineers. All rights reserved.

622218 CCN Activity Analysis of Low-Hygroscopicity Aerosols Using the Aerodynamic Aerosol Classifier (AAC)

Abstract:

In this paper we describe the methodology for Cloud Condensation Nuclei (CCN) activity analysis of aerosol particles involving an Aerodynamic Aerosol Classifier (AAC). The AAC is a novel instrument that size-selects aerosol particles based on their mechanical mobility. While moving through a rotating AAC column an aerosol particle is subjected to a centrifugal force that varies according to particle mass and is balanced by the drag force acting on it. The drag force defines the mechanical mobility of the particle, which is related to its relaxation time within the AAC column. The relaxation time depends on the rotational speed of the AAC column and determines the "aerodynamic diameter" for which the aerosol particles can be size-selected/classified. So far, previous studies have examined the use of the AAC for numerous applications. These include calibration of Optical Particle Counter (OPC), determination of particle mass, effective density and dynamic shape factor, and separation of particles from a polydisperse stream. In addition to this, theoretical models to represent the classifying characteristics of an AAC have also been studied in vast detail. However, the utility of an AAC for CCN activity analysis of aerosols has not yet been explored. Traditionally, a Differential Mobility Analyzer (DMA) is used for aerosol classification in a CCN analysis setup. A DMA size-selects/classifies particles based on their electrical mobility. This requires the particles to acquire a unit charge before being subjected to an external electrostatic field while they move along the DMA column. Despite the utility of a DMA-based setup, there are issues related to particle multiple charging artifacts that can introduce uncertainties in CCN measurements and subsequent CCN activity predictions. Substituting the DMA with an AAC can help with eliminating these uncertainties as classification using an AAC does not require particle charging. This can improve classification for aerosol particles and has the potential to then improve the CCN activity predictions, particularly of low-hygroscopicity species. In this work, we present the results for CCN analysis of a moderately aqueous soluble and weakly hygroscopic organic aerosol (sucrose) using an AAC-based experimental setup. We show how the uncertainties associated with aerosol particle sizes can be removed with the help of extensive transfer function analysis. Previously, the AAC transfer function and uncertainties associated with particle relaxation times have been combined for expressing the resolution for the AAC classification. Since the particle relaxation times are translated into aerodynamic diameters, their uncertainties will also carry over. We extend the transfer function analysis to examine the variability AAC resolution in terms of the aerosol aerodynamic diameter instead of relaxation time. This facilitates quantification of the size-dependent uncertainties in the particle measurements corresponding to specific aerodynamic diameter of aerosols being classified using the AAC. We conclude that this is effective for reducing the disparities in size-resolved aerosol measurement, therefore increasing the accuracy in the predictions related to CCN activity and droplet nucleation.

Copyright © American Institute of Chemical Engineers. All rights reserved.

624924 Investigating Wildfire Smoke Plume Transport in the Western U.S. Using Satellite Remote Sensing and Atmospheric Models

Abstract:

In recent decades, there has been a significant increase in both the size and number of wildfires in the western U.S. The mountains create complex atmospheric flows that lead to terrain-induced wildfires from dry, downslope winds, especially during drought conditions. Wildfire smoke plumes can be transported several hundreds of miles downwind, adversely impacting populations living near or downwind of the fires. To investigate the health outcomes associated with smoke exposure and implement warning systems to protect public health the spatiotemporal distribution of smoke plume concentrations is required. Air quality monitors do not specifically quantify air pollution concentrations from smoke plumes because the measurements are for the total ambient air pollution concentrations (i.e., all air pollution sources). The monitoring data can be supplemented with information from satellite remote sensing products, wildfire emissions inventories, dispersion models, and chemical transport models to characterize the increase in air pollution concentrations due to wildfire smoke. Many of these tools have uncertainties in the western U.S. because of the mountainous terrain, elevation changes, heterogeneous land surface, and/or aerosol mixtures associated with unique air pollution sources. Results of ongoing air quality modeling efforts to estimate and forecast wildfire smoke plume transport in the western U.S will be presented. Including recent advancements in modeling smoke plume injection height and use of satellite remote sensing products to investigate fire behavior and smoke plume transport.

Copyright © American Institute of Chemical Engineers. All rights reserved.

632975 High Performance Air Assisted Flare Design to Handle Purge Flow Conditions

Abstract:

One of the biggest challenges in the gas flaring process in petrochemical plants is the purge flow conditions. During normal operating conditions, the flare system receives much lower waste gases flow than the maximum expected amount of waste gases during the sudden shutdown or upset in the whole system. Therefore, waste gases at purge flow conditions combust with poor efficiency because of the lack of the required mixing energy between the waste gases and the ambient air. In this study, our research group designed, built, and tested a new air assisted flare of an industrial size that is capable of overcoming this challenge by providing the required momentum and air to the waste gases flow, thus ensuring an efficient combustion process for waste gases and a better flare performance. The new design included a radial slot jet at the end of the flare stack from which inward radial air was injected to adjust the exit area available for the waste gases flow, and increasing the waste gases flow rate. The new air assisted flare design also provided a protection for the flare tip from the radiation heat; the flame heat could cause damage to the flare tip by cooling it the used air. Five different experiments from two groups were carried out to test the new designs under the low flow conditions. The first group included tests without using the new design while the second group included the same test conditions using the new design. The obtained new design results were compared to the tests without the new design. The results showed a significant improvement in the combustion efficiency and in the flare performance when the new design was employed.

Copyright © American Institute of Chemical Engineers. All rights reserved.

624057 Computational Study of Simultaneous Removal of Nitrous Oxide and Carbon Mono Oxide from Flue Gases

Abstract:

N_2O have very noxious effects on environment as green hose gas in stratosphere and ozone depleting catalyst in lithosphere and CO is a fatal gas. In this study we used spin polarized density functional theory calculations to investigate the possible action of the Si-coordinated nitrogen doped ($Si@N_4C_{54}$) fullerene as potent catalyst for N_2O reduction and CO oxidation. Our computed results suggest that Si atom is strongly trapped in the N_4-cavity of the N_4C_{54} with larger negative adsorption energy of -185.91 kcal/mol. The N_2O molecule dissociates into O* and N_2, where O* remains adsorb over the $Si@N_4C_{54}$ surface. The carbon monoxide on the surface of the catalyst reacts with oxygen forming CO_2 and the energy barrier for the elimination of the adsorbed O* by CO molecule is 10.90 kcal/mol. The catalytic reduction of N_2O and oxidation of CO take place in the presence of O_2 molecule. This process can be used effectively for the mitigation of N_2O and CO pollution.

Copyright © American Institute of Chemical Engineers. All rights reserved.

631090 Determination of Particulate Matter in Dental Clinics with and without the Use of Air Purifiers

Abstract:

Introduction

In light of the SARS-CoV-2 virus pandemic, the Occupational Safety and Health Administration of the US Department of Labor has classified dentistry as one of the very high risk professions for the virus transmission due to the aerosols produced during various dental treatments [1]. Most of the instruments used in standard dental treatments produce aerosols [2, 3]. It has been found that the use of ultrasound equipment in dental treatments can produce aerosols that spread out to distances of up to 2 meters from the patient and that can remain in the air for time periods ranging from 35 minutes to several hours [4]. The production of aerosols and saliva droplets can cause infection via air transmissible microbes. Recent studies have shown that the SARS-CoV-2 virus can be transferred by particulates [5] and a positive correlation has been found between airborne particulates $PM_{2.5}$ and the transmission of other viruses, including the influenza virus [6]. Another study found that SARS-CoV-2 occurs more frequently in airborne particulates with diameters of 0.25-1.00 μm and >2.5 μm [7]. Zhao et al. [8] reported that the use of air purification systems with HEPA and F6 class filters during dental treatments can reduce aerosol concentrations by 54-83%. Another study carried out in a dental clinic found that air purifiers can reduce aerosol concentrations by 80-90% therefore reducing pathogen exposure for both employees and patients [9]. The purpose of this study was to measure the number and concentration of airborne particulates occurring in a dental clinic while performing dental procedures, with and without the simultaneous use of air purifier systems. To the best of our knowledge, no other studies comparing air purification units have been conducted in dental clinics.

Methodology

A Mini Laser Aerosol Spectrometer - IAQ-11R (by GRIMM Aerosol) automatic data recorder was used to determine particulate matter in real time. The Mini Laser captures airborne particulates with diameters of 0.25-32.0 μm and classifies them into 31 size channels. Samples were taken once per minute. The air purification systems applied and compared were the Winix Zero Pro, Aerte, Airocide, Rohnson Pure Air, and VacStation by Eighteeth (Extraoral Dental Vacuum System) systems. Baseline concentrations of airborne particles were initially recorded before air purifier operation. The quantity of airborne particulates was then measured during dental procedures, both with and without the simultaneous operation of the air purifying systems. Statistical analysis of the results was performed using IBM SPSS Statistics Version 25. Results were compared by applying one-way ANOVA analysis of variance and the Tukey test was used to determine statistically significant differences between particulate concentrations ($p \leq 0.05$ was considered statistically significant).

Results

Results showed that the use of air purifiers during dental work can reduce concentrations of particulate matter. Compared to the baseline values, concentrations of particulate matter recorded during dental procedures with operating air purifiers got reduced by 77% for $PM_{0.25}$, 81% for $PM_{0.28}$, 82% for $PM_{0.3}$, and 82% for $PM_{0.35}$. However, under the same conditions but while the treatment area was being aired by open windows alone, particulate concentrations were reduced by 53% for PM , 64% for PM , 72% for PM , and 77% for PM . Given that airborne particles with

0.25 0.28 0.3 0.35

diameters <1 μm are the ones which can enter lung alveoli and potentially harm human respiratory systems [10], use of air purifiers reduces concentrations of particles of this size, thereby improving air quality and reducing potential risk of airborne viral transmission. Continuous operation of air purifiers combined with airing (with open windows) contributes decisively to maintain low particulate levels in and around treatment areas. During dental procedures carried out with the simultaneous operation of air purifying systems, reductions in recorded airborne particulates ranged from 13-69% for PM_1, 27-67% for $PM_{2.5}$, and 19-23% for PM_{10}. Experiments conducted with air purifiers that do not possess filters did not show reduced particulate concentrations. The majority of particulate matter captured had diameters of 0.25-0.30 μm, 0.5 μm, and 1.0-4.0 μm, while particles with diameters of 0.65 μm and >5.0 μm were the least commonly observed in all experiments.

Conclusions

The results of this study show a statistically significant difference between the concentrations of particulate matter recorded during dental procedures carried out with and without the simultaneous operation of air purifiers. There is a high risk of SARS-CoV-2 virus contamination in dental clinics due to the aerosols emitted by the use of common dental instruments during standard treatments. Considering the necessity of dental treatments even during the SARS-CoV-2 pandemic, the use of air purifiers is highly recommended as an effective and economic means of reducing virus transmission, especially in small dental clinics / treatment areas.

References

1.Administration, U.S.D.o.L.O.S.a.H. *Dentistry Workers and Employers*. 2020; Available from: https://www.osha.gov/SLTC/covid-19/dentistry.html

2.Leggat, P.A. and U. Kedjarune, *Bacterial aerosols in the dental clinic: a review.* Int Dent J, 2001. **51**(1): p. 39-44.

3.Graetz, C., et al., *Spatter contamination in dental practices--how can it be prevented?* Revista medico-chirurgicala a Societatii de Medici si Naturalisti din Iasi, 2014. **118 4**: p. 1122-34.

4.Miller, R.L., *Characteristics of Blood-Containing Aerosols Generated by Common Powered Dental Instruments.* American Industrial Hygiene Association Journal, 1995. **56**(7): p. 670-676.

5.Guo, Z.D., et al., *Aerosol and Surface Distribution of Severe Acute Respiratory Syndrome Coronavirus 2 in Hospital Wards, Wuhan, China, 2020.* Emerg Infect Dis, 2020. **26**(7): p. 1583-1591

6.Su, W., et al., *The short-term effects of air pollutants on influenza-like illness in Jinan, China.* BMC Public Health, 2019. **19**(1): p. 1319.

7.Liu, Y., et al., *Aerodynamic analysis of SARS-CoV-2 in two Wuhan hospitals.* Nature, 2020. **582**(7813): p. 557-560.

8.Zhao, B., N. An, and C. Chen, *Using an air purifier as a supplementary protective measure in dental clinics during the coronavirus disease 2019 (COVID-19) pandemic.* Infection Control & Hospital Epidemiology, 2020: p. 1-2.

9.Chen, C., et al., *The effectiveness of an air cleaner in controlling droplet/aerosol particle dispersion emitted from a patient's mouth in the indoor environment of dental clinics.* Journal of the Royal

Society, Interface, 2010. **7**(48): p. 1105-1118.

10.Thomas, R.J., *Particle size and pathogenicity in the respiratory tract.* Virulence, 2013. **4**(8): p. 847-858.

Copyright © American Institute of Chemical Engineers. All rights reserved.

632065 Aerosol Generation during Routine Otolaryngology Procedures and the Effect of Hepa Filtration on Aerosol Concentrations

Abstract:

The SARS COVID pandemic has highlighted safety concerns for healthcare workers exposed to potentially aerosol-borne communicable diseases. Healthcare workers working in otolaryngology may be particularly at risk due to the prevalence of the SARS CoV-2 virus in the nares. In response, measures like use of HEPA filters and increased turnover times have been implemented to safeguard against these aerosolized particles. However, prospective investigations have not clearly defined aerosolization or how HEPA filtration affects in-clinic aerosol concentrations. The purpose of this study was to determine whether increased aerosol concentrations are associated with patients undergoing nasal endoscopy and to determine the effectiveness of HEPA filtration in reducing aerosol concentrations in the clinic.

Patients were prospectively enrolled in a cross-sectional study at a tertiary medical center. Demographic information, as well as clinic room characteristics, such as room size and air exchange with and without a HEPA filter, were measured and recorded. Clinic personnel also recorded activities in the clinic room, which were subsequently aligned with aerosol concentrations. A scanning mobility particle sizer, aerosol particle sizer, and GRIMM aerosol monitor measured aerosolized particles ranging from 14.3 nm to 34 µm. Associations between aerosol concentration, nasal endoscopy, and use of HEPA filter were assessed. Room turnover times needed to achieve baseline aerosol concentration were also calculated.

A total of 48 patients met inclusion criteria (29 with nasal endoscopy and 19 without (control)). The two patient cohorts exhibited no significant differences in age or gender. Aerosol concentrations among the nasal endoscopy cohort were greater than those in the control group ($p<0.05$). The coughing/sneezing event, followed by the patient/provider discussion events resulted in the highest aerosol concentrations. Aerosol concentrations returned to baseline within 8.76 minutes without the use of a HEPA filter versus. 4.79 minutes with a HEPA filter ($p=<0.01$). This translated into an air change exchange per hour value of 2.14 and 4.17, with and without HEPA filtration, respectively.

Copyright © American Institute of Chemical Engineers. All rights reserved.

632326 Characteristics of Electrospun Nanofiber Mats and Their Aerosol Filtration Performances As Respiratory Personal Protective Equipment

Abstract:

During the COVID-19 pandemic, respiratory personal protective equipment (PPE) has played a crucial role in controlling and minimizing exposure to respiratory droplets that can transmit the SARS-CoV-2 virus. Conventional filter media for PPE have been produced using meltblown processes, wherein polymer microfibers such as polypropylene are spun to nonwoven fabrics by extrusion. Moreover, these meltblown fabrics are followed by electrostatic charging to be capable of filtering out small particles while providing comfortable breathability. However, the meltblown process is a capital-intensive process carried out at high temperatures in dedicated facilities. Also, the important step of electrostatic charging of the fibers is kept as a proprietary process by conventional PPE manufacturers. These factors have resulted in meltblown fabrics becoming the major bottleneck for respiratory PPE production during the COVID-19 crisis. Along with the critical shortage in respiratory PPE due to the overwhelmed supply chain, the need for alternative production approaches and material options for respiratory filter media has been highlighted.

Electrospinning is a process capable of producing sub-micron diameter fibers with control over the morphology and chemical composition. It has found utility in multiple industries, including HEPA filtration, medical textiles, industrial catalysis, and drug delivery. Market forces have already driven this process due to its low start-up costs and scalability; therefore, it is an excellent approach for respiratory PPE production during a pandemic. To leverage this capability during a time of crisis, performance, process, and materials data are necessary so that electrospun fabric manufacturers can meet the specifications for N95 respirators, as well as responding to surge in demand and managing future outbreaks. Currently, there is limited understanding of and knowledge about the performance of electrospun filter media and required characteristics as respiratory PPE.

Here, we study the correlation between characteristics of electrospun filter media and their performance. The relevant characteristics include the type of polymer material, thickness of fabric, fiber diameter, porosity, and pore size distribution. The evaluation of performance is based on filtration efficiency and pressure drop for breathability. Testing protocols that approximate the National Institute for Occupational Safety and Health (NIOSH) standard for N95 masks are followed. The structure-property relationships for efficient filtration of particles down to 0.07-0.1 micron (i.e., the size of the SARS-CoV2 virus) are presented in this study. In contrast to meltblown fabrics, the filtration efficiency of electrospun nanofibers remains stable over time, as they capture the particles primarily through mechanical mechanisms, although induced electrostatics arising from intrinsic polarity in some polymers may also contribute. Among a wide range of achievable electrospun filter media, we identify specific characteristics of electrospun filter media that are suitable for respiratory PPE and meet the requirements for the existing PPE standards, such as NIOSH N95 masks. Finally, the electrospinning recipes are provided for immediate transfer to the production floor in the industry.

Copyright © American Institute of Chemical Engineers. All rights reserved.

625260 Effect of Polycyclic Aromatic Hydrocarbons on Alzheimer's Disease: Insights from Molecular Dynamics Simulations

Abstract:

Alzheimer's Disease is a neurodegenerative disorder, related to the development of amyloid plaques in human brain due to the aggregation of amyloid beta (Aβ) peptides. Polycyclic aromatic hydrocarbons (PAHs) are environmental pollutants, produced from human activities, such as cooking and smoking, and associated with increased risk for Alzheimer's Disease. In-vitro studies showed an increased aggregation kinetics of $A\beta_{40}$ peptide in the presence of PAHs, such as phenanthrene, pyrene and benzo[a]pyrene (B[a]P). In addition, recent in-vivo experiments revealed an enhanced formation of amyloid plaques and amyloid burden percentage in the brain of animals due to the exposure to B[a]P. Furthermore, according to literature, exposure to B[a]P may lead to the decline of the neurobehavioral function in human organism. While in-vivo and in-vitro experiments show an enhanced aggregation of Aβ peptides and formation of plaques in the presence of PAHs, the molecular interactions between peptides and PAHs are obscure. In this work, we performed a molecular dynamics study via GROMACS software aimed to investigate the structure of amyloid beta peptides and their aggregation in the presence of PAHs. According to the results of the performed simulations, strong interactions between amyloid beta peptides and B[a]P molecules enhanced the aggregation kinetics of the peptides. Moreover, presence of PAHs increased amounts of beta sheets in the aggregated structures of Aβpeptides.

References:

[1] I. Moreno-Gonzalez, L. D. Estrada, E. Sanchez-Mejias, and C. Soto, "Smoking exacerbates amyloid pathology in a mouse model of Alzheimer's disease," (in eng), *Nat Commun,* vol. 4, p. 1495, 2013, doi: 10.1038/ncomms2494.

[2] C. Wallin *et al.*, "Alzheimer's disease and cigarette smoke components: effects of nicotine, PAHs, and Cd(II), Cr(III), Pb(II), Pb(IV) ions on amyloid-β peptide aggregation," (in eng), *Sci Rep,* vol. 7, no. 1, p. 14423, 10 2017, doi: 10.1038/s41598-017-13759-5.

[3] C. Wallin *et al.*, "Mercury and Alzheimer's Disease: Hg(II) Ions Display Specific Binding to the Amyloid-β Peptide and Hinder Its Fibrillization," (in eng), *Biomolecules,* vol. 10, no. 1, 12 2019, doi: 10.3390/biom10010044.

[4] D. Gao, M. Wu, C. Wang, Y. Wang, and Z. Zuo, "Chronic exposure to low benzo[a]pyrene level causes neurodegenerative disease-like syndromes in zebrafish (Danio rerio)," (in eng), *Aquat Toxicol,* vol. 167, pp. 200-8, Oct 2015, doi: 10.1016/j.aquatox.2015.08.013.

[5] D. Gao, C. Wang, Z. Xi, Y. Zhou, Y. Wang, and Z. Zuo, "Early-Life Benzo[a]Pyrene Exposure Causes Neurodegenerative Syndromes in Adult Zebrafish (Danio rerio) and the Mechanism Involved," (in eng), *Toxicol Sci,* vol. 157, no. 1, pp. 74-84, 05 2017, doi: 10.1093/toxsci/kfx028.

[6] D. Liu *et al.*, "Benzo(a)pyrene exposure induced neuronal loss, plaque deposition, and cognitive decline in APP/PS1 mice," (in eng), *J Neuroinflammation,* vol. 17, no. 1, p. 258, Aug 2020, doi: 10.1186/s12974-020-01925-y.

[7] Q. Niu, H. Zhang, X. Li, and M. Li, "Benzo[a]pyrene-induced neurobehavioral function and neurotransmitter alterations in coke oven workers.," *Occupational and environmental medicine,* vol. 67, no. 7, pp. 444-8, 2010, doi: doi:10.1136/oem.2009.047969.

Copyright © American Institute of Chemical Engineers. All rights reserved.

623798 Implementation of a New Defoamer in the Defense Waste Processing Facility to Reduce Batch Processing Time and Improve Safety

Abstract:

The legacy of the waste from the production of nuclear weapons materials is the biggest environmental challenge in the state of South Carolina, and one of the biggest in the United States. Although great progress has been made in processing and vitrifying high level waste at the Savannah River Site, approximately 35 million gallons remains to be processed. In this paper, we discuss the new defoamer that has been developed for use in processing high level radioactive waste. The use of the new defoamer will allow higher evaporation rates, leading to as much as a 50% decrease in processing time per batch.

The defoamer is novel because it is effective in the harsh conditions needed for processing high-level radioactive waste, including high temperatures, a wide pH range (4 to 13), strong oxidizing and reducing agents, and the presence of a myriad of metal catalysts. The defoamer has replaced an old antifoam agent that was chemically unstable at non-neutral pH and decomposed to form three flammable gases. Because the old antifoam agent decomposed quickly, processing changes were needed which resulted in increasing processing time, including lowering acid addition rates and evaporation rates.

The new defoamer is not only more effective at controlling foam but, as a result, will shorten the batch processing time and lead to safer and more efficient processing by eliminating the production of flammable gases. Testing demonstrated that the new defoamer would meet all the requirements needed for foam mitigation during DWPF processing. The new defoamer will shorten the mission lifetime, allowing the vitrification facility to be shut down several years early, potentially saving the federal government hundreds of millions of dollars in operating costs.

This paper will discuss the implementation of the new defoamer in the Defense Waste Processing Facility (DWPF). Use of the new defoamer has demonstrated that excessive foaming and foamovers have been eliminated allowing DWPF to increase acid addition rate and steam flow rate to maximum rates permitted by procedures and safety requirements.

Copyright © American Institute of Chemical Engineers. All rights reserved.

633803 Water Sorption/Desorption Characteristics in Electrorefiner Salt-Occluded Zeolites

Abstract:

Electrorefiners currently operating in the Fuel Conditioning Facility of Idaho National Laboratory contain radioactive, actinide-containing molten chloride salt. The primary components of this salt are LiCl and KCl in a eutectic composition. The salt also contains UCl_3, $PuCl_3$, and numerous fission products from processing irradiated sodium-bonded metal fuel from the Experimental Breeder Reactor-II. It has been proposed to occlude the salt into a zeolite such as zeolite-4A or Na-Y zeolite before eventual conversion into ceramic waste forms for permanent disposal. In this work, we studied the characteristics of the salt-occluded zeolite for hydration and if such hydration is reversible. Zeolite-4A with 20 wt% LiCl-KCl was found to absorb up to about 10 wt% water when in contact with 40°C air at 20% relative humidity. As is the case with zeolite-4A without any occluded salt, essentially all of the moisture can be desorbed by heating the salt occluded zeolite-4A to 500°C under flowing argon. However, the dehydration is accompanied by a small amount of HCl gas formation due to hydrolysis of the salt. HCl generation was measured continuously while the salt occluded zeolite was heated to 500°C using an off-gas sparged auto-titrator. The amount of HCl generated ranged from about 0.3 to 0.5% of the total Cl contained in the salt for any given sample of salt occluded zeolite. The effect of temperature ramp rate and maximum temperature on the amount and rate of HCl generated was studied and will be reported.

Copyright © American Institute of Chemical Engineers. All rights reserved.

631169 Performance Assessment for the Low Activity Waste and Intermediate Level Vaults (LAWV and ILV) at the Savannah River Site

Abstract:

Low Activity Waste (LAW) generated at the Savannah River Site (SRS) that exceeds disposal limits for burial in ground trenches is being disposed of in the LAW Vault (LAWV) located in E-Area. The LAWV is a large, above-grade, reinforced concrete structure 643 ft long, 145 ft wide, and 27 ft high at the roof crest. It is divided into 12 cells designed to contain more than 12,000 B-25 boxes. Intermediate level waste generated at SRS is being disposed of in the Intermediate Level Vault (ILV) also located in E-Area. The ILV is a largely below-grade, reinforced concrete vault with a footprint of 279 ft by 48 ft and consists of two modules: the Intermediate Level Tritium (ILT) module, which contains two 26-ft deep cells, and the Intermediate Level Non-Tritium (ILNT) module, which contains seven 28.4-ft deep cells. At the end of radwaste-generating operations at SRS, both LAWV and ILV will be closed, sealed, and covered with an earthen closure cap for long-term disposition.

Groundwater models have been developed for the ILV and LAWV using the PORFLOW software package with the purpose of evaluating dose to members of the public following site closure and thereby producing waste disposal limits. Modeling results are presented and discussed in this paper.

Copyright © American Institute of Chemical Engineers. All rights reserved.

632329 Performance Assessment for Naval Reactor Component Disposal Areas (NRCDA) at the Savannah River Site

Abstract:

Naval Reactor (NR) waste generated by the reactor components from the U.S. Navy is comprised of highly radioactive components consisting of activated corrosion-resistant metal alloy contained within welded thick-walled steel casks, and auxiliary equipment primarily contaminated with lower levels of Activated Corrosion Products residing on the metal surfaces and contained within thinner-walled bolted containers. The NR waste has been stored and disposed in the two Naval Reactor Component Disposal Areas (NRCDAs), 643-7E and 643-26E, at the Savannah River Site (SRS).

Groundwater models have been developed using the PORFLOW software package for the two NRCDAs with the purpose of evaluating dose impacts and producing disposal limits for the SRS E-Area Low-Level Facility (LLWF). Modeling results are presented and discussed in this paper.

Copyright © American Institute of Chemical Engineers. All rights reserved.

627761 Continuous-Flow Centrifugal Solid-Liquid Separation for the Recovery of Rare-Earth Elements Containing Particles from Phosphoric Acid Sludge

Abstract:

Phosphoric acid sludge contains concentrated phosphoric acid (up to 54% P_2O_5) and solid precipitates, including rare earth elements (REEs) at concentrations near 2,200 ppm. Low cost recovery of valuable P_2O_5 and simultaneous separation of solid particles from the phosphoric acid sludge could be an economically feasible approach to recovering REEs, while increasing the production of phosphoric acid. The phosphoric acid sludge, however, is a complicated fluid stream that cannot be simply separated by traditional separation technologies such as batch centrifugation, decantation, settling, or filtration because of its high viscosity and high solids content (e.g., 30–40%). Guided by a force balance model, a highly efficient solid/liquid separation method was demonstrated in this study, using a continuous-flow centrifugal contactor. A high shear is typically used to increase the contact surface between two liquid phases, thus enhancing the interphase mass transfer rate, and a centrifugal force is exerted on the fluids, inducing phase separation at the exit. By introducing the sludge directly through the phase separation region, we demonstrated sludge recovery and studied the effects of process parameters on solid particle capture with the goal to optimize the separation process. Three contactors in series yielded 94% solid recovery and 98% recovery of the liquid component. After saturation, the slurry was withdrawn and the accumulated solids in the contactor were flushed out. Recovered solid particles contained 1,500–2,895 ppm of REEs, which can be recovered after further processing.

Copyright © American Institute of Chemical Engineers. All rights reserved.

Effective Adsorption of Chromium from Tannery Wastewater Using Green Synthesis Nano-Zero Valent Iron (GT-nZVI)

Ahmed S. Mahmoud[1], M. S. Mahmoud[2], Ahmed M. Noureldin[2], Robert W. Peters[3], and Mohamed K. Mostafa[4]

[1]*Scientific Research Development unit, Egyptian Russian University (ERU), Badr, Egypt*

[2] *Sanitary and Environmental Institute (SEI), Housing and Building National Research Center (HBRC), Giza, Egypt.*

[3]*Department of Civil, Construction and Environmental Engineering, University of Alabama at Birmingham, Birmingham, AL, USA*

[4]*Faculty of Engineering and Technology, Badr University in Cairo (BUC), Cairo, Egypt*

Abstract

Nowadays, Egypt built the biggest industrial area for tanning industries; this industry produces a vast amount of chromium in the tannery wastewater effluent. This study successfully prepared and characterized GT-nZVI by reducing agents extracted from soft black tea. The prepared GT-nZVI was characterized using x-ray diffraction (XRD), scanning electron microscopy (SEM), and energy dispersive analysis of x-rays (EDAX) analysis. The chromium removal efficiency reached 97.1% and 73.7% from an initial chromium concentration of 1.0 and 9.0 mg/L, respectively. The optimum removal efficiencies were achieved using 0.6 g/L GT-nZVI at acidic pH 5 after 30 min of shaking at a rate of 250 rpm. The equilibrium isotherm and kinetic results indicated that the adsorption of chromium onto GT-nZVI agrees with the Koble-Corrigan model with the lowest summation of errors of 0.394 and a Pseudo-Second-Order adsorption with the lowest summation of errors of 0.07. An artificial neural network (ANNs) was built to describe the factors affecting the adsorption process by 6 inputs, 4 hidden layers, and 1 output. The results obtained indicated that the initial concentration and GT-nZVI dose are the most significant variables that influence the adsorption process. The response surface methodology (RSM) equation was developed to allow the prediction of chromium removal efficiency at different operating parameters. Finally, this study strongly supports using GT-nZVI for chromium removal from tannery effluents.

1. Introduction

Water is the most traditional polar solvent with the capacity to exceed 75% of total earth capacity (Weingärtner and Franck, 2005, Mahmoud et al., 2021b, Farag et al., 2018). It is responsible for humans, plant, and animals nutrition and it is necessary for development especially in developing countries so, it considers as a principal source of life (Pahl-Wostl et al., 2008, Mostafa and Peters, 2015). Because of water polarity, it has multiple usages and can dissolve most of the Inorganic and Organic compounds and cause water and wastewater contamination (Schwarzenbach et al., 2010, Jason A. Heberling, 2018). Increasing water consumption, by increasing world population according to the estimates of the United Nations Organization the world's population October 2011exceeding than seven billion, increases global awareness about it to save water by decreasing consumption and water recycling (Dudgeon et al., 2006, Kennedy, 2011, Carpenter et al., 2011). Recently, water is becoming scarcer globally with evidence that it will become rarer. The water scarcity problem is the main reason for many wars in that world, especially in the Middle East (Beaumont, 1994, Dolatyar and Gray, 2000, Amery, 2002, Barnaby, 2009, Hamdy et al., 2019). There are different types of water resources include surface water such as rivers, seas and oceans, and groundwater (Mostafa and Peters, 2016a, Mostafa and Peters, 2016b). About 44% of the world population using groundwater as the second source of drinking water, such as China, the United States of America (USA), and Egypt (Shiklomanov, 1991, Pimentel et al., 2004, Taylor et al., 2013). In China, more than one-third of total water is demanded and some people using groundwater as the main source of drinking water (Oki and Kanae, 2006, Mostafa et al., 2017, SaryEl-deen et al., 2017). In the United States, Nearly 50% of Californian's people using groundwater as the main source of drinking water and agriculture and extract more than 14 billion gallons per day so it required to organize from scientists and government to overcome groundwater contaminations(Barlow and Clarke, 2017, Mostafa and Peters, 2017).

Wastewater is the most observable environmental problem among many in urban areas (Ngoc and Schnitzer, 2009, El-Shafei et al., 2016). The increasing world population increases water consumption and produces a huge amount of wastewater every day. There is an expectation of increasing the demand for sanitation by 2 billion will live in towns and cities, mainly in developing countries within 20 years (Langergraber and Muellegger, 2005). More than 90% of developing country's sewage is discharged untreated and polluting lakes, rivers, and coastal areas (Islam and Tanaka, 2004, Liu and Diamond, 2005). Conventional sanitation techniques, based on water-wasting technology, flush toilets, aren't a suitable solution for industrial and developing countries (Weng, 2002, Del Porto, 2005, Diaz and Barkdoll, 2006). The sanitation systems were built and designed on the premise that human excreta are a waste suitable for disposal into the environment and it can assimilate this waste (Esrey and Andersson, 1999, Langergraber and Muellegger, 2005).

The tanning industry is one of the most important industries all over the world. Throughout ancient history, the Ancient Egyptians employed the tanning process to cover the local needs of leather, belts, footwear, and drums (Forbes, 1956, Mahmoud et al., 2021b). By increasing the world population, increased the requirement of different leather products, leading to creating large tanneries projects such as Al Robaiky region, Egypt (Frosch and Gallopoulos, 1989). In Egypt, tanning businesses are the most significant source of national income (Nelson and Rosenberg, 1993, M Abdeldayem et al., 2020). The total number of tannery projects in Egypt is about 300 and most of them using the chromium tanning technique (Hafez et al., 2002, Walsh and O'Halloran, 1996). The tanning process using the chromium salts technique is effective, fast, stable, and low cost during the operation and production processes (Gupta et al., 2001, Sreeram and Ramasami, 2003). More than 65% of used chromium salts react with leathers and the chromium leftover remains in the effluent (Karabay, 2008, Kurniawan et al., 2011). So, tannery wastewater is recognized as highly contaminated industrial effluent including heavy metals, toxic chemicals, chloride, lime with high dissolved and suspended salts, and other pollutants (Chowdhury et al., 2013).

Several treatment technologies, like filtration, electrocoagulation, chemical precipitation, ion exchange, bioremediation, and adsorption, were tested for chromium removal from aqueous solutions and real wastewater ((Morsy et al., 2020)El-Shafei et al., 2016). Additionally, many papers have been published addressing the utilization of nano materials, especially iron nano particles, which is a promising material for the removal of heavy metals from wastewater due to its high reactivity and surface area (Li et al., 2020). This study attempts to use low-cost green synthesis of nano-Zero Valent Iron (GT-nZVI) sorbent material for chromum removal from aqueous solutions and real tannery wasteater. Batch adsorption experiments were conducted using GT-nZVI by adapting the operating conditions, viz., solution pH, GT-nZVI dosage, stirring rate , and adsorption time. Response surface methodology (RSM) and artificial neural network

(ANN) were also applied to find the correlation between the experimental factors and chromium removal efficiency

2. Material and Methods

2.1. Preparation of GT-nZVI

The extraction of reducing agents from soft black tea as well as reduction of metal cation from its soluble salt was adapted from previous studies (Mahmoud et al., 2019a, Mahmoud et al., 2020a).

2.2. Batch studies

When green synthesized nZVI (GT-nZVI) are placed in standard solutions of chromium (Cr) and agitated for an adequate time, the adsorption process occurs. Aqueous solutions (100 mL) were used and the adsorption capacity was determined at different environmental conditions (pH, Dose, Time, Stirring rates, and Concentrations). After filtration, the withdrawn samples were analyzed by atomic absorption spectroscopy according to the standard method for examination of water and wastewater 23 edition 2013 (Baird, 2017). Then the adsorption capacity and isothermal studies were conducted using adsorption isotherms (Mahmoud et al., 2020a, Mahmoud and Mahmoud, 2021, Rabie S. Farag and Peters, 2019). The percentage of removal efficiency was calculated using Eq. 1 (Mahmoud et al., 2021a, Rabie S. Farag and Peters, 2018). The amount of sorbed contaminants were calculated using Eq. 2 (Maha M. Elshfai and Elsaid, 2021, Mahmoud et al., 2021a, Rabie S. Farag and Peters, 2018). All experiments were performed in triplicate and all results were within lab measurements (Mahmoud et al., 2020b, Mahmoud et al., 2019b).

$$Sorption\ (\%) = \left(\frac{C_0 - C_e}{C_0}\right) \times 100 \quad (1)$$

where C_o is the initial concentration (mg/L) and C_e is the equilibrium concentration in solution (mg/L).

$$Q_e(mg/g) = \frac{(C_0 - C_e)V}{m} \quad (2)$$

where q_e is the equilibrium adsorption capacity (mg/g), V is the volume of aqueous solution (L), and m is the dry weight of the adsorbent (mg) (Mahmoud et al., 2018).

Table 1. Batch experiments at different operating parameters.

Effect	Dose (g)	Contact time (min)	pH	Stirring rate (rpm)	Initial concentrations (mg/L)
Effect of pH	0.6	30	3-11	250	5
Effect of adsorbent dose	0.1-1.0	30	5	250	5
Effect of contact time	0.6	10-50	5	250	5
Effect of stirring rate	0.6	30	5	100-300	5
Effect of concentration	0.6	30	5	250	1-9

2.3. Neural network structure

An artificial intelligence network using the backpropagation technique was familiar to approximation the adsorption of chrome onto GT-nZVI, the importance of each operating condition pH, dose, time, stirring rate, and concentrations. Also, to predict the corrected relation between predicted values and true measurements. The ANN model combined the input covariable (pH, dose, time, stirring rate, and concentrations) with hidden layers (weight and bias), and the final output removal layer. The estimated ANN structure for DY 50 removal using nano bentonite was expressed as 5 – 3 – 1 as shown in Figure 1 (Mahmoud et al., 2020a, Mahmoud et al., 2019a).

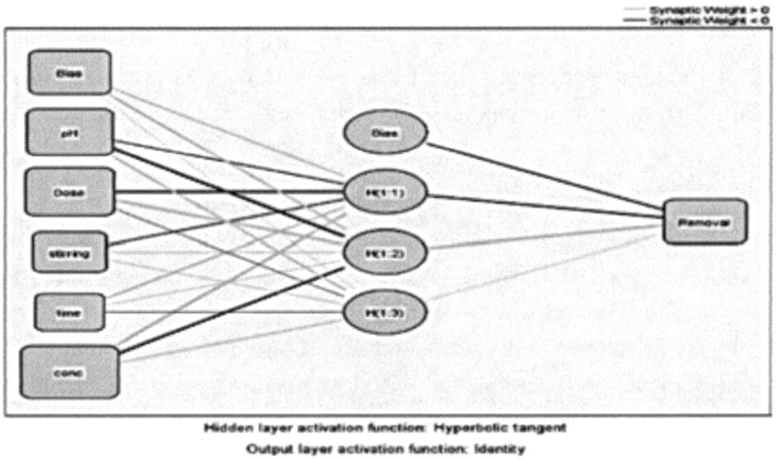

Figure 1. ANN of 5 – 3 – 1 structure used for the prediction of Cr removal efficiency.

2.4. Response surface methodology

A linear equation was performed to determine the connection between operating parameters on chromium removal efficiency (Eq. 3) and the significant parameter was used to predict the final RSM relation equation (Mahmoud and Mahmoud, 2021, Mahmoud et al., 2021a, Maha M. Elshfai and Elsaid, 2021).

$$Y = \beta_0 + \beta_1 x_1 + \beta_2 x_2 + \beta_3 x_3 + \beta_4 x_4 + \beta_5 x_5 \quad (3)$$

where, Y is the predicted percent of chromium adsorption onto GT-nZVI (%); x_1 is pH (2 – 10); x_2 is the adsorbent dose (0.1 – 1.0 g/L); x_3 is stirring rate (100 – 300 rpm); x_4 is contact time (10 – 50min); x_5 is chromium concentrations ranged between (1 to 9 mg/L); $\beta 0$ is the model intercept; β_1, β_2, β_3, β_4, and β_5 are the linear coefficients of x_1, x_2, x_3, x_4, and x_5, respectively (Rabie S. Farag and Peters, 2019, Mahmoud et al., 2017).

3. Result and dissections

3.1. Characterization

The XRD pattern as shown in Figure (2a) shows two peaks at $2\theta = 44.58°$ and $64.99°$ indicated the formation of alpha iron structure 110 and 200 planes. The XRD analysis also verified that there are other iron oxides, carbide, iron sodium phosphate, magnesium iron hydride, iron-chromium oxide, iron (III) borate, and ferrimolybdite formations. Figure (2b) shows the SEM image of the prepared GT-nZVI particles before any treatment process. The GT-nZVI showed an irregular surface structure with an average size of about 83 nm. Also, many pores were observed, which help to mass transfer of soluble low molecular weight organic compounds to the inner iron nanoparticles. Figure (2c) shows the EDAX analysis for the selected nanoparticles at 20 mm, indicated the formation of nano-iron particles with outer oxide and carbon and oxygen layers and other elements which come from the extraction solution.

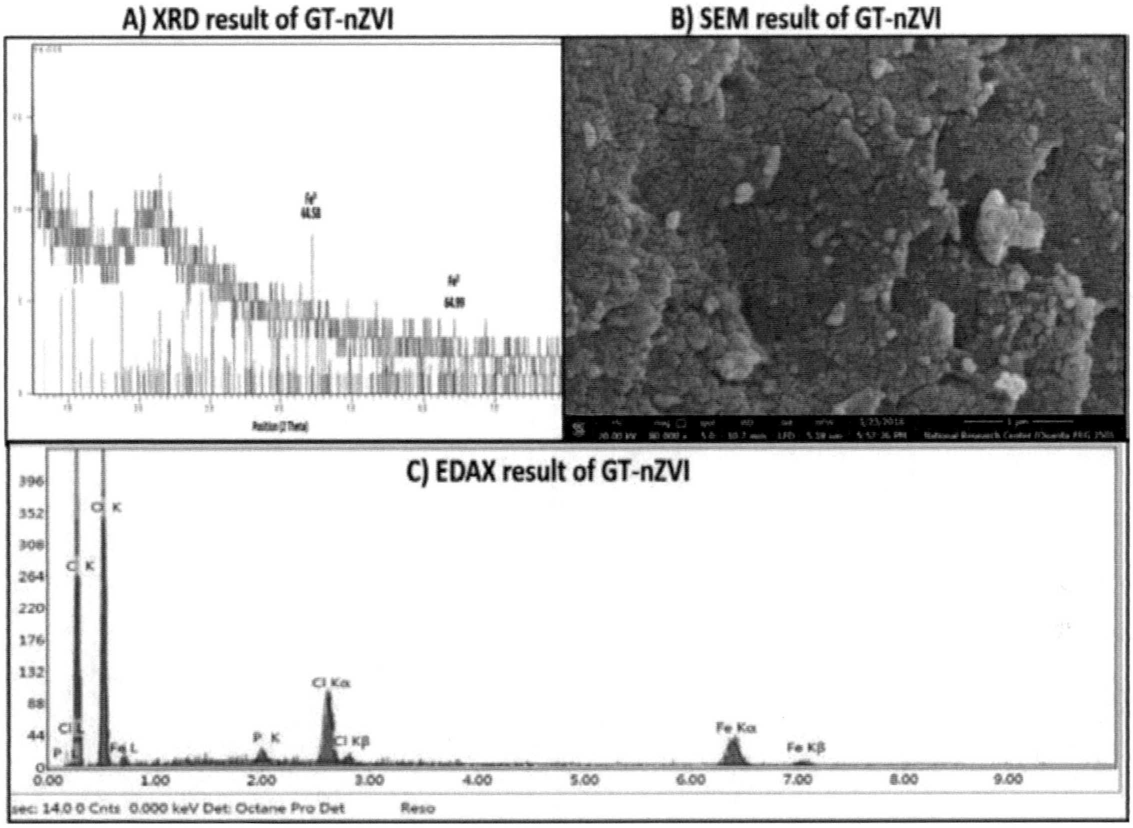

Figure 2. Characterization of GT-nZVI A) XRD of the freshly prepared GT-ZVI, B) SEM of GT-nZVI and C) EDAX results of GT-nZVI.

3.2. Effect of operating parameter

3.2.1. Effect of pH

The pH value is an important parameter affecting chemical and biological reactions in wastewaters (Zhang and Frankenberger, 2003). The obtained results indicated that the most effective pH was 5 showing a removal percentage of 89.6% from the initial chrome concentration of 5 mg/L as shown in Figure 3a. Different studies conducted with the removal of chromium metal ions from aqueous solution using different sorbent materials need high acidic solutions (Gao et al., 2008, Malkoc and Nuhoglu, 2007, Garg et al., 2004, Verma et al., 2006). These phenomena may be explained by that at low pH the surface of rice straw will be surrounded by high quantities of hydronium ions (H^+), which will promote the approach of negatively charged $Cr_2O_7{}^{2-}$. The more positive the surface charge of the biomass, the faster the removal rate of Cr (VI) in the rice straw solution, since the binding of anionic Cr (VI) ion species is enhanced. Therefore, solution pH is of great importance for Cr (VI) removal by different sorbent materials (Park et al., 2005).

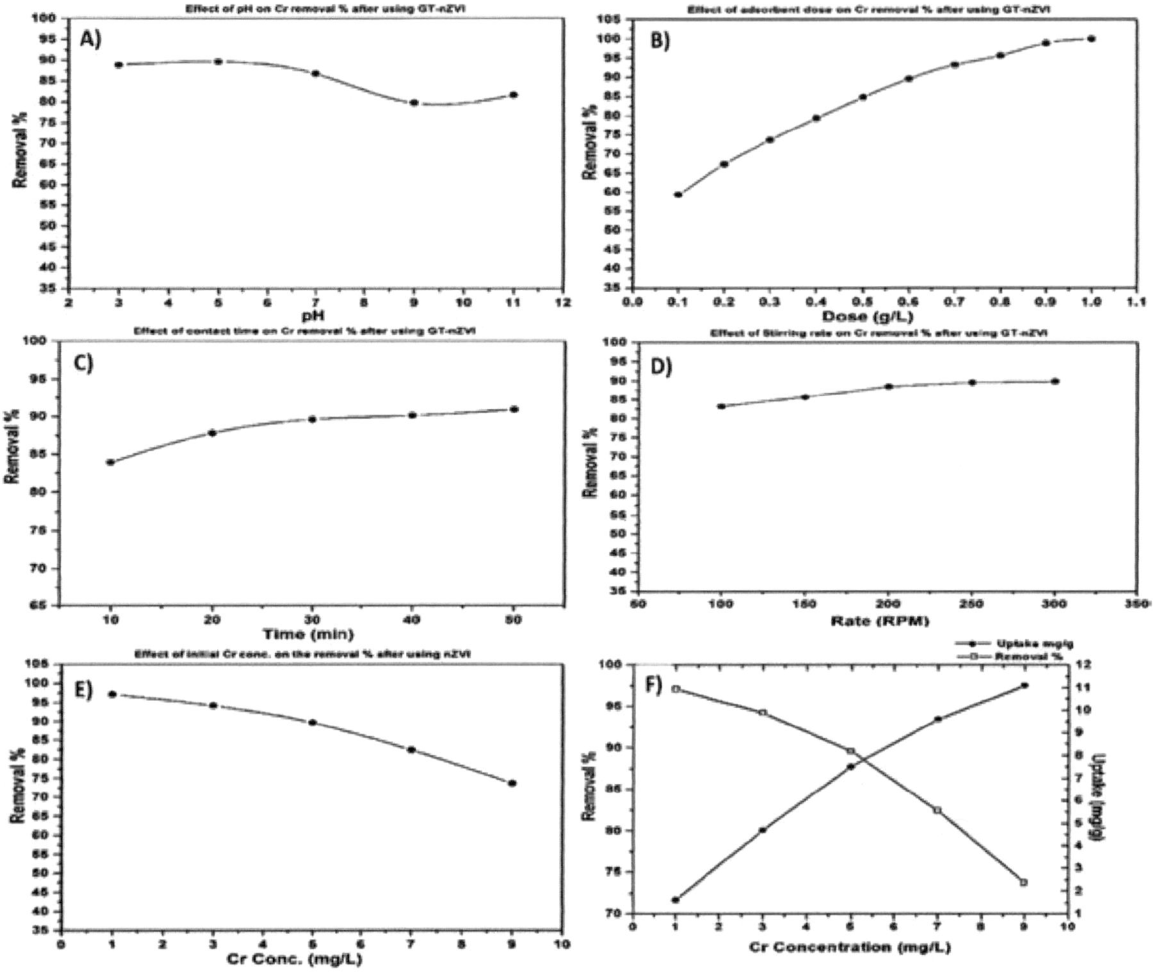

Figure 3. Effect of operating parameter for chromium removal: A) pH effect, B) dose-effect, C) Time effect, D) stirring rate effect, E) concentration effect and F) relation between uptake and removal percent.

3.2.2. Effect of dose

Figure 3b shows the effect of adsorbent dose on chromium removal efficiency after using GT-nZVI as sorbent material. The results showed that the minimum effective adsorbent dose for best removal efficiency is 0.6g/L for showing removal efficiency 89.6. The removal efficiency was increased with dose due to increasing the vacant site for adsorption and free electrons for the degradation Process (Mahmoud, 2017).

3.2.3. Effect of contact time

Figure 3c shows the effect of contact time on chrome removal using at different times namely (10, 20, 30, 40, and 50min). showing that the effective time for best removal efficiency is 30 min. Table 2 explains experimental Qt which is important for kinetic model estimations. Previous studies for Cr removal indicated the effect of contact time for chrome removal using different sorbent materials at pH2, shaking speed 100 rpm and the obtained results indicated that effective time was 10 min (Memon et al., 2009). Another study indicated that the effective time to remove 50 mg/L from chrome solution at pH 2 and stirring rate 250rpm was 60 min (Garg et al., 2009).

Table 2. Effect of contact time on chrome adsorption by using GT-nZVI.

Time (min)	Initial chrome concentration	Final chrome concentration after GT- nZVI	
min	mg/L	mg/L	Qt (mg/g)
10	5	0.805	6.99
20	5	0.61	7.32
30	5	0.52	7.47
40	5	0.495	7.51
50	5	0.455	7.58

3.2.4. Effect of stirring rate

Figure 3d shows the effect of stirring rate on chrome removal using GT-nZVI at different stirring rates namely (100, 150, 200, 250, and 300 rpm). The obtained results indicated that the optimum stirring rate was 250 rpm. Two types of adsorption occur in GT-nZVI surface physical and chemical adsorption. Chemical adsorption depends on the charge between the adsorbed molecule and the sorbent charge. Physical adsorption depends on vacant sites on the surface. Chemical adsorption is more stable than physical adsorption and mainly the stirring rate effect cannot decrease its stability (Ho and McKay, 1998). The amount of adsorbed molecule in physical adsorption equals the summation of adsorbed materials for all individual active sites. The mechanism of the adsorption process depends on the binding energy between adsorbed molecules and sorbents. The adsorption energy decreased gradually until vanishes with the complete adsorption process (Zeldowitsch, 1934). So, the little increase of removal percent according to stirring rate relative to chemisorption reaction and this in agreement with kinetic data (Ho and McKay, 1999).

3.2.5. Effect of initial concentration

Figure 3e shows the effect of initial concentration on chrome removal using GT-nZVI at different concentrations namely (1, 3, 5, 7, and 9 mg/L). The obtained results indicated that the effective chrome concentration was 5 mg/L as shown in Figure 3f.

3.3. Artificial intelligence neural networks (ANNs)

ANNs were trained using the back-propagation algorithm for chrome removal by using GT-nZVI. Network training is a process by which the connection weight and bias on the ANN are adapted through a continuous process of simulation by the environment in which the network is embedded as shown Figure 1 with Rescaling Method for Covariates (Standardized), Activation Function (Hyperbolic tangent) and The Sum of Squares Error were existing in Table 3. GT-nZVI, the sample was trained 15 times and tested 9 times without excluding any result with a total number of 24 runs using network structure 6-4-1. The results indicated that the most significant operating parameter is the effect of concentration 100% followed by the effect of dose 73.1%. The ANN results indicated that there is a small deviation between the predictive and actual results. Also, the residual percent between actual removal % and predicted removal % was between (-6 to +4).

Table 3. Neural network results for chrome removal using GT-nZVI.

		ANN model summaries	
Training	Sum of Squares Error		.197
	Relative Error		.028
	Stopping Rule Used		1 consecutive step(s) with no decrease in error
Testing	Sum of Squares Error		1.846
	Relative Error		.329
		Independent Variable Importance	
		Importance	**Normalized importance**
	pH	.237	65.2%
	Dose	.266	73.1%
	Stirring rate	.094	25.8%
	Contact time	.040	10.9%
	Concentration	.364	100.0%

3.4. Response Surface Methodology (RSM)

Table 4 describes the positive linear effect of the all-independent variable "pH", "dose", "time", "stirring rate", and "initial concentration" on chrome removal were observed to be significant ($p< 0.05$). The coefficient of determination between measured data and simulated results (R^2) and adjusted R^2 existed in Table 4. The significant result of the linear RSM model highly significant results agrees with the obtained ANN model to describe the most significant operating parameters for chrome removal using GT- nZVI. By applying Equation 3, the removal equation 4 is obtained. The removal equation can be used to control the removal percent by minimizing or maximizing the operating variables using all obtained results not restricted to the optimum conditions.

$$Y = 77.208 + 1.306\ x_1 + 33.551\ x_2 + 0.027\ x_3 + 0.170\ x_4 - 3.0\ x_5 \qquad (4)$$

where, Y is the predicted response of different wastewater contaminants removal efficiency (%); $x_1$is pH (3 – 11); x_2 is adsorbent dose (0.1 – 1.0 g) dry dose and (0.5 – 5 g) wet dose; x_3 is stirring rate (10 – 300 rpm); x_4 is contact time (10 – 50min); x_5 is concentration (1 – 9 mg/L); β_0 is the model intercept; β_1, β_2, β_3, β_4 and β_5 are the linear coefficients of x_1, x_2, x_3, x_4 and x_5 respectively.

Table 4. *t*-statistics and p-values for coefficients of a linear regression model after using GT-nZVI.

				RSM model (Enter Method)			
Model Summary	R			0.970			
	R^2			0.940			
	Adjusted R^2			0.924			
	Std. Error of the Estimate			1.729			
ANOVA	F			56.733			
	Model sig.			00.001			
Coefficients		Term	Estimate	Standard Error	t-Ratio	Prob> |t|	Effect*
	Constant	β0	77.208	3.897	19.812	0.001	Significant
	pH	β1	-1.306	.234	-5.587	0.001	Significant
	Dose	β2	33.551	3.447	9.733	0.001	Significant

Stirring rate	β3	.027	.009	2.877	.010	Significant
Time	β4	.170	.055	3.109	.006	Significant
Concentration	β5	-30.001	.273	-10.97	0.001	Significant

4. Conclusion

Effect of chrome (Cr) removal was studied for initial Cr concentrations (1, 3, 5, 7, and 9 mg/L) and the removal percentages were (97.1, 94.2, 89.6, 82.4, and 73.7%), respectively after using GT-nZVI after applying the optimum conditions at pH 5, dosages 0.6 g, contact time 30 min, and stirring rate 250 rpm. The adsorption isotherms were studied by applying nonlinear equations of Redlich- Peterson, Hill, Sips, Khan, Toth, Koble-Corrigan, Javanovic, Freundlich, and Langmuir models. The obtained results indicated that GT-nZVI preferred Koble–Corrigan isotherm model with the lowest summation of errors 0.394. The kinetic analysis was carried by practicing nonlinear equations of Pseudo First Order (P.F.O), Pseudo Second Order (P.S.O), Elovich, Avrami, and Intraparticle kinetic models. The obtained results indicated that GT-nZVI preferred the Pseudo Second Order mechanism with the lowest summation of errors equal to 0.07. ANN results indicated that GT-nZVI was able to describe the removal of chrome with the maximum residual variations between (-6, +4 %). RSM results indicated that the operating conditions are significant for chrome removal using GT-nZVI.

References

AMERY, H. A. 2002. Water wars in the Middle East: a looming threat. *Geographical Journal,* 168, 313-323.

BAIRD, R. B. 2017. Standard methods for the examination of water and wastewater, 23rd. Water Environment Federation, American Public Health Association, American ....

BARLOW, M. & CLARKE, T. 2017. *Blue gold: the battle against corporate theft of the world's water*, Routledge.

BARNABY, W. 2009. Do nations go to war over water? *Nature,* 458, 282.

BEAUMONT, P. 1994. The myth of water wars and the future of irrigated agriculture in the Middle East. *International journal of water resources development,* 10, 9-21.

CARPENTER, S. R., STANLEY, E. H. & VANDER ZANDEN, M. J. 2011. State of the world's freshwater ecosystems: physical, chemical, and biological changes. *Annual review of Environment and Resources,* 36, 75-99.

CHOWDHURY, M., MOSTAFA, M., BISWAS, T. K. & SAHA, A. K. 2013. Treatment of leather industrial effluents by filtration and coagulation processes. *Water Resources and Industry,* 3, 11-22.

DEL PORTO, D. 2005. Urban and industrial watersheds and ecological sanitation: two sustainable strategies for on-site urban water management. *Water Crisis: Myth or Reality?*, 285.

DIAZ, J. & BARKDOLL, B. Comparison of wastewater treatment in developed and developing countries. World Environmental and Water Resource Congress 2006: Examining the Confluence of Environmental and Water Concerns, 2006. 1-10.

DOLATYAR, M. & GRAY, T. S. 2000. The politics of water scarcity in the Middle East. *Environmental Politics,* 9, 65-88.

DUDGEON, D., ARTHINGTON, A. H., GESSNER, M. O., KAWABATA, Z.-I., KNOWLER, D. J., LÉVÊQUE, C., NAIMAN, R. J., PRIEUR-RICHARD, A.-H., SOTO, D. & STIASSNY, M. L. 2006. Freshwater biodiversity: importance, threats, status and conservation challenges. *Biological reviews,* 81, 163-182.

EL-SHAFEI, M., MAHMOUD, A., MOSTAFA, M. & PETERS, R. Effects of entrapped nZVI in alginate polymer on BTEX removal. AIChE Annual Meeting. San Francisco, CA, 2016. 13-18.

ESREY, S. & ANDERSSON, I. Environmental Sanitation from an Eco-Systems Approach. Proceedings of the Workshop on Ecological Sanitation. Mexico, 1999.

FARAG, R. S., ELSHFAI, M. M. & MAHMOUD, A. S. 2018. Adsorption and kinetic studies using nano zero valent iron (nZVI) in the removal of chemical oxygen demand from aqueous solution with response surface methodology and artificial neural network approach. *Journal of Environment & Biotechnology Research,* 7, 12-22.

FORBES, R. J. 1956. *Studies in Ancient Technology Vol V*, Brill Archive.

FROSCH, R. A. & GALLOPOULOS, N. E. 1989. Strategies for manufacturing. *Scientific American,* 261, 144-153.

GAO, H., LIU, Y., ZENG, G., XU, W., LI, T. & XIA, W. 2008. Characterization of Cr (VI) removal from aqueous solutions by a surplus agricultural waste—rice straw. *Journal of Hazardous Materials,* 150, 446-452.

GARG, U. K., KAUR, M., SUD, D. & GARG, V. 2009. Removal of hexavalent chromium from aqueous solution by adsorption on treated sugarcane bagasse using response surface methodological approach. *Desalination,* 249, 475-479.

GARG, V., GUPTA, R., KUMAR, R. & GUPTA, R. 2004. Adsorption of chromium from aqueous solution on treated sawdust. *Bioresource technology,* 92, 79-81.

GUPTA, V. K., GUPTA, M. & SHARMA, S. 2001. Process development for the removal of lead and chromium from aqueous solutions using red mud—an aluminium industry waste. *Water research,* 35, 1125-1134.

HAFEZ, A., EL-MANHARAWY, M. & KHEDR, M. 2002. RO membrane removal of unreacted chromium from spent tanning effluent. A pilot-scale study, Part 2. *Desalination,* 144, 237-242.

HAMDY, A., MOSTAFA, M. K. & NASR, M. 2019. Techno-economic estimation of electroplating wastewater treatment using zero-valent iron nanoparticles: batch optimization, continuous feed, and scaling up studies. *Environmental Science and Pollution Research,* 26, 25372-25385.

HO, Y.-S. & MCKAY, G. 1999. Pseudo-second order model for sorption processes. *Process biochemistry,* 34, 451-465.

HO, Y. & MCKAY, G. 1998. A comparison of chemisorption kinetic models applied to pollutant removal on various sorbents. *Process safety and environmental protection,* 76, 332-340.

ISLAM, M. S. & TANAKA, M. 2004. Impacts of pollution on coastal and marine ecosystems including coastal and marine fisheries and approach for management: a review and synthesis. *Marine pollution bulletin,* 48, 624-649.

JASON A. HEBERLING, Y. G. A., AHMED S. MAHMOUD, MOHAMED K. MOSTAFA, ROBERT W. PETERS. AOP Performance at Wastewater Treatment Plants. ". Proc. 2018 Annual AIChE Meeting, Pittsburgh, PA, (October 28 – November 2). 2018.

KARABAY, S. 2008. *Waste management in leather industry.* DEÜ Fen Bilimleri Enstitüsü.

KENNEDY, P. 2011. *Preparing for the twenty-first century*, Vintage.

KURNIAWAN, A., SISNANDY, V. O. A., TRILESTARI, K., SUNARSO, J., INDRASWATI, N. & ISMADJI, S. 2011. Performance of durian shell waste as high capacity biosorbent for Cr (VI) removal from synthetic wastewater. *Ecological Engineering,* 37, 940-947.

LANGERGRABER, G. & MUELLEGGER, E. 2005. Ecological Sanitation—a way to solve global sanitation problems? *Environment international,* 31, 433-444.

LIU, J. & DIAMOND, J. 2005. China's environment in a globalizing world. *Nature,* 435, 1179.

M ABDELDAYEM, O., ELDAGHAR, O., K MOSTAFA, M., M HABASHY, M., HASSAN, A. A., MAHMOUD, H., M MORSY, K., ABDELRADY, A. & PETERS, R. W. 2020. Mitigation plan and water harvesting of flashflood in arid rural communities using modelling approach: A case study in Afouna village, Egypt. *Water,* 12, 2565.

MAHA M. ELSHFAI, A. S. M. & ELSAID, M. A. Comparative Studies of using nZVI and Entrapped nZVI in Alginate Biopolymer (Ag/ nZVI) for Aqueous Phosphate Removal. 12th international conference on nano-technology for green and sustainable construction, 2021. https://www.researchgate.net/publication …, 21.

MAHMOUD, A. S. 2017. Study the Removal of Most Hazardous Pollutants from Aqueous Solutions Using Nanoparticles. http://lis.cl.cu.edu.eg/search*ara.

MAHMOUD, A. S., FARAG, R. S. & ELSHFAI, M. M. 2020a. Reduction of organic matter from municipal wastewater at low cost using green synthesis nano iron extracted from black tea: Artificial intelligence with regression analysis. *Egyptian Journal of Petroleum,* 29**,** 9-20.

MAHMOUD, A. S., FARAG, R. S., ELSHFAI, M. M., MOHAMED, L. A. & RAGHEB, S. M. 2019a. Nano zero-valent aluminum (nZVAl) preparation, characterization, and application for the removal of soluble organic matter with artificial intelligence, isotherm study, and kinetic analysis. *Air, Soil and Water Research,* 12**,** 1178622119878707.

MAHMOUD, A. S., ISMAIL, A., MOSTAFA, M. K., MAHMOUD, M., ALI, W. & SHAWKY, A. M. 2020b. Isotherm and kinetic studies for heptachlor removal from aqueous solution using Fe/Cu nanoparticles, artificial intelligence, and regression analysis. *Separation Science and Technology,* 55**,** 684-696.

MAHMOUD, A. S., MOHAMED, N. Y., MOSTAFA, M. K. & MAHMOUD, M. S. 2021a. Effective Chromium Adsorption From Aqueous Solutions and Tannery Wastewater Using Bimetallic Fe/Cu Nanoparticles: Response Surface Methodology and Artificial Neural Network. *Air, Soil and Water Research,* 14**,** 11786221211028162.

MAHMOUD, A. S., MOSTAFA, M. K. & ABDEL-GAWAD, S. A. 2018. Artificial intelligence for the removal of benzene, toluene, ethyl benzene and xylene (BTEX) from aqueous solutions using iron nanoparticles. *Water Supply,* 18**,** 1650-1663.

MAHMOUD, A. S., MOSTAFA, M. K. & NASR, M. 2019b. Regression model, artificial intelligence, and cost estimation for phosphate adsorption using encapsulated nanoscale zero-valent iron. *Separation Science and Technology,* 54**,** 13-26.

MAHMOUD, A. S., MOSTAFA, M. K. & PETERS, R. W. 2021b. A prototype of textile wastewater treatment using coagulation and adsorption by Fe/Cu nanoparticles: Techno-economic and scaling-up studies. *Nanomaterials and Nanotechnology,* 11**,** 18479804211041181.

MAHMOUD, A. S., SARYEL-DEEN, R. A., MOSTAFA, M. K. & PETERS, R. W. Artificial Intelligence for Organochlorine Pesticides Removal from Aqueous Solutions using Entrapped nZVI in Alginate Biopolymer. 2017 Annual AIChE Meeting; Minneapolis, MN, October 29 - November 3, 2017, 2017.

MAHMOUD, M. & MAHMOUD, A. S. 2021. Wastewater treatment using nano bimetallic iron/copper, adsorption isotherm, kinetic studies, and artificial intelligence neural networks. *Emergent Materials***,** 1-9.

MALKOC, E. & NUHOGLU, Y. 2007. Potential of tea factory waste for chromium (VI) removal from aqueous solutions: Thermodynamic and kinetic studies. *Separation and purification technology,* 54**,** 291-298.

MEMON, J. R., MEMON, S. Q., BHANGER, M. I., EL-TURKI, A., HALLAM, K. R. & ALLEN, G. C. 2009. Banana peel: a green and economical sorbent for the selective removal of Cr (VI) from industrial wastewater. *Colloids and surfaces B: Biointerfaces,* 70**,** 232-237.

MORSY, K. M., MOSTAFA, M. K., ABDALLA, K. Z. & GALAL, M. M. 2020. Life cycle assessment of upgrading primary wastewater treatment plants to secondary treatment including a circular economy approach. *Air, Soil and Water Research,* 13**,** 1178622120935857.

MOSTAFA, M. & PETERS, R. 2017. Reuse paint wastewater in the manufacture of cement bricks and tiles. *Journal of Material Cycles & Waste Management,* 19.

MOSTAFA, M. & PETERS, R. W. 2016a. A comprehensive assessment of water quality at the Rosetta branch of the Nile River, Egypt. *Journal of Civil Engineering and Architecture,* 10**,** 513-529.

MOSTAFA, M. K., MAHMOUD, A. S., SARYEL-DEEN, R. A. & PETERS, R. W. Application of Entrapped Nano Zero Valent Iron into Cellulose Acetate Membranes for Domestic Wastewater Treatment. 2017 Annual AIChE Meeting; Minneapolis, MN, October 29 - November 3, 2017, 2017.

MOSTAFA, M. K. & PETERS, R. W. 2015. Use river pollutant modeling to simulate and predict the change in the Damietta Branch water quality before and after construction of the Ethiopian Dam. *Journal of Environmental Protection,* 6**,** 935.

MOSTAFA, M. K. & PETERS, R. W. 2016b. Improve effluent water quality at Abu-Rawash wastewater treatment plant with the application of coagulants. *Water and Environment Journal,* 30**,** 88-95.

NELSON, R. R. & ROSENBERG, N. 1993. Technical innovation and national systems. *National innovation systems: A comparative analysis,* 322.

NGOC, U. N. & SCHNITZER, H. 2009. Sustainable solutions for solid waste management in Southeast Asian countries. *Waste management,* 29**,** 1982-1995.

OKI, T. & KANAE, S. 2006. Global hydrological cycles and world water resources. *science,* 313**,** 1068-1072.

PAHL-WOSTL, C., MOSTERT, E. & TÀBARA, D. 2008. The growing importance of social learning in water resources management and sustainability science. *Ecology and Society,* 13**,** 24.

PARK, D., YUN, Y.-S. & PARK, J. M. 2005. Studies on hexavalent chromium biosorption by chemically-treated biomass of Ecklonia sp. *Chemosphere,* 60**,** 1356-1364.

PIMENTEL, D., BERGER, B., FILIBERTO, D., NEWTON, M., WOLFE, B., KARABINAKIS, E., CLARK, S., POON, E., ABBETT, E. & NANDAGOPAL, S. 2004. Water resources: agricultural and environmental issues. *BioScience,* 54**,** 909-918.

RABIE S. FARAG, M. M. E.-S., AHMED S. MAHMOUD, MOHAMED K. MOSTAFA, & PETERS, R. W. Green Synthesis of Nano Iron Carbide: Preparation, Characterization and Application for Removal of Phosphate from Aqueous Solutions. Proc. 2018 Annual AIChE Meeting, Pittsburgh, PA, (October 28 – November 2). 2018.

RABIE S. FARAG, M. M. E., AHMED S. MAHMOUD, MOHAMED K. MOSTAFA, AHMED KARAM, & PETERS, R. W. Study the Degradation and Adsorption Processes of Organic Matters from Domestic Wastewater using Chemically Prepared and Green Synthesized Nano Zero-Valent Iron. 2019 Annual AIChE Meeting; Orlando, FL, November 10 - 15, 2019, 2019. https://www.researchgate.net/publication ….

SARYEL-DEEN, R. A., MAHMOUD, A. S., MAHMOUD, M., MOSTAFA, M. K. & PETERS, R. W. Adsorption and Kinetic Studies of using Entrapped Sewage Sludge Ash in the Removal of Chemical Oxygen Demand from Domestic Wastewater, with Artificial Intelligence Approach. 2017 Annual AIChE Meeting, 2017.

SCHWARZENBACH, R. P., EGLI, T., HOFSTETTER, T. B., VON GUNTEN, U. & WEHRLI, B. 2010. Global water pollution and human health. *Annual Review of Environment and Resources,* 35**,** 109-136.

SHIKLOMANOV, I. A. The world's water resources. International symposium to commemorate the, 1991. 93-105.

SREERAM, K. & RAMASAMI, T. 2003. Sustaining tanning process through conservation, recovery and better utilization of chromium. *Resources, conservation and recycling,* 38**,** 185-212.

TAYLOR, R. G., SCANLON, B., DÖLL, P., RODELL, M., VAN BEEK, R., WADA, Y., LONGUEVERGNE, L., LEBLANC, M., FAMIGLIETTI, J. S. & EDMUNDS, M. 2013. Ground water and climate change. *Nature climate change,* 3**,** 322.

VERMA, A., CHAKRABORTY, S. & BASU, J. 2006. Adsorption study of hexavalent chromium using tamarind hull-based adsorbents. *Separation and Purification Technology,* 50**,** 336-341.

WALSH, A. R. & O'HALLORAN, J. 1996. Chromium speciation in tannery effluent—II. Speciation in the effluent and in a receiving estuary. *Water Research,* 30**,** 2401-2412.

WEINGÄRTNER, H. & FRANCK, E. U. 2005. Supercritical water as a solvent. *Angewandte Chemie International Edition,* 44, 2672-2692.
WENG, C. N. A critical review of Malaysia's accomplishment on water resources management under agenda 21. Proceedings of International Conference on Environmental Management, Centre for Graduate Studies, National University of Malaysia (UKM), Bangi, 2002. 315-331.
ZELDOWITSCH, J. 1934. Adsorption site energy distribution. *Acta phys. chim. URSS,* 1, 961-973.
ZHANG, Y. & FRANKENBERGER, W. T. 2003. Factors affecting removal of selenate in agricultural drainage water utilizing rice straw. *Science of the total environment,* 305, 207-216.

629405 Enhanced Biodegradability Assessment of Total Petroleum Hydrocarbon By Implementing a Novel Bioaugmentation Strategy of Indigenous Bacterial Consortium

Abstract:

Petroleum industry is the of the main sectors of the world's energy production. A huge amount of petroleum sludge generates during crude oil processing. Being significant in the amount and toxicity of petroleum sludge cause a serious issue to the environment and human health worldwide. The current research intends to isolation and identification of petroleum hydrocarbon degrading bacterial strains from the habitat petroleum refinery sludge. Biodegradation of total petroleum hydrocarbon of petroleum refinery sludge in liquid media by a bacterial consortium having pure culture of four bacterium was performed. The four bacterial strains isolated from the petroleum sludge are *Dietzia* sp. IRB191, *Dietzia* sp. IRB192, *Staphylococcus* sp. BSM19 and *Stenotrophomonas* sp. IRB19. Implementation of novel bioaugmentation strategy was carried out to enhance the biodegradation potential of total petroleum hydrocarbons. Total petroleum hydrocarbon present in the sludge was determined by liquid-liquid solvent extraction. Further, the petroleum fractions were determined by column chromatography. The characterization of oil sludge was performed by ICP-AES, GC-FID, FTIR, SEM analysis. The oil contents asphaltene (50%), and the maltene fraction comprised of 52 ± 3.5% aliphatic, 39 ± 2.5% aromatics, and 9 ± 1% polar fractions, respectively. The ICP-AES analysis depicted that petroleum sludge contains 11 heavy metals (Pb, Cu, Ni, Zn, Fe, Hg, Cd, Mn, As, Cr, and Co) and six metals (Be, Na, Mg, K, Ca, Al). The consortium degrades 77 ± 2.5 %of TPH, whereas the pure culture showed 59-65 % in 2% (w/v) of petroleum sludge in mineral salt media. The degradation of TPH showed a better fit to first order kinetics. This result specified that the developed microbial consortium is suitable for bioremediation process in contaminated oily sludge sites. Considering the high production, toxicity and carcinogenic potential of PRS towards humans, the biotreatment of TPH employing biodegradation is an efficient ecofriendly strategy.

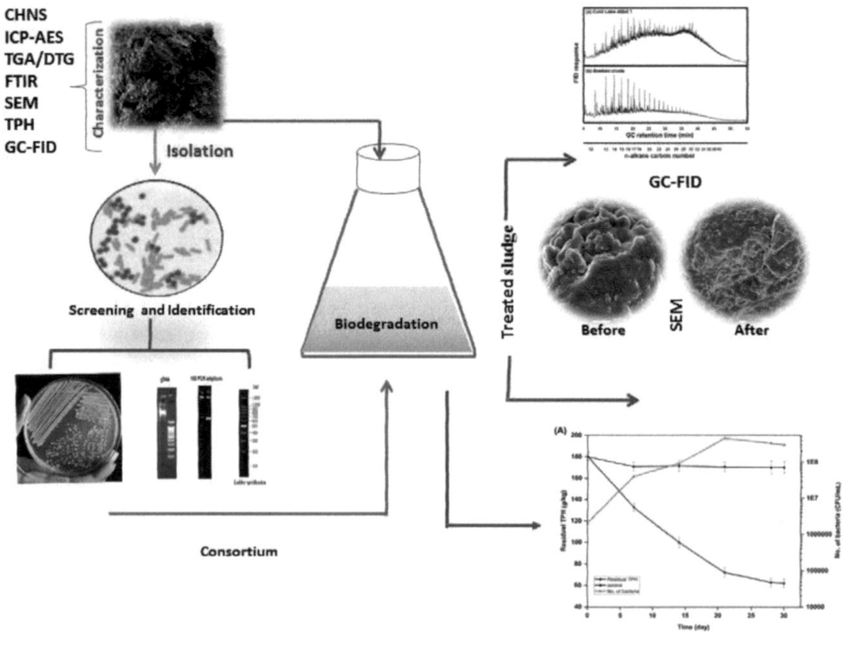

Copyright © American Institute of Chemical Engineers. All rights reserved.

632981 Evaluation of Catalyzed Pretreatment Processes of Cocoa Biomass Residues for Further Pyrolysis Applications.

Abstract:

Cocoa is a commonly cultivated fruit for the world's chocolate industry. Colombian regions of Orinoquía and Andina are the ones with the highest cocoa production in the country. However, out of 100% of cocoa biomass production, only 20% is destined for producing chocolate. During the last decade, the areas planted with cocoa have increased, thus generating a high amount of biomass residue per unit of chocolate produced. Cocoa farmers use this agricultural waste to fertilize the same crop and animal feeding. But these uses generate problems such as the proliferation of pathogens that affect the cocoa crop and unbalanced animal feeding due to low digestibility. Likewise, zinc sulfate and ferric sulfate catalysts are used for dehydrating, fragmenting, and depolymerizing lignocellulosic materials. This feedstock preprocessing step improves the pyrolysis process to obtain bio-oil. Therefore, this research aims to study the use of these catalysts for pretreating cocoa biomass residues and enhance their usability as a valuable pyrolysis feedstock. Also, the carbonization yield and the synthesis for sulfates were observed. Finally, during the result analysis, the ferric sulfate catalyst showed a high decrease in activation energy and temperature peaks among the zinc sulfate catalyst. Moreover, the carbonization yield was estimated in a range of 30-35%. This experimental analysis represents a new advancement to employ cocoa biomass as a valuable raw material for producing biofuels. Besides, this can help minimize the proliferation of pests in crops, reducing the use of pesticides, and increasing the economic sustainability of farmers.

Copyright © American Institute of Chemical Engineers. All rights reserved.

624639 Lead and Sulfate Ions Removal from Aqueous Solutions with Electrodialysis Process

Abstract:

The pollution incurred by heavy metals is a serious and escalating environmental problem, as a result of the continuous increase of industrial activities and technological development. Lead exists in wastewater from various sources, including processes for the production of batteries, pigments, plastics, and electronics, whereas its removal has been a major concern to scientists until this day. Sulfuric acid is also contained in the effluents of battery production processes and efficient treatment is required to avoid the discharge of water with low pH. Conventional treatment of wastewater with the abovementioned properties includes ion exchange (Petruzzelli et al., 1999), electrochemical reduction (Lin et al., 1999), adsorption (Runtti et al., 2016), and ion precipitation (Vu et al., 2019). However, these technologies may exhibit incomplete ion removal, high startup costs and operation time and so forth. Electrodialysis (ED) is a promising process as it requires little pretreatment and produces clean and reusable streams. ED is an electrochemical separation process in which ions are transferred through ion exchange membranes by means of a direct current (DC) voltage. The solution to be treated is pumped and circulated through the ED unit, that consists of a series of alternating parallel anion exchange membranes (AEM) and cation exchange membranes (CEM) fixed between two electrically charged electrodes, the anode and the cathode (Fig. 1a).

In published literature, results are available regarding the separation of lead ions from wastewater using ED, but in small scale systems of rather low active membrane area and small volumes of treated wastewater. The group of Gherasim (2014) used a batch ED system of 10 cell pairs to remove lead ions from aqueous solutions at 10 Volts, 70 L/h flow rate and 1000 mg/L feed concentration. In another study (Mohammadi et al., 2004), the effect of process parameters on lead separation from wastewater was examined with the use of a 1-cell-pair system, at 0.25 L/h, 30 Volts and 1000 mg/L. For sulfate ions removal by ED, research is rather limited. Amor and coworkers (1998) investigated the removal of fluoride and other ions, including sulfate, from brackish water with electrodialysis, with the aid of a system composed of 10 cell pairs, at 15 Volts, 180 L/h and 730 mg/L sulfate ions concentration. In another work (Lee et al., 2003), the recovery of ammonium and sulfate ions from waste was investigated with an ED system of 10 cell pairs, at 180 L/h, 30 Volts and 730 mg/L ammonium sulfate concentration. To the authors' knowledge, there is no work in literature concerning the removal of sulfate or the simultaneous removal of sulfate and lead ions by ED at larger scale under realistic operating conditions.

The experimental results of a pilot-scale investigation of sulfate ions removal from aqueous solutions by electrodialysis are presented in this study. For the scope of the work, an ED pilot plant composed of 68 cell pairs of cation- and anion-exchange membranes of 5.17 m^2 effective membrane area and maximum nominal feed flow rate 350 L/h, operating in batch mode, was used. The capability of the ED plant to remove sulfate ions was initially optimized using an experimental design based on central composite design (CCD) coupled with response surface methodology (RSM). This aimed to evaluate the effect of key process parameters like the applied voltage in the stack (U_{st}), the product (diluate) to concentrate ratio (*r in percentage*), and the sulphate concentration ($[SO_4^{2-}]$). The best operating conditions are applied for the pilot testing of the simultaneous removal of lead and sulfates from simulated battery manufacturing industry effluents.

The pilot unit (Fig. 1b) is equipped with an electrodialysis four-chamber ED stack and is designed to operate also as a salt splitting (EDBM) and salt metathesis (EDM) unit. For the scope of the standard ED tests, only two out of the four circuits of the cell are in use with one of the cell systems being the diluate (where the ions are removed) and the other one being the concentrate (in which the ions are collected). Feed solutions were prepared by dissolving H_2SO_4 in demineralized water. The electrolyte rinse solution was Na_2SO_4 60 g/L. The experiments were carried in batch recirculation mode at ambient temperature with diluate/concentrate tank volumes of 35 L. Samples from dilute and concentrate compartments were obtained at specific time intervals and the sulfate concentrations were measured by means of Ion Chromatography (IC). Three optimization criteria were considered: a) maximization of the sulphate separation (*SP* in percentage), b) minimization of the energy consumption (*EC*) (not including the energy for pumping), and c) maximization of current efficiency (*CE* in percentage), defined as the ratio between the electricity consumed in the cell for the separation of the sulfate ions and the total electricity supplied.

Based on the CCD, a total of seventeen (17) experiments were conducted to determine the optimum values of the three independent variables. The experiments were conducted by recirculating synthetic sulfate solutions at flow rate 150 L/h, and they were randomly performed to minimize the effect of systematic errors. The three independent variables varied over three levels (maximum, minimum and midpoint levels) at determined ranges chosen based on preliminary experiments (data not shown here) and background knowledge (U_{st}: 5, 10 or 15 Volts, r: 40%, 65% or 90%, $[SO_4^{2-}]$: 1000, 1500 or 2000 mg/L). It is noted that the flow rate did not have any significant effect in the sulfate ions separation, whereas the value of 150 L/h was proven the optimum due to the lower current density monitored in the system and therefore the lower energy consumption.

The obtained contour (Fig. 1c) and response surface plots (Fig. 1d) assisted in identifying the interrelationship between the variables U_{st}, r and $[SO_4^{2-}]$ and providing a three-dimensional view of the changes in the three performance responses (*SP*, *EC*, *CE*) for different combinations of independent variables. These RSM plots were functions of two independent variables by maintaining the third variable at a constant value and helped determine the optimum experimental conditions for maximum response. As verified by the ANOVA analysis, *SP* is linearly correlated with the variable r, with U_{st} and $[SO_4^{2-}]$ having marginal effect on the sulfate ions separation. As expected, the higher the U_{st} the larger the EC. A similar conclusion is drawn for variable r and the interaction factor $U_{st}\bullet r$, that correlates positively to EV values. On the contrary, r correlated negatively to the current efficiency, since CE was higher at lower diluate recoveries. A number of 99 solutions of optimum U_{st}, r and $[SO_4^{2-}]$ values were proposed by the statistical software yielding maximum *SP*, *CE* and minimum *EC*. Among the solutions proposed, the first solution, presenting a rather good desirability of 0.712, was selected as the optimum one. This solution required U_{st}=15 V, r=90% for $[SO_4^{2-}]$=2000 mg/L which is representative of the sulfate ions content in lead-acid battery wastewater effluents. Under these conditions, a set of experiments was conducted aiming to validate the correlation models (Fig. 1e). Under the optimum operating conditions, sulphate ions were removed by approx. 65% within 30 min, with a current efficiency of 33% and a recorded 0.5 kW/m^3 consumption. High efficiency is also achieved for experiments that involved the simultaneous removal of lead and sulfate ions.

Acknowledgements

This research has been co-financed by the European Union and Greek national funds through the Operational Program Competitiveness, Entrepreneurship and Innovation, under the call RESEARCH–CREATE–INNOVATE (project code: T1EDK-02677).

References

Amor Z., Malki S., Taky M., Beriou B., Mameri N., Elmidaoui A., Optimization of fluoride removal from brackish water by electrodialysis, 1998, Desalination 120, 263-271.

Gherasim C., Krivcik J., Mikulasek P., Investigation of batch electrodialysis process for removal of lead ions from aqueous solutions, 2014, Chemical Engineering Journal 256, 324-334.

Lee H., Oh S., Moon S., Recovery of ammonium sulfate from fermentation waste by electrodialysis, 2003, Water Research 37, 1091-1099.

Lin S.W., Navarro R.M.F., An innovative method for removing Hg2+ and Pb2+ in ppm concentrations from aqueous media, 1999, Chemosphere, 39, 1809–1817.

Mohammadi T., Razmi A., Sadezadeh M., Effect of operating parameters on Pb^{2+} separation from wastewater using electrodialysis, 2004, Desalination 167, 379-385.

Petruzzelli D., Pagano M., Tiravanti G., Passino R., Lead removal and recovery from battery wastewaters by natural zeolite clinoptilolite, 1999, Solvent Extraction and Ion Exchange 17, 677–694.

Runtti H., Luukkonen T., Niskanen M., Tuomikoski S., Kangas T., Tynjälä P., Tolonen E.T., Sarkkinen M., Kemppainen K., Rämö J., Sulphate removal over barium-modified blast-furnace-slag geopolymer, 2016, Journal of Hazardous Materials 317, 373–384.

Vu H.H.T., Gu S., Thriveni T., Khan M.D., Tuan L.Q., Ahn J.W., Sustainable treatment for sulfate and lead removal from battery wastewater, 2019, Sustainability 11, 3497.

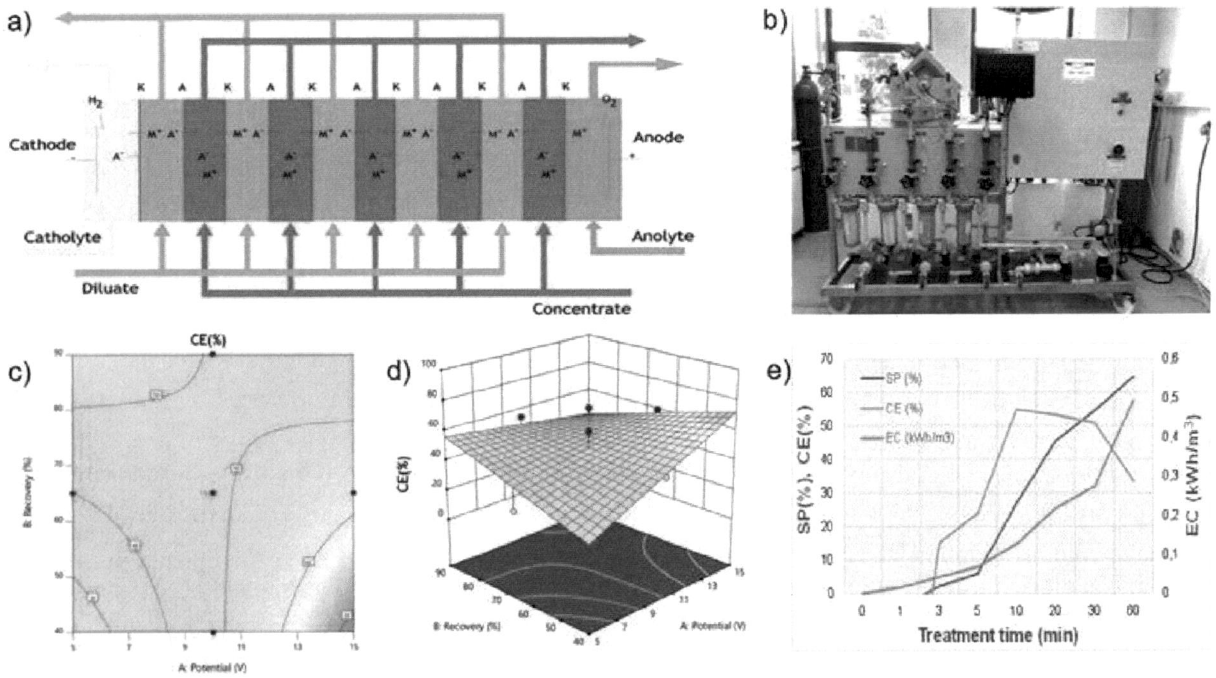

Figure 1: a) Functional setup of a standard ED stack, b) ED pilot plant at CPERI/CERTH, contour plot, d) 3D surface plot, e) variation of process responses with electrodialysis time und the optimum operating conditions.

Copyright © American Institute of Chemical Engineers. All rights reserved.

623993 A Recycling System for Wastewater from Oil Refineries Using Hydrogen and Photocatalysis

Abstract:

In oil refineries, water is used for many different purposes and a substantial amount of wastewater is produced. The traditional wastewater treatment process in oil refineries, which is comprised of primary, secondary, and tertiary treatment, the organic sulfur, organic nitrogen, and other organic compounds are not fully removed. A treatment system for the wastewater from oil refineries to promote the recycling of industrial wastewater has been proposed. In addition to the traditional wastewater treatment process, the process to remove organic sulfur and the process to decompose organic compounds in wastewater have been implemented in the proposed system. To realize the proposed system, the removal of organic sulfur using hydrogen by simulation and photochemical reactions by experiments were studied. In the simulation of the removal of organic sulfur using hydrogen, oil refinery plant data which contain the structures of all components were analyzed. The results showed that organic sulfur was removed in the process, and the conversion ratio highly depended on the chemical structure of the components. As a representative of the remaining organic compound in the wastewater from the oil refinery, phenol, a compound toxic even in low concentrations, was chosen. In the experiment, the decomposition of phenol using UV and TiO2 was investigated. The experimental results show phenol was successfully decomposed. The proposed process may improve the treatment of industrial wastewater and contribute to the conservation of water resources.

Copyright © American Institute of Chemical Engineers. All rights reserved.

627592 The Impact of Trace Ammonia Impurities on the Utilization of Renewable Natural Gas

Abstract:

Renewable natural gas (RNG), produced from biogas upgrading, is an important alternative to fossil fuels for power generation. Unfortunately, RNG contains several trace contaminants. One such contaminant is NH_3, of particular concern for RNG produced from farming operations. The presence of NH_3 in RNG, particularly if it is injected into the natural gas (NG) pipeline network, can have serious consequences. These can range from damage to components of the NG infrastructure, to corrosion of analytical and end-user equipment and added pollutant formation during combustion. The research in this paper addresses all such impacts.

In the area of pollutant emissions, three common end-user equipment were tested: An internal combustion engine, a household tankless water heater and a wall furnace. They were all operated with NG injected with NH_3 concentrations ranging from 0 to 500 ppm_v. Emissions studied included unburned hydrocarbons (UHC), CO, and nitrogen oxides (NO_X), the latter being of particular concern. The presence of NH_3 resulted in increased NO_X emissions for all equipment. For some of the equipment, the relationship between the amount of NO_X formed and the NH_3 concentration in the RNG was direct and quantitative, meaning that for every molecule of NH_3 fed to the equipment, one molecule of NO_X was produced. For others, a lower quantity of NO_X was formed to that corresponding to complete conversion, underscoring the impact that the prevailing conditions in the combustion chamber has on the pollutant formation processes.

We also investigated NH_3-induced corrosion in materials relevant to the NG infrastructure with the goal to identify various (bulk and surface) species potentially forming, their rates of formation, and their dependence on the operating conditions. The tested materials included components made of steel and brass, and a variety of polymer gaskets. They were exposed to both dry and humidified (RH>90%) NH_3/ NG mixtures with 0-500 ppm_v of NH_3 for periods, under each concentration, ranging from 400-1200 hours. NH_3-induced changes in the materials were identified via direct weight change measurements and optical observations, SEM-EDS, ICP-OES and microhardness testing. For the metal components, corrosion phenomena were observed, particularly for the higher NH_3 concentrations under humidified conditions.

We also studied the impacts of NH_3 in the NG may have on Gas Chromatographs (GC) used to analyze the gas for customer billing. Two issues of concern were: (i) interference in the measurements due to the presence of NH_3; (ii) corrosion of GC internals due to its long-term exposure to NH_3. Two different experiments were conducted to evaluate each of these concerns. In both experiments, the GC was gradually exposed to NG containing various NH_3 concentrations ranging from 30 ppm_v to 200 ppm_v. For experiments probing interference, the duration of experiments was rather short, typically, consisting of 50-70 measurements (each measurement lasting 5 min., 14 sec) for each NH_3 concentration studied. For the experiments to study potential instrument corrosion, each cycle of experiments for a given NH_3 concentration lasted about a week. Two different experimental protocols were utilized: During the early phase of the study, for the first group of experiments, the instrument was calibrated at the beginning of the experimental run using a standard calibration gas, but for the whole series of experiments it was not calibrated again. The reason for doing that was to observe any cumulative effects that continuous exposure to NH_3 may have on the instrument. In the second group of experiments, we followed the normal operation protocol for the instrument when used in the field by re-calibrating the instrument daily. For both the interference and the corrosion experiments, the presence of NH_3 has a detrimental impact, more so, as expected, for the long-term (corrosion) experiments rather than the short-term (interference) experiments. The impact was significantly more pronounced for the experiments without re-calibration, during which at some point, and before the planned completion of experiments, the GC stopped functioning all together. The daily re-calibration using standard gas had a good benefit, with the NH_3 impact on the instrument still being present but not to the extent encountered in the runs without re-calibration.

The results of the study are concerning, indicating the need for thorough purification of the RNG prior to its potential injection into the NG network.

Copyright © American Institute of Chemical Engineers. All rights reserved.

623588 Impact of Co-Processing Bio-Crude with Petroleum Stream on Hydrotreating Catalyst Deactivation

Abstract:

Co-processing of bio-crudes, from hydrothermal liquefaction (HTL) of biomass or wet wastes, in current refinery process such as hydrotreating with petroleum streams has great potential to produce green transportation fuels in the near term. However, one of the risks is the possible faster deactivation of the hydrotreating catalysts caused by co-processing HTL bio-crude, especially considering some problematic characteristics of HTL bio-crudes such as high content of nitrogen containing species, oxygenates, and inorganics. They can potentially lead to the poisoning of catalyst active sites by strongly adsorbing molecules like nitrogen compounds and the blocking of the pore structure or fouling the catalyst active sites by coke formation from oxygenates or inorganic deposition. In this research, we developed an understanding of the deactivation mechanism of a $NiMo/Al_2O_3$ hydrotreating catalyst for co-processing a wastewater sludge HTL bio-crude with straight run diesel or vacuum gas oil. Spent catalysts after testing petroleum feed and co-processing different bio-crudes have been collected after >300 hours' time on stream (TOS) testing. These catalysts have been characterized by various techniques such as physisorption, chemisorption, elemental analysis, ICP-OES, TGA-MS, and SEM-EDX for their chemical and physical properties. Their catalytic activity has also been evaluated by model compound testing including hydrodesulfurization (HDS) of dibenzothiophene (DBT) and hydrogenation (HYD) of cyclohexene (CHE) to quantitively determine the concentration of different active sites. The results indicate more carbonaceous species and inorganic deposits (such as Fe and Si) on the external surface of spent catalysts at the top of the reactor when co-processing untreated bio-crude compared with pretreated bio-crude, diesel only or fresh ones, which leads to severe loss of catalyst active sites. A clear understanding of the deactivation behavior particularly caused by bio-crudes is highly desirable and critical to improve hydrotreating catalyst/process to enable co-processing renewable feedstocks and incorporate biogenic carbon in fuels.

Copyright © American Institute of Chemical Engineers. All rights reserved.

628561 Integrating Thermal Analysis and Reaction Modeling for Rational Design of Pyrolytic Processes to Remediate Soils Contaminated with Polyaromatic Hydrocarbons

Abstract:

Background and Motivation

There is a pressing need for reliable, broadly applicable and sustainable remediation of soils and sediments impacted by polyaromatic hydrocarbons (PAHs) and other persistent hydrophobic hazardous pollutants. Whereas numerous site remediation approaches have been developed, many of these technologies are marginally cost-effective or unreliable. For example, bioremediation (including landfarming) is relatively slow and often difficult to accomplish and has the potential to generate more toxic PAH byproducts [1].

We recently demonstrated that pyrolytic treatment of contaminated soils under carefully selected operating conditions can rapidly and reliably remove heavy hydrocarbons (including PAHs) and completely eliminate toxicity to human lung cells, while preserving the fertility of the soil to facilitate ecosystem restoration and re-greening efforts [2,3]. The more volatile hydrocarbons desorb as temperature of the contaminated soil is raised to 350°C. When pyrolysis temperatures rise above 350°C, however, heavier hydrocarbons (including PAHs) undergo a cascade of pyrolytic reactions resulting in the release of hydrogen and C_1-C_4 hydrocarbons. The final product of these reactions is a thin layer of carbonaceous material (coke) that coats the particles of the contaminated soil and drastically reduces its toxicity. The advantages of pyrolytic treatment over other thermal treatment methods (like incineration) include significantly lower energy requirements and preservation of soil fertility due to retention of some organic carbon and water holding capacity.

While pyrolytic treatment can be very effective in removing hydrocarbons from contaminated soils to levels well below regulatory standards, the treated soils may exhibit significant differences in fertility, as determined by germination and biomass production metrics. Moreover, heating contaminated soils in air at temperatures above 600°C (incineration) either drastically reduced or eliminated the fertility of treated soils [4]. Our earlier work also revealed potential tradeoffs between pyrolytic treatment intensity (and associated energy consumption), soil detoxification efficacy, and soil fertility restoration. For example, treating a contaminated soil in an isothermal reactor at 420°C for 30 min removed 99.9% for total petroleum hydrocarbons and 94.5% for PAHs, while also ***restoring its fertility to virtually clean soil levels*** [4]. When the same contaminated soil was treated at 470°C for 15 and 30 min, however, its fertility dropped to 51% and 39% respectively of the clean soil level, with the latter fertility value being only marginally higher than that of the contaminated soil [ibid.]. But not all soil/contaminant mixtures showed the same behavior. Working with another soil/contaminant mixture, for example, we did not observe a soil fertility "maximum" with increasing treatment temperature [1].

The fact that different soil/hydrocarbon mixtures respond differently to pyrolytic treatment underscores the need for deeper understanding of the processes taking place as contaminated soils are heated in anoxic atmospheres. Specifically, we need to distinguish between (a) thermally induced changes to soil components that affect soil fertility, and (b) contaminant desorption and pyrolysis, which ultimately determines the rate and extent of their removal.

While the primary objective of soil remediation is regulatory compliance, a deeper mechanistic understanding of the interactions between contaminants and soil components will allow us to optimize pyrolytic treatment to achieve multiple objectives like detoxification and soil fertility restoration in addition to contaminant removal. Since the optimal treatment conditions are system-specific, the design of remediation processes should be based on a systematic study of (a) the properties of the soil (e.g., texture and porosity, clay type and content, presence of catalytic metals) and the contamination scenario (type and concentration of contaminants, interactions between contaminants and metal sites, age and weathering of contamination). However, the operating conditions for current remediation processes are often selected using empirical approaches that do not address the complexity of these systems. This underscores the need for novel and comprehensive testing protocols that quantify soil/contaminant interactions, followed by robust data analysis and system modeling that will ultimately lead to the development of guidelines for selecting cost-effective operating conditions.

Integrating Thermal Analysis and Reaction Engineering to Optimize Pyrolytic Remediation

To address this need, we will present a novel approach that integrates advanced analytical measurements and mathematical modeling to rationally design a pyrolytic remediation process for specific soil/contaminant systems. We first use thermogravimetry with online infrared and mass spectrometry (TG-IR and TG-MS) to model the thermal transformations occurring when clean soil is heated in an anoxic atmosphere. By analyzing the patterns of water and carbon dioxide release (the two major products of clean soil pyrolysis), we can determine the temperature ranges over which (a) the clay components of soil release the water held between their aluminosilicate 2D sheets, (b) soil organic matter (SOM) pyrolyzes, or (c) carbonate minerals decompose to release carbon dioxide. We can then obtain the kinetics of soil decomposition by deconvolving the multipeak patterns of water and carbon dioxide release. Deconvolution is achieved by assuming that the multipeak patterns are weighted sums of the decomposition rates of a number of pseudo-components. These pseudo-components are actually mixtures of individual species (clays, SOM, carbonates) each decomposing to release either water or carbon dioxide with different activation energies that are distributed with a Gaussian probability density function (PDF). The full soil decomposition kinetics are obtained by a distributed activation energy (DAE) approach that allows us to estimate (a) the rate constants, (b) the means and variances of each activation energy PDF and (c) the weight fraction of each pseudo-component by solving a minimization problem. We will demonstrate that predictions from a DAE model with four or five pseudo-components can very accurately match the experimental thermogravimetry data [5].

Once we have obtained the full kinetics of the thermal transformations of clean soil, TG-IR and TG-MS are again used to determine how contaminant desorption and pyrolysis affects the decomposition pattern of the contaminated soil. We will demonstrate that contaminant mixtures consisting of light (LH) and heavy hydrocarbons (HH) superimpose two extra peaks to the multipeak decomposition pattern of clean soil. Thus, the kinetics for the pyrolytic treatment of contaminated soil can now be obtained by adding two additional pseudo-components (denoted here by LH and HH) to the four or five pseudo-components used to describe the kinetics of soil pyrolysis. Again, the pre-exponential factors, the distribution of activation energies and the weight fractions of the final number of pseudo-components are obtained by formulating and solving a DAE minimization problem.

The full kinetics are then used to formulate and solve a reactor model that predicts the conversion of the contaminants (including PAHs) and decomposition of soil components as they are heated in a continuous pyrolysis reactor. We first used an isothermal reactor model to compare our model predictions to experimental data obtained with a pilot-scale continuous rotary kiln reactor operating under flowing nitrogen. This electrically heated reactor treated two contaminated soils with different hydrocarbon contents at three temperatures (370, 420 and 470°C) and three residence times (15, 30 and 60 min). We will present data showing that the isothermal model provides very accurate predictions for both (a) the residual amount of total petroleum hydrocarbons (TPH) in the treated soil and (b) the amount of PAHs remaining at the exit of the isothermal reactor.

Since commercial pyrolysis reactors will operate under non-isothermal conditions, we extended the model to simulate the operation of non-isothermal rotary kiln reactors with co-current and counter-current flow of the nitrogen sweep gas. We will present results to demonstrate how the key operating parameters (solids flow rate, loading, wall temperature, direction and flow rate of sweep gas etc.) affect the removal efficiency of contaminants.

Significance

This study presents the development of a novel framework that integrates thermal analysis methods and reactor modeling to design, scale up and optimize pyrolytic remediation processes. It consists of:

- Workflows of thermal analysis tests (TG-IR, TG-MS) that (a) identify the temperatures where key soil mineral and contaminant transformations occur, and (b) quantify the gaseous products of these transformations.
- A systematic method for deconvolving the complex decomposition patterns of contaminated soils to obtain the full pyrolysis kinetics. (DAE model for clean soil followed by a full DAE analysis of contaminated soil).
- Reactor models for selecting process conditions and guiding multi-objective optimization with a rigorous decision model.

This novel framework has the potential to become an essential tool for remediation professionals by allowing them to scale up and optimize pyrolytic treatment processes for a broad spectrum of soil/contaminant systems using a rigorous methodology that does not rely on "black box" approaches.

References

1. Chibwe, L.; Geier, M. C.; Nakamura, J.; Tanguay, R. L.; Aitken, M. D.; Simonich, S. L.,. *Environmental Science & Technology* **2015**, 49, (23), 13889-98.
2. Vidonish, J. E.; Zygourakis, K.; Masiello, C. A.; Gao, X.; Mathieu, J.; Alvarez, P. J.,*Environmental Science & Technology* **2016,** *50*, (5), 2498-2506.
3. Vidonish, J. E.; Alvarez, P. J. J.; Zygourakis, K.,*Industrial & Engineering Chemistry Research* **2018,** *57*, (10), 3489-3500.
4. Song, W.; Vidonish, J. E.; Kamath, R.; Yu, P. F.; Chu, C.; Moorthy, B.; Gao, B. Y.; Zygourakis, K.; Alvarez, P. J. J.,*Environmental Science & Technology* **2019,** *53*, (4), 2045-2053.
5. Gao, Y., Zygourakis, K., *Industrial & Engineering Chemistry Research*, **2019**, 58, 10829-10843.

Copyright © American Institute of Chemical Engineers. All rights reserved.

621078 Valorizing Phosphogypsum As a Set Retarder in Portland Cement: Investigation on Its Heavy Metals Release into Buffered Aqueous Media

Abstract:

Phosphogypsum is a by-product in phosphate fertilizers and phosphoric acid industries. It is valorized as a set retarder in Portland cement replacing the gypsum rocks. In this industry, influence of phosphogypsum contents on concrete and Portland cement properties was investigated. Phosphogypsum was mixed with Portland cement at different mass contents extended from 1% to 20% wt/wt. Setting time, specific area of Portland cement and response to compression strength were followed. Chemical characteristics and heavy metals release into buffered acidic and alkaline aqueous solutions were also investigated. Obtained results showed that less than 8% of phosphogypsum is the best mass content to enhance cement quality without altering significantly the mechanical properties. The highest compressive strength applied to concrete was recorded after 28 days under saturated humidity conditioning. Conversely, the increase of phosphogypsum content in cement over than 8% resulted in a decrease of mechanical strengths of the corresponding concrete. Metal leaching experiments were conducted on different mixture of concrete mortars immersed in buffered acid solutions. Leaching results have showed that highest concentration of chromium, zinc, copper and lead are transferred to aqueous media during the first three days of impregnation. Zinc has a particular behavior. It was entrapped in concrete mortars before being transferred into aqueous solutions. In addition, metal release was significantly enhanced when phosphogypsum contents increased in the

Copyright © American Institute of Chemical Engineers. All rights reserved.

Biogas Production by co-digestion of Goat Manure with Rice Husk

Mahmoud Soliman[1], Harjinder Kaur[1], Raghava R. Kommalapati[1, 2]*

[1] Center of Energy and Environmental Sustainability, Prairie view A&M University 77446
[2] Department of Civil & Environmental Engineering, Prairie View A&M University 77446
**Corresponding Author; rrkommalapti@pvamu.edu*

Abstract

Anaerobic digestion (AD) is the primary technique used in aiding and limiting odors in waste management systems, reducing the organic strength in waste streams, and is a major player in producing a high energy source biogas. The AD is a process of degradation of organic materials by the microorganisms that operate in an oxygen-free environment that produces methane (CH_4) and carbon dioxide (CO_2) and small amounts of water vapors and other gases which is also called biogas. Biogas is a combination of methane, and carbon dioxide, which is produced from dairy waste by anaerobic digestion (AD). Manures have the potential to contaminate drinking water, affect air quality, and foster pathogenic species that can negatively affect human health. The objective of this study is to produce biomethane from anaerobic co-digestion of goat manure (GM) with rice husk (RH). The GM was collected from International Goat Research Center, PVAMU. The RH was collected from Gulf Pacific Co. The inoculum used in this study was a sludge that is obtained from the Prairie View wastewater treatment plant. The trace nutrient or known as the anaerobic medium was prepared according to Moody et al. (2009). The analysis of the substrates (GM, and RH) was conducted following APHA method 2540. After preparing the substrates, inoculum, and anaerobic medium nutrients, a triplicate of 250 mL serum bottles (Fisher Scientific, Pittsburgh, PA) was prepared for ratios of GM: RH varied from 0-100% were prepared with an organic loading rate of 10g Total solids (dried at 105°C). The theoretical maximum biomethane yields (TMYs), Biodegradability (BD), Synergetic effect index, and Carbon to Nitrogen ratio of GM and RH were calculated. After 63 days, it is determined that the highest three yielded ratios were 100% GM, 90% GM, and 30% GM produced 267.6 mL/gVS, 252.6 mL/gVS, and 239.3 mL/gVS, respectively. The BD of the mono-digestions of GM and RH was 78.36% and 18.99% respectively. The highest SEI of 105.45% was detected in the GM co-digestion with 20% GM. Daily and Cumulative BMP show a positive trend toward 30% GM, and 90% GM as high producers of BMP.

1 Introduction

Anaerobic digestion (AD) is the primary technique used in aiding and limiting odors in waste management systems, reducing the organic strength in waste streams, and is a major player in producing a high energy source biogas. The AD is a process of degradation of organic materials by the microorganisms that operate in an oxygen-free environment that produces methane (CH_4) and carbon dioxide (CO_2) and small amounts of water vapors and other gases which is also called biogas. Biogas is a very promising substitute as a renewable source of energy that contributes to fewer greenhouse emissions. Biogas can be used as a fuel that can be used in daily life needs such as: generating electricity from burned biogas, conversion of biogas to methanol by liquefication, chemical feedstocks, and can be used as compressed biogas [1, 2]. AD supplements are not only an option to use an alternative source of energy but also an alternative direction to put into good use the organic waste, as well reduce the releasing of greenhouse gas emissions from landfills [3,4]. California is the first leading state of largest dairy in the USA with approximately 1.7 million cows that produce more than 20,000 pounds of milk per cow each year, which generates close to 3.6 million dry tons of manure [5]. With this huge amount of manure being exerted on the environment, it must be properly managed to minimize air pollution and emissions.

Manures have the potential to contaminate drinking water, affect air quality, and foster pathogenic species that can negatively affect human health [5]. Biogas is a combination of methane, and carbon dioxide, which is produced from dairy waste by anaerobic digestion (AD). The AD is a natural process that decomposes organic material in an oxygen-free environment. Greenhouse gases that are produced from the agriculture sector are due to the stored piles of manures in lagoons which are left untreated. Anaerobic digesters play the role of allowing the dairies to produce and collect biogas that can be used as a renewable energy source. Most farms use anaerobic digesters to burn the biogas as a source of renewable energy to generate electricity. Around 14.6 billion ft^3 of methane is potentially produced in California by cows annually [5]. Co-digestion of goat manures with different biomass is possible such as rice waste, food waste, and other animal waste. The U.S. Environmental Protection Agency (EPA) estimates that 15% of total GHG emissions from agriculture can be attributed to manure management [10]. The AD does not only provide an alternative source of energy but also initiates the solution of reducing greenhouse gas emissions. A total of 287 anaerobic digesters were operating at different animal farms across the United States of America in 2019 [14, 15, 16]. Rice Husk (RH) is known to have a high content of inorganic compounds; therefore, it goes through thermo-chemical processes to be converted into energy [6]. The RH is considered among the top agricultural waste in the United States of America and is a very attractive feedstock for energy production because of its wide availability around the globe especially for the top countries known for rice production. The Food and Agriculture Organization (FAO) of the United States has estimated roughly rice production of 760 million tons in 2017 [9]. The agriculture sector accounts for about 10% of all total U.S. greenhouse gas emissions in the year 2019 [2]. Manure contains major elements which are important for plant growth such as Nitrogen (N), Phosphorus (P), Potassium (K), and micronutrients. When mixed with organic carbon, it aids in the growth of a healthy plant. A quarter of the emissions such as methane gas (CH_4) is a natural process that livestock produce as per their normal digestive process. It is, therefore, necessary to understand how emissions are being released to the environment with known knowledge of the harmful effects it may cause to human health and the environment. Animal feeding operations (AFOs) which release animal waste to the environment are known as CAFOs (reference).The main goal of USDA's towards operation sites is that owners collaborating to help in minimizing the potential of air and water pollution sourced from storage facilities, land areas, and confinement areas [17]. As the problem that many agencies and farm owners struggle with the most is finding an efficient way to minimize methane emissions. The proposed solution in this study is to use animal manure and rice waste to produce biogas through the (AD) process. The AD has a wide area of advantages which makes it very profitable and desired in such farms with high animal waste amounts such as diversified farm revenue, rural economic growth, conservation of agricultural land, energy independence, sustainable food production, and farm-community relationships [18].

This work is part of a large project studying different Agricultural residues and food waste (different C/N) to be co-digested with goat manure to produce biomethane.

The objectives of this study were to determine the best mixing ratio of GM and RH for maximum biomethane recovery by running lab scale biochemical methane potential assays (BMP) and the efficacy the AD process.

2 Materials and Methods

2.1 Chemicals, substrates, and inoculum,

The analytical grade chemicals used in this study were purchased from Thermo Fisher Scientific Inc. (Waltham, MA, USA) or from HACH Co. (Loveland, Colorado, USA).

The GM was collected from International Goat Research Center, PVAMU. The collected goat manure went through a drying process in an oven (Thermo Scientific oven) at 80 °C for 24 hours. Mechanical pretreatment of the goat manure was done using a mechanized grinder (LHB dual refiner Model 6SM14A) to reduce its size into fine particles. The uniform size particles were obtained using a 2 mm sieve. The dried GM was then transferred inside a Ziplock bag and stored inside a laboratory-grade refrigerator at 4°C for future use. The RH was collected from Gulf Pacific Co. (Houston, Tx), inspected for any impurities that might contain rocks, rice hulls, and other impurities and passed through the willy mill to grind into smaller portions where particle size is homogenous. Then the portions it was blended using a Roswell Home blender (1400-Watt model #RHPB-18001(YB-P26)) to minimize the particle size into smaller particle size. The rice residues were sieved (2 mm) to obtain the homogenous size particles. The processed rice residues were then sealed in a Ziplock bag and were stored inside a laboratory-grade refrigerator at 4°C for future use.

The inoculum used in this study was a sludge that is obtained from the Prairie View wastewater treatment plant (operating at mesophilic temperature) and was de-gassed for a week. The digester function at a temperature of 36±1 °C. The inoculum was degassed every day for a week or until there were no signs of any significant biogas produced.

The trace nutrient or known as the anaerobic medium to aid the microbial biodegradation of the substrates was prepared according to Moody et al. [13].

2.2 Substrate Characterization

The analysis of the substrates (GM, and RH) was conducted following APHA method 2540 [23] by using Linderberg Blue M electric furnace (Fisher Scientific, Pittsburgh, PA). The elemental composition of Carbon (C), Hydrogen (H), Nitrogen (N), Sulfur (S), and oxygen (O) of the pure feedstocks were determined using Flash 2000 Organic Elemental Analyzer (Thermo Fisher Scientific, Cambridge, UK). The standard used for determining the elemental composition was BBOT (C, H, N, and S; 72.5%, 6.1%, 6.5%, and 7.5%, respectively).

2.3 Biochemical Methane Potential (BMP) assays

An amount of 20 mL of anaerobic medium, 20 mL of inoculum, and the rest is filled with Deionized (DI) Water to satisfy a total volume of 100 mL as described. There were total of 36 batch reactor serum bottles including blank, mono-digestions of GM and RH and the co-digestions for the varied ratios of (GM/RH; 100:0, 90:10, 80:20, 70:30, 60:40, 50:50, 40:60, 30:70, 20:80, 10:90, and 100:0) in triplicates. The organic loading rate of 10 g total solids (dried at 105°C). The blanks contained inoculum, anaerobic medium nutrient, and DI water without the presence of the substrates. The inoculum to substrate ratio (ISR) of 2.6 was maintained in all mono-digestions and co-digestions. The alkalinity was adjusted by adding 20, 18, 16, 14, 12, 10, 8, 4 and 2 mL of 50,000 ppm sodium bicarbonate solution in RH:GM ratios of 100:10, 90:10, 80:20, 70:30, 60:40, 30:70, 20:80, and 10:90, respectively. The pH of each bioreactor was kept within the optimum range of 7.0 [19] The serum bottles were then purged with compressed nitrogen for 3 minutes and sealed with rubber plug stoppers and secured with aluminum top crimp seals (Fisher Scientific, Pittsburgh, PA), and stored in an incubator at 36±1 °C or the duration of bedding time of 63 days.

The biomethane was collected and measured by using the water displacement method where a 3-gallon container was filled with 1 gallon of DI Water and the pH of the DI water was adjusted to 10:30 with 0.5 M NaOH solution. The NaOH solution was added to the container with a 1000 mL inverted plastic graduated cylinder in the plastic container filled with DI water as described in a previous study [1]. The readings were taken by recording the initial and final values and then subtracting them. The reactors were shaken daily before biomethane reading to eliminate the formation of dead zones [20]. The collection of biomethane was recorded daily until the production of biomethane is zero or less than 1% of the last 3 days of recording [20].

2.4 Effluent Characterization

The 10 mL effluent in glass tubes was centrifuged (Thermo Scientific, Waltham, MA) for 15 minutes at 5500 rpm. The supernatant was tested and utilized using HACH test kits for Volatile Acids (HACH TNT 872), Nitrate (HACH TNT 835), and Ammonia (HACH TNT 832) of the co-digestion of GM and RH by utilizing DRB 200 reactor and DR 3900 spectrophotometer.

2.5 Theoretical Maximum biomethane yield (TMY)

The theoretical maximum biomethane yields (TMYs) of GM and RH were determined by Boyle's equation (Eq. 1), which is a modification of the original Buswell Muller's original equation (Eq. 2) after including Nitrogen (N) and Sulfur (S) to obtain Ammonia (NH_3) and sulfur dioxide (SO_2) fractions [21, 22]. From the elemental formula of the pure feedstocks (GM, RH) in Table 1, the theoretical maximum methane yield (TMY) of GM and RH were calculated and identified using Boyle's equation (Eq. 1).

$$C_aH_bO_cN_dS_e + \left(a - \frac{b}{4} - \frac{c}{2} + \frac{3d}{4} + \frac{e}{2}\right) H_2O \rightarrow \left(\frac{a}{2} + \frac{b}{8} - \frac{c}{4} - \frac{3d}{8} - \frac{e}{4}\right) CH_4 + \left(\frac{a}{2} - \frac{b}{4} + \frac{c}{4} + \frac{3d}{8} + \frac{e}{4}\right) CO_2 + dNH_3 + eH_2S \quad \text{(Eq. 1)}$$

$$\text{TMY (mL } CH_4/g_{VS}) = \frac{22400 \times [(\frac{a}{2})+(\frac{b}{8})-(\frac{c}{4})-(\frac{3d}{8})-(\frac{e}{4})]}{12a+b+16c+14d+32e} \quad \text{(Eq. 2)}$$

Table 1. Characterization of goat manure (GM) and rice husk (RH).

Parameter and unit	GM	RH	Inoculum
Moisture content (%)	9.52	3.70	96.00
TS (%)	90.48	96.29	3.90
VS (%)	74.63	78.75	1.94
Ash (%)	10.00	18.00	1.90
N (%)	2.22	0.52	-
C (%)	34.77	35.91	-
H (%)	4.90	5.01	-
O (%)	57.59	58.52	-
S (%)	0.45	0.01	-
C/N	15.66	69.05	-
Empirical Formula	$C_{202.23}H_{342.77}O_{251.43}N_{11.10}S_{1.00}$	$C_{299.05}H_{497.62}O_{365.82}N_{3.74}S_{0.05}$	-

2.6 Biodegradability (BD)

Biodegradability (BD) is the volatile fraction of each ratio that is converted to biomethane during anaerobic digestion. The (BD) was calculated for each of the mixing ratios by dividing the experimental biomethane yield (EMY) by the theoretical methane yield (TMY) multiplied by 100 as shown below (Eq. 3) as described by Li et al. [24].

$$BD(\%) = \frac{EMY}{TMY} \times 100 \qquad \text{(Eq. 3)}$$

EMY= Experimental Methane Yield
TMY= Theoretical Methane Yield

2.7 Synergetic Effect Index (SEI)

The Synergetic effect index (SEI) was calculated for the anaerobic digestion (Eq. 4) as described by Li et al. [24].

$$SEI(\%) = \frac{EMY_{CO} - (EMY_1 \times X_1 + X_2 \times EMY_2)}{(X_1 \times EMY_1 + X_2 \times EMY_2)} \times 100 \qquad \text{(Eq. 4)}$$

EMY_{CO} = The biomethane potential of co-digestion
EMY_1 = The biomethane potential of GM
EMY_2 = The biomethane potential of RH
X_1 = The volatile solid fraction of GM
X_2 = The volatile solid fraction of RH

3 Results and Discussion

3.1 Characteristics of GM and RH

Table 1 presented characteristics of GM, and RH. The values align within optimum range results to be considered and suitable for anaerobic degradation. The C/N ratio of GM was 16:1, which falls within the optimum range for anaerobic digestion. On the other hand, the C/N ratio for RH was 69:1. RH has a slighter low C/N ratio, than the previously reported range of 97-99 [6]. GM & RH C/N ratios 16:1 & 69:1, respectively falls within the optimum [7]. The VS/TS ratio of GM calculated showed a high ratio of 82%, while RH also showed a high VS/TS ratio of 82% which indicates a low ash content of 18%.

3.2 Daily biomethane production

The daily data was acquired to generate a daily biomethane chart as shown in figure 1. After 63 days, it is determined that the highest three yielded ratios were 100% GM, 90% GM, and 30% GM. The biomethane production was active from the first day yielding 30% as the highest ratio for the first 4 days, from day 5 to day 63 there was a minimal change of increase in the 30% GM. The second-highest yielding ratio was 20% GM which was leading second from day 1 to day 4 then stabilized with a minimal increase till day 63. The total sum of the biomethane produced over a period of 63 days for the various ratios from (100:0) of (GM: RH) to (0:100) (GM: RH) are 267.6 mL/gVS, 90% GM (90:10) is 252.6 mL/gVS, and 30% GM (30:70) is 239.3 mL/gVS. Figure 1 shows the daily biomethane production yield for 63 continuous days.

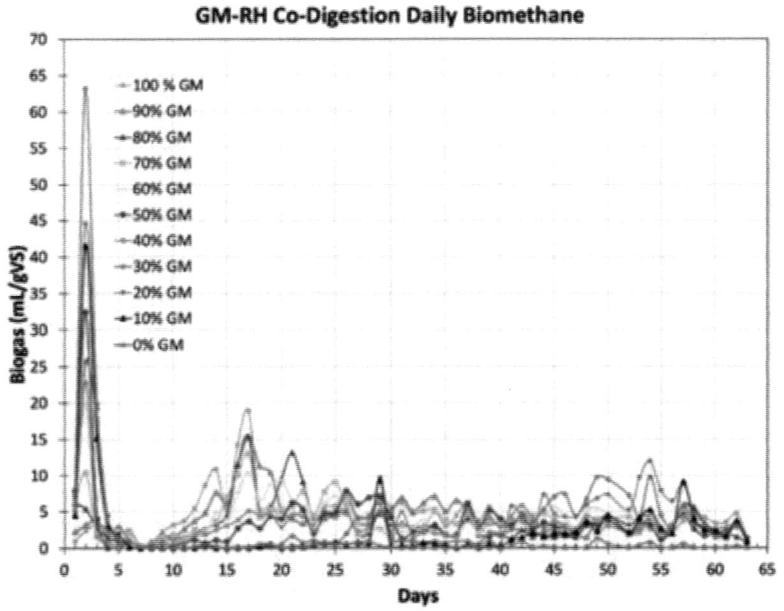

Figure 1. Daily Biomethane production yield for 63 continuous days.

3.3 Cumulative biomethane production

A cumulative data was generated for 63 days of daily data recording, it is determined that the highest three yielded ratios were 100% GM, 90% GM, and 30% GM. The biomethane production was active from the first day yielding 30% as the highest ratio for the first 18 days, between day 6 and day 18 there was a minimal change of increase in the 30% GM. The second highest yielded ratio was 20% GM which was leading second from day 3 to day 15 then stabilized with a minimal increase till day 28. Leading thirdly, 30% GM with being the highest ratio on the productivity of biogas since day 1 leading first till day 18, when 100% GM surpassed to be leading till day 63. The cumulative biomethane production for the mixing ratio from (100:0) of (GM: RH) to (0:100) (GM: RH) was recorded after 63 days of daily data recordings. The cumulative biomethane production for 100% GM (100:0) is 267.6 mL/gVS, 90% GM (90:10) is 252.6 mL/gVS, and 30% GM (30:70) is 239.3 mL/gVS. Figure 2. shows the cumulative biomethane yield for 63 continuous days.

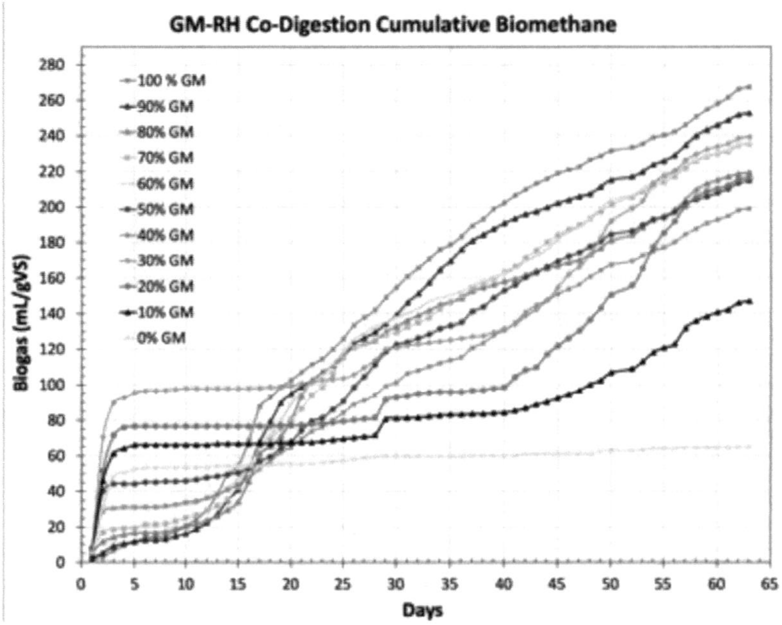

Figure 2. Cumulative biomethane production yield for 63 continuous days.

3.4 Biodegradability (BD) and Synergetic Effect Index (SEI)

The biodegradability (BD) and synergetic index effect (SEI) were calculated using equation 3 and equation 4 of the anaerobic digestions for each ratio as shown in table 2. The BD of the mono-digestions of GM and RH was 78.36% and 18.99% respectively. The BD decreased as the proportion of RH in the co-digestion was increased. The highest SEI of 105.45% was detected in the GM co-digestion with 20% GM.

Table 2. Biodegradability (BD), synergetic effect index (SEI), and Carbon to Nitrogen ratio of goat manure (GM) and rice husk (RH) anaerobic co-digestions.

GM: RH	BD(%)[c]	SEI(%)[d]	C/N
100:0	78.36	-	18.22
90:10	72.47	2.14	20.45
80:20	63.26	-3.40	23.01
70:30	68.25	13.82	26.01
60:40	68.82	26.58	29.54
50:50	62.95	29.22	33.77
40:60	58.69	36.50	38.94
30:70	70.91	90.57	45.38
20:80	64.44	105.45	53.65
10:90	43.99	72.84	64.63
0:100	18.99	-	79.92

[c] Biodegradability, [d] Synergetic Effect Index

3.5 Effluent Characterization

The volatile acids, nitrate, and ammonia of the co-digestion of GM and RH were tested using HACH kits utilizing DRB 200 reactor and DR 3900 spectrophotometer. volatile acids, nitrate, and ammonia of the co-digestion of GM and RH are shown below in figures 3, 4, and 5. The volatile acids recorded for 100% GM, 90% GM, 80% GM, 70% GM, 60% GM, 50% GM, 40% GM, 30% GM, 20% GM, 10% GM, and 0% GM are 735.67, 952.67, 979.00, 1506.00, 1146.00, 1772.67, 1679.33, 2921.67, 2892.67, 3299.33, 1351.67 and 568.67 mg/L which all of them falls within recorded range as shown in figure 3.

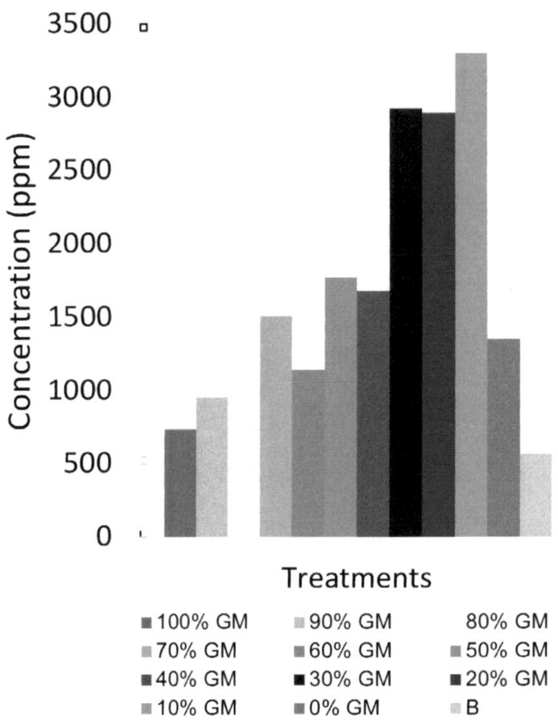

Figure 3. Volatile acid concentration in effluents.

The Nitrate recorded for 100% GM, 90% GM, 80% GM, 70% GM, 60% GM, 50% GM, 40% GM, 30% GM, 20% GM, 10% GM, and 0% GM are 13.60, 16.73, 17.03, 20.37, 48.80, 24.40, 21.97, 17.87, 18.93, 11.75, 13.12, and 1.44 as shown in figure 4.

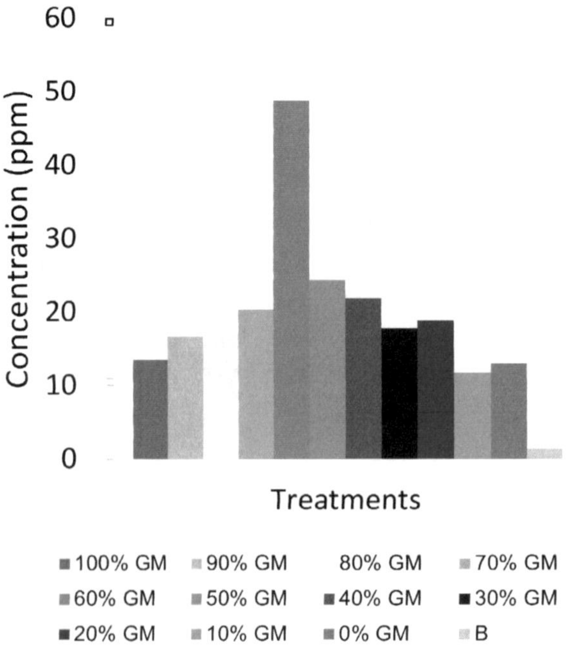

Figure 4. Nitrate concentration in effluents.

The ammonia recorded for 100% GM, 90% GM, 80% GM, 70% GM, 60% GM, 50% GM, 40% GM, 30% GM, 20% GM, 10% GM, and 0% GM are 47.29, 55.79, 36.36, 36.36, 20.96, 14.21, 16.17, 6.27, 19.29, 1.97, 11.34, 42.26, and 0.59 ppm as shown in figure 5.

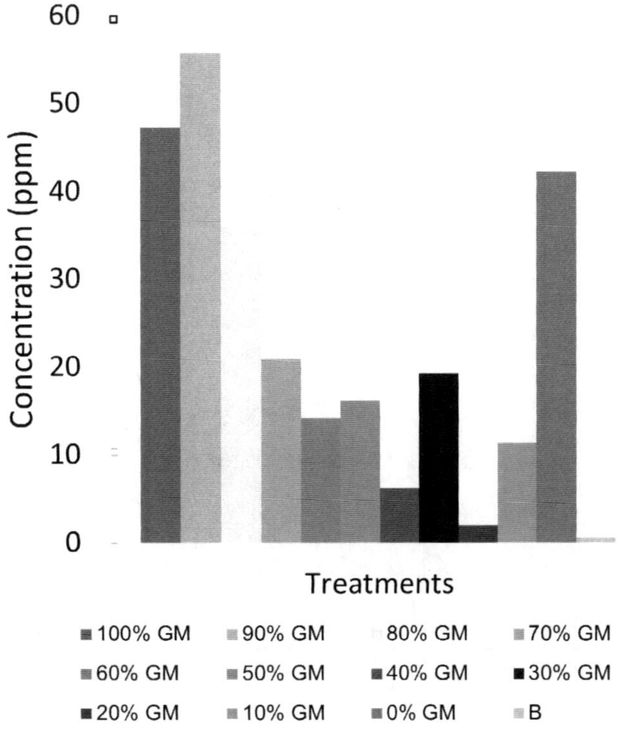

Figure 5. Ammonia concentration in effluents.

The TMY, EMY and the elemental formulas for various mixing ratios of (GM:RH) 100% - 0% were calculated using equations 2 and 3 as shown in Table 2. The TMY of 100% GM, 90% GM, 80% GM, 70% GM, 60% GM, 50% GM, 40% GM, 30% GM, 20% GM, 10% GM, and 0% GM are 341.50 mL/gVS, 348.57 mL/gVS, 346.67 mL/gVS, 344.78 mL/gVS, 342.92 mL/gVS, 341.08 mL/gVS, 339.26 mL/gVS, 337.45 mL/gVS, 335.67 mL/gVS, 333.90 mL/gVS, and 340.69 mL/gVS. The EMY of 100% GM, 90% GM, 80% GM, 70% GM, 60% GM, 50% GM, 40% GM, 30% GM, 20% GM, 10% GM, and 0% GM are 267.6 mL/gVS, 252.6 mL/gVS, 219.3 mL/gVS, 235.3 mL/gVS, 236.0 mL/gVS, 214.7 mL/gVS, 199.1 mL/gVS, 239.3 mL/gVS, 216.3 mL/gVS, 146.9 mL/gVS, and 64.7 mL/gVS.

Table 2. Theoretical maximum methane yield, experimental biomethane potential, and elemental formula of goat manure (GM) and rice husk (RH) anaerobic digestions.

GM: RH	Elemental formula	TMY[a] (mL/g_{vs})	EMY[b] (mL/g_{vs})
100:0	$C_{202.23} H_{342.77} O_{251.43} N_{11.10} S_{1.00}$	341.50	267.6
90:10	$C_{211.91} H_{358.25} O_{262.87} N_{10.36} S_{0.91}$	348.57	252.6
80:20	$C_{221.60} H_{373.72} O_{274.31} N_{9.63} S_{0.81}$	346.67	219.3
70:30	$C_{231.28} H_{389.18} O_{285.75} N_{8.89} S_{0.72}$	344.78	235.3
60:40	$C_{240.96} H_{404.65} O_{297.19} N_{8.16} S_{0.62}$	342.92	236.0
50:50	$C_{250.64} H_{420.12} O_{308.63} N_{7.42} S_{0.53}$	341.08	214.7
40:60	$C_{260.32} H_{435.59} O_{320.07} N_{6.69} S_{0.43}$	339.26	199.1
30:70	$C_{270.01} H_{451.06} O_{331.51} N_{5.95} S_{0.34}$	337.45	239.3
20:80	$C_{279.69} H_{466.53} O_{342.94} N_{5.24} S_{0.24}$	335.67	216.3
10:90	$C_{289.37} H_{481.99} O_{354.38} N_{4.48} S_{0.15}$	333.90	146.9
0:100	$C_{299.05} H_{497.62} O_{365.82} N_{3.74} S_{0.05}$	340.69	64.7

[a] Theoretical biomethane potential, [b] Experimental biomethane potential

4.0 Conclusions

Daily and cumulative BMP show a positive trend toward 30% GM, and 90% GM as high producers of BMP. The BD of the mono-digestions of GM and RH was 78.36% and 18.99% respectively. The highest SEI of 105.45% was detected in the GM co-digestion with 20% GM. Future work will involve implementation of a 2L bioreactor designed to improve the efficiency and biogas production and life cycle assessment (LCA) of the AD process. The kinetics parameters will be assessed utilizing different models.

5.0 Funding/Acknowledgments

This work is supported by the USDA-CBG program (Award No. 2018-38821-27750), and partial support received from the NSF CREST Center for Energy and Environmental Sustainability (CEES) at Prairie View A&M University (NSF Award #1914692). Special thanks to Dr. Hongbo Du, for his support and to Dr. Vamsi Botlaguduru for his contribution while he worked with us as a post-doctoral researcher at PVAMU."

References

[1] Roubaud A, Favrat D. Improving performances of a lean burn cogeneration biogas engine equipped with combustion prechambers. Fuel 2005;84(16):2001e7.

[2] Ghosh S, Henry MP, Sajjad A, Mensinger MC, Arora JL. Pilot- scale gasification of municipal solid wastes by high-rate and two-phase anaerobic digestion (TPAD). Water Sci Technol 2000;41(3):101e10.

[3] Mata-Alvarez J, Mace S, Llabres P. Anaerobic digestion of organic solid wastes. An overview of research achievements and perspectives. Bioresour Technol 2000;74(1):3e16.

[4] Frigon JC, Guiot SR. Biomethane production from starch and lignocellulosic crops: a comparative review. Biofuel Bioprod Biorefin Biofpr 2010;4(4):447e58.

[5] Krich, Ken, et al. "Biomethane from Dairy Waste - Sustainable Conservation." Biomethane from Dairy Wastes, July 2005.

[6] Contreras, L. M., Schelle, H., Sebrango, C. R., & Pereda, I. (2012). Methane potential and biodegradability of rice straw, rice husk and rice residues from the drying process. Water Science and Technology, 65(6), 1142–1149. https://doi.org/10.2166/wst.2012.951

[7] Kalra, M. S., & Panwar, J. S. (1986). Anaerobic digestion of rice crop residues. Agricultural wastes, 17(4), 263-269.

[8] Zhang, T., Liu, L., Song, Z., Ren, G., Feng, Y., Han, X., & Yang, G. (2013). Biogas production by co-digestion of goat manure with three crop residues. PloS one, 8(6), e66845.

[9] Y. Li et al. (2013) Comparison of methane production potential, biodegradability, and kinetics of different organic substrates

[10] Environmental Protection Agency, https://www.epa.gov/ghgemissions/sources-greenhouse-gas-emissions.

[11] Shankman, Sabrina. "What Is Nitrous Oxide and Why Is It a Climate Threat?" *Inside Climate News*, 7 Dec. 2020, https://insideclimatenews.org/news/11092019/nitrous-oxide-climate-pollutant-explainer-greenhouse-gas-agriculture-livestock/.

[12] Food and Agriculture Organization of the United Nations (FAO), et al. "Enteric Fermentation." *Climate & Clean Air Coalition*, 1 Jan. 1970, https://www.ccacoalition.org/en/activity/enteric-fermentation.

[13] Moody L, Burns R, Wu-haan W, Spajic R (2009) Use of biochemical methane potential (BMP) assays for predicting and enhancing anaerobic digester performance. In: Proceedings of the 44th Croatian and the 4th international symposium on agriculture, Opatija, Croatia, 16–20 February 2009

[14] USEPA (2019) Livestock anaerobic digester database AgSTAR:Biogas recovery in the agriculture sector. https:// www. epa. gov/agstar/ lives tock- anaer obic- diges ter- datab ase. Accessed 5 Jan 2020

[15] Guo M, Song W, Buhain J (2015) Bioenergy and biofuels: history,status, and perspective. Renew Sustain Energy Rev 42:712–725

[16] Moriarty K, Milbrandt A, Lewis J, Schwab A (2017) Bioenergy industry status report

[17] Natural Resources Conservation Service. NRCS. (n.d.). Retrieved November 20, 2021, from https://www.nrcs.usda.gov/wps/portal/nrcs/main/national/plantsanimals/livestock/afo/.

[18] Environmental Protection Agency. (n.d.). The Benefits of Anaerobic Digestion. EPA. Retrieved November 20, 2021, from https://www.epa.gov/agstar/benefits-anaerobic-digestion.

[19] Raposo, F., Fernández-Cegrí, V., De la Rubia, M. A., Borja, R., Béline, F., Cavinato, C., Demirer, G., Fernández, B., Fernández-Polanco, M., Frigon, J. C., Ganesh, R., Kaparaju, P., Koubova, J., Méndez, R., Menin, G., Peene, A., Scherer, P., Torrijos, M., Uellendahl, H., … de Wilde, V. (2011). Biochemical methane potential (BMP) of solid organic substrates: Evaluation of anaerobic biodegradability using data from an international interlaboratory study. Journal of Chemical Technology & Biotechnology, 86(8), 1088–1098.https://doi.org/10.1002/jctb.2622

[20] Holliger, C., Astals, S., de Laclos, H. F., Hafner, S. D., Koch, K., & Weinrich, S. (2020). Towards a standardization of biomethane potential tests: A commentary. Water Science and Technology, 83(1), 247–250. https://doi.org/10.2166/wst.2020.569

[21] Boyle, W.C. "Energy Recovery from Sanitary Landfills - A Review." Microbial Energy Conversion, 1977, pp. 119–138., https://doi.org/10.1016/b978-0-08-021791-8.50019-6.

[22] Buswell, A. M., and H. F. Mueller. "Mechanism of Methane Fermentation." Industrial & Engineering Chemistry, vol. 44, no. 3, 1952, pp. 550–552., https://doi.org/10.1021/ie50507a033.

[23] APHA (2005) Standard methods for the examination of water and wastewater, 20th Edn. American Public Health Association, Washington.

[24] Richards, Brian K., et al. "Methods for Kinetic Analysis of Methane Fermentation in High Solids Biomass Digesters." *Biomass and Bioenergy*, vol. 1, no. 2, 1991, pp. 65–73., https://doi.org/10.1016/0961-9534(91)90028-b.

623239 Advances in Treated Industrial Water Quality for Reuse – Life-Cycle Perspective

Abstract:

Tapas K. Das[1] and Murilo D.M. Innocentini[2]

[1]Department of Civil Engineering, School of Engineering, Saint Martin's University

Lacey, WA 98503 USA; Email: tdas@stmartin.edu

[2]Department of Chemical Engineering, University of Ribeirão (UNAERP), São Paulo, Brazil, Email: muriloinnocentini@yahoo.com.br

Abstract

The increasing scarcity of water coupled with escalating cost of fresh water and its treatment has prompted industry to think of water conservation, reuse, and recycling. Incorporating advanced technologies such as reverse osmosis, enhanced biological nutrient removal for phosphorus and nitrogen, membrane bioreactor, nanofiltration, ion exchange, and so on, would result in reclamation and reuse of water and less environmental damage, but to what degree, and with what trade-offs? To answer these questions, this paper will present a number of life cycle assessments (LCA) on industrial and wastewater systems, with a focus on short-term and/or long-term effects and/or benefits of reusing and recycling waters on economics (life cycle cost), energy and materials recovery using the Pinch Analysis, public health, environmental protection, clean water act, water quality standards, and water-ecology.

This presentation will also highlight some of the recent advanced treatment technologies included (1) enhanced biological nutrient removal for phosphorus and nitrogen, and (2) chemical addition with filtration for phosphorus removal. These technologies appeared to remove some pharmaceutical and personal care products (PPCP) chemicals (analytes) from wastewater, due to extended biological contact time, nutrient conversion and removal, and using advanced or tertiary filtration methods and additional disinfection of secondary effluent produce high quality reclaimed water which is used in industries and some part of US for other beneficial uses. Few selected stream-lined LCA case studies will be presented on water reuse/recycling practices toward zero-liquid discharge plants in pharmaceutical, food processing, pulp and paper, power generating industries, ultraviolet disinfection for wastewater reuse, agricultural and land applications of reclaimed water.

References:

Das, T. K.(2020). "*Industrial Environmental Management: Engineering, Science and Policy*", John Wiley and Sons, Hoboken, NJ.

Das, T. K., ed. (2005). "Toward Zero Discharge: Innovative Methodology and Technologies for Process Pollution Prevention," John Wiley and Sons, Hoboken, NJ.

Copyright © American Institute of Chemical Engineers. All rights reserved.

629577 Production of Jet Fuel from Municipal Solid Waste (MSW): Process Modeling, Techno-Economic Feasibility, and Life Cycle Analysis

Abstract:

The transportation sector is a leading contributor to greenhouse gas (GHG) emissions, accounting for approximately 28% of the total U.S. GHGs emissions, 9% of which comes from aircraft operations [1]. Jet-fuel derived from renewable resources such as municipal solid waste (MSW) and/or woody biomass offers the potential to reduce the aviation sector's GHG emissions. In 2018, about 292.4 million tons of MSW was generated. Of this total, about 44% was recovered via recycling, composting, and energy generation, and more than 50% of MSW was landfilled [2]. The HHV value of the organic fraction of MSW is similar to other biomass feedstocks and it has the potential to be utilized as a feedstock with current gasification technology. Moreover, it is readily available at little to no cost [3,4].

In this work, a rigorous plant-wide process model is developed in Aspen Plus for the production of jet fuel from MSW via ethanol synthesis through syngas fermentation technology [5]. Cycloalkane-rich jet fuel is produced through a selective pathway developed by Pacific Northwest National Laboratory, where ethanol is directly converted to long chain ketones. The ketone mixture is cyclized to branched cyclohexanones ranging from C_9 to C_{15}, followed by conversion to cycloalkane rich jet fuel product through hydrodeoxygenation and hydrogenation. To study the economic feasibility of the process, a techno-economic (TEA) model is developed using material and energy balance information from the process model. A discounted cash flow economic model is used to estimate the minimum fuel selling price (MFSP) of the conceptual process at scale. Sensitivity studies are also performed to study the effect of key process and economic parameters such as feedstock (MSW vs. woody biomass), product yield, plant scale, and internal rate of return. Life cycle analysis is also performed to determine the total GHGs emissions associated with the fuels from the process.

References

[1] Agency USEP. Sources of Greenhouse Gas Emissions. United States Environ Prot Agency 2018.

[2] United States Environmental Protection Agency. Facts and Figures about Materials, Waste and Recycling 2021.

[3] Jones SB, Zhu Y. Techno-economic analysis for the thermochemical conversion of lignocellulosic biomass to gasoline via methanol-to-gasoline (MTG) process. US Dep Energy 2009;PNNL-18483:1–46.

[4] Valkenburg C, Walton CW, Thompson BL, Gerber MA, Jones SB, Stevens DJ. Municipal Solid Waste (MSW) to Liquid Fuels Synthesis, Volume 1: Availability of Feedstock and Technology. Richland, WA (United States): 2008. https://doi.org/10.2172/962858.

[5] Köpke M, Mihalcea C, Bromley JC, Simpson SD. Fermentative production of ethanol from carbon monoxide. Curr Opin Biotechnol 2011;22. https://doi.org/10.1016/j.copbio.2011.01.005.

Copyright © American Institute of Chemical Engineers. All rights reserved.

AUTHOR INDEX